Advances in
ORGANOMETALLIC CHEMISTRY

VOLUME 16

Molecular Rearrangements

CONTRIBUTORS TO THIS VOLUME

Arlene Courtney

John J. Eisch

John Evans

J. W. Faller

Erling Grovenstein, Jr.

E. Alexander Hill

Thomas J. Katz

John P. Oliver

M. T. Reetz

Minoru Tsutsui

Robert West

Advances in Organometallic Chemistry

EDITED BY

F. G. A. STONE

DEPARTMENT OF INORGANIC CHEMISTRY
THE UNIVERSITY
BRISTOL, ENGLAND

ROBERT WEST

DEPARTMENT OF CHEMISTRY
UNIVERSITY OF WISCONSIN
MADISON, WISCONSIN

Molecular Rearrangements

VOLUME 16

1977

ACADEMIC PRESS   New York · San Francisco · London
A Subsidiary of Harcourt Brace Jovanovich, Publishers

ACADEMIC PRESS, INC.
111 Fifth Avenue, New York, New York 10003

United Kingdom Edition published by
ACADEMIC PRESS, INC. (LONDON) LTD.
24/28 Oval Road, London NW1

LIBRARY OF CONGRESS CATALOG CARD NUMBER: 64–16030

ISBN 0–12–031116–X

PRINTED IN THE UNITED STATES OF AMERICA

Contents

1,2-Anionic Rearrangements of Organosilicon and Germanium Compounds

ROBERT WEST

Dyotropic Rearrangements and Related σ–σ Exchange Processes

M. T. REETZ

Rearrangements of Unsaturated Organoboron and Organoaluminum Compounds

JOHN J. EISCH

Contents

Rearrangements of Organoaluminum Compounds and Their Group III Analogs

JOHN P. OLIVER

Organomagnesium Rearrangements

E. ALEXANDER HILL

Aryl Migrations in Organometallic Compounds of the Alkali Metals

ERLING GROVENSTEIN, JR.

Fluxional and Nonrigid Behavior of Transition Metal Organometallic π-Complexes

J. W. FALLER

σ–π Rearrangements of Organotransition Metal Compounds

MINORU TSUTSUI and ARLENE COURTNEY

The Olefin Metathesis Reaction

THOMAS J. KATZ

Molecular Rearrangements in Polynuclear Transition Metal Complexes

JOHN EVANS

List of Contributors

Numbers in parentheses indicate the pages on which the authors' contributions begin.

ARLENE COURTNEY (241), *Department of Chemistry, Texas A & M University, College Station, Texas*

JOHN J. EISCH (67), *Department of Chemistry, State University of New York at Binghamton, Binghamton, New York*

JOHN EVANS (319), *Department of Chemistry, The University, Southampton, United Kingdom*

J. W. FALLER (211), *Department of Chemistry, Yale University, New Haven, Connecticut*

ERLING GROVENSTEIN, JR. (167), *School of Chemistry, Georgia Institute of Technology, Atlanta, Georgia*

E. ALEXANDER HILL (131), *Department of Chemistry, University of Wisconsin-Milwaukee, Milwaukee, Wisconsin*

THOMAS J. KATZ (283), *Department of Chemistry, Columbia University, New York, New York*

JOHN P. OLIVER (111), *Department of Chemistry, Wayne State University, Detroit, Michigan*

M. T. REETZ (33), *Fachbereich Chemie der Universität, Lahnberge, Marburg, West Germany*

MINORU TSUTSUI (241), *Department of Chemistry, Texas A & M University, College Station, Texas*

ROBERT WEST (1), *Department of Chemistry, University of Wisconsin-Madison, Madison, Wisconsin*

Preface

This volume of *Advances in Organometallic Chemistry* represents a new departure in that it is concerned exclusively with molecular rearrangements of organometallic compounds. The idea for such a volume arose from the symposium, "Organometallic Rearrangements in the Main Group and Transition Metals," held at the Inorganic Chemistry Division of the National Meeting of the American Chemical Society in September 1976. Most of the contributors to this volume were participants in the symposium, which was organized by Professors John Eisch and Minoru Tsutsui. At our request, Professor Eisch has written the Introduction to the volume.

F. G. A. STONE
ROBERT WEST

Introduction

From the very beginning of modern chemistry molecular rearrangements have exerted a strong, constant influence on the development of structural theory. Indeed, modern organic chemistry is widely considered to have been born when Wöhler in 1828 observed the rearrangement of ammonium cyanate into urea, thereby rendering the quietus to the theory of vital force in organic synthesis. Yet of even wider significance was the discovery of silver fulminate by Wöhler in 1823 and of silver cyanate by Liebig in 1824, whose identical compositions stimulated Berzelius to introduce the concept of isomers in 1820. With the realization that a given number of atoms could be arranged in different ways to give isomeric molecules with different properties, chemistry had been placed on the highroad to rapid, rational development.

Throughout the succeeding 150 years, the configurational changes observed in a reacting molecule have served to point up inadequacies of existing theory and to provide indispensable insight into reaction mechanisms. Once the theory of van't Hoff and Le Bel (1874) had clarified how the bonds about carbon are arrayed in space, the stage was set to understand molecular rearrangements, such as cis, trans-isomerization (van't Hoff, 1874), prototropic shifts (Laar, 1885), and the inversion at saturated carbon (Walden, 1893). The applicability of these structural concepts to organometallic and inorganic chemistry was demonstrated early in this century, mainly through the efforts of Werner, Pope, and Kipping.

The chemistry of organometallic compounds not only displays all of the configurational variety of carbon compounds, but is complicated as well by the varying coordination number, oxidation state, and electronic configuration of the metal. As a consequence, organometallics often undergo reaction via one or more metastable intermediates, which frequently can be detected or even isolated. Thus, products of kinetically controlled pathways abound in organometallic chemistry and it becomes the imposing task of the research chemist to understand and to control the rearrangement pathways of such molecular intermediates.

In choosing contributors both to the symposium and to this volume, the co-chairmen, Professor Minoru Tsutsui and I and the Editors of the series have sought to include significant areas of current research, while maintaining a considerable breadth of coverage. Thus, among the chapters treating main group organometallics, we have selected one dealing with carbon skeletal rearrangements of alkali metal derivatives by E. Grovenstein, followed by those on such processes in Grignard reagents by E. A. Hill, on Group III compounds by J. P. Oliver and by myself, and on Group IV compounds by R. West and by M. Reetz. With transition metal derivatives, the treatments have tended to stress general properties, rather than specific periodic families of metals: $\sigma-\pi$ rearrangements by M. Tsutsui, and A. Courtney fluxional behavior by J. W. Faller, olefin metathesis by T. J. Katz, and rearrangements in metal clusters by J. Evans.

In its juxtaposition of main group and transition metal chemistry, this volume is consistent with the aim of this series, to serve the needs of chemists working in both of these areas by stressing the many similarities which exist between main group and transition organometallics. Advances in reaction mechanisms have uncovered common underlying features in hydrometallation, carbometallation, hydrogen–metal exchanges, and transalkylation common to organometallic compounds from all parts of the periodic table. We hope that this volume will contribute to increased awareness of these similarities by workers in various areas of organometallic chemistry.

JOHN J. EISCH

Advances in

ORGANOMETALLIC CHEMISTRY

VOLUME 16

Molecular Rearrangements

1,2-Anionic Rearrangements of Organosilicon and Germanium Compounds

ROBERT WEST

Department of Chemistry
University of Wisconsin-Madison
Madison, Wisconsin

I

INTRODUCTION

A. *Mobility of Silicon in Anionic Rearrangements*

In 1964 it was discovered that organosilicon groups migrate from one nitrogen atom to the other in anions of silicon-substituted hydrazines (*1, 20*). Subsequent investigations established that the anionic rearrangement of organosilyl hydrazines takes place with unprecedented speed.

1

The rate of migration of R_3Si groups in the hydrazine rearrangement is at least 10^{12} more rapid than for phenyl or methyl groups (*21, 22*). Now, 12 years after the initial discovery, we know that extremely high mobility is characteristic for silicon in anionic rearrangements generally. This fact has led to the development of an extensive new area of organometallic chemistry. Equally important, organosilicon substituents are excellent probes for the study of anionic rearrangement reactions because of their high mobility. So far three completely new reactions, the N→N, O→N, (*23*), and S→C (*32*) 1,2-rearrangements, have been discovered through the use of organosilicon substituents.

Why is silicon (and germanium) so mobile in anionic rearrangements? When a carbon substituent migrates to an anionic site a simple [1,2]-sigmatropic shift is disallowed by the Woodward–Hoffmann rules (*14, 31*). Accordingly the known examples are slow reactions, usually requiring prolonged heating and proceeding by complex mechanisms involving radical pair formation and recombination (*10, 11, 13, 14, 36*). But when an organosilicon group migrates, rearrangement can occur without these symmetry restrictions because pentacoordinate transition states are possible (*24, 33*) [Eq. (1)]. Migration can therefore be very rapid indeed. Moreover, organosilyl anionic rearrangements are typically clean, intramolecular, and quantitative.

$$
\begin{array}{ccc}
\underset{\underset{Si}{|}}{\overset{R}{\diagdown}}X{-}Y\overset{R'}{\diagup}{}^{-} & \rightleftharpoons \left[\underset{\underset{Si}{|}}{\overset{R}{\diagdown}}X{\cdots}Y\overset{R'}{\diagup}\right]^{-} & \rightleftharpoons \overset{R}{\diagdown}X{-}Y\underset{\underset{Si}{|}}{\overset{R'}{\diagup}}{}^{-}
\end{array}
\qquad (1)
$$

B. *Types of Anionic Rearrangements*

In an anionic rearrangement the migrating group moves from one atom X to another atom Y which bears a negative charge:

$$R{-}X{-}\bar{Y} \overset{a}{\longrightarrow} \bar{X}{-}Y{-}R \qquad (2)$$

The discussion in this chapter will be limited to cases in which X and Y are directly bonded to each other, that is, to 1,2-anionic rearrangements. These are by far the best studied and in many ways the most interesting anionic rearrangements.

The two classical anionic rearrangements are the Wittig rearrangement (*8*), wherein migration takes place from oxygen to anionic carbon as in

Eq. (3):

$$PhCH_2-OR \xrightarrow{RLi} Ph\bar{C}H-OR \xrightarrow{\quad \Delta \quad} PhCH-O^-$$
$$\underset{R}{|}$$

$$(3)$$

and the Stevens rearrangement, where migration is from nitrogen to carbon (16).

$$ArCOCH_2\overset{+}{N}Me_3 \xrightarrow[\Delta]{OH^-} ArCO\bar{C}H-\overset{+}{N}Me_3 \xrightarrow{\quad \Delta \quad} ArCOCH-NMe_2$$
$$\underset{Me}{|}$$

$$(4)$$

Both of these are examples of *heteroatomic* anionic rearrangements, that is, ones where X and Y are different. *Homoatomic* anionic rearrangements, where X = Y, are also possible. The one example known before the discovery of organometallic migrating groups was the C—C anionic rearrangement (11, 36), reviewed by Grovenstein elsewhere in this volume:

$$Ar_3C-CH_2Cl \xrightarrow{M} Ar_3C-\bar{C}H_2 \xrightarrow{\quad \Delta \quad} Ar_2\bar{C}-CH_2Ar$$

$$(5)$$

II

THE HYDRAZINE (N→N) REARRANGEMENT

A. *Anionic Rearrangement of Silylhydrazines*

The hydrazine rearrangement was the first known anionic rearrangement of organosilicon compounds[1] and has been studied more thoroughly than any other. A previous review, now somewhat out-of-date (24), covered the early experimental results on this reaction.

The first case came to light accidentally in an attempt to dimethylate 1,1-bis(trimethylsilyl)hydrazine with *n*-butyllithium followed by methyl iodide. In this reaction almost equal amounts of the 1,1 and 1,2 isomers are produced [Eq. (6)]. It was soon

$$(Me_3Si)_2NNH_2 \xrightarrow[2\ CH_3I]{2\ n-BuLi} (Me_3Si)_2NN(CH_3)_2 + Me_3SiN\underset{CH_3}{\overset{|}{\quad\quad}}N(SiMe_3)_2$$
$$\underset{CH_3\quad CH_3}{|\quad\quad\quad|}$$

$$(6)$$

---

[1] The α-silylcarbinol rearrangement, discussed in Section IV,B, was discovered by Brook and his students as early as 1958, but was not recognized as an anionic rearrangement until later (5).

established that this isomerization was due to anionic rearrangement (*1, 20, 21*).

For detailed study of this reaction, the compound 1,2-bis-(trimethylsilyl)methylhydrazine (**1**) proved to be quite useful. This substance can be prepared almost isomerically pure from methylhydrazine and trimethylchlorosilane (*18*), as shown in Eq. (7). When **1**,

$$3 \text{ MeNH—NH}_2 + 2 \text{ Me}_3\text{SiCl} \longrightarrow \underset{\underset{\text{CH}_3 \quad (\mathbf{1})}{|}}{\text{Me}_3\text{SiN—NSiMe}_3} + 2 \text{ MeN}_2\text{H}_4{}^+\text{Cl}^- \qquad (7)$$

or its 1,1-isomer (**2**), is treated with one equivalent of *n*-butyllithium, the same mixture of anions is obtained. This may be derivatized with methyl iodide or reprotonated with pyrrole. In either case a nearly equal mixture of the two possible isomers is produced (*20*).

$$\underset{\underset{\text{CH}_3 \quad \text{CH}_3}{| \quad |}}{\text{Me}_3\text{SiN——NSiMe}_3} + (\text{CH}_3)_2\text{N—N}(\text{SiMe}_3)_2 \ (1:1)$$

50%   50%   (8)

$$\uparrow_{\text{CH}_3\text{I}}$$

$$\underset{\underset{\text{CH}_3}{|}}{\text{Me}_3\overset{-}{\text{SiN}}\text{—NHSiMe}_3} \xrightarrow{\text{RLi}} \underset{\underset{\text{CH}_3}{|}}{\text{Me}_3\text{SiN—}\overset{-}{\text{N}}\text{SiMe}_3} \xleftarrow{\ \ p\ \ } \text{CH}_3\overset{-}{\text{N}}\text{—N}(\text{SiMe}_3)_2 \xleftarrow{\text{RLi}} \text{CH}_3\text{NH—N}(\text{SiMe}_3)_2$$

(1) (46%)   pyrrole   (54%) (2)

\+

Unlike previously known anionic rearrangements these reactions proceed rapidly, even at or below room temperature. Moreover, complete lithiation and reprotonation are unnecessary for the rearrangement of 1 to 2. Treatment of either isomer with a *trace* of *n*-butyllithium rapidly converts it to a 1:1 mixture of the two isomers (*25*). This reaction is therefore a *catalytic anionic rearrangement*. Such reactions with organic migrating groups are still unknown.

B. *The Catalyzed Hydrazine Rearrangement*

1. *Studies of the Reaction Mechanism*

The kinetics of the catalyzed rearrangement have been thoroughly investigated (*21*). Crossover experiments show that it is fully intramolecular. Reaction rates were followed by nuclear magnetic resonance (NMR) spectroscopy. The methylsilyl resonances for the different isomers are usually well resolved, so the rate was obtained by tracing through the methylsilyl region repeatedly until no further change was observed.

Because the alkyllithium compound used as "catalyst" is consumed within the time of mixing, the actual catalyst for most of the reaction is hydrazide anion. Letting one isomer by AH and the other BH, the reaction sequence can be summarized as in Eq. (9). The equilibration

$$\text{AH} + \text{B}^- \underset{k_{-1}}{\overset{k_1}{\rightleftharpoons}} \text{A}^- + \text{BH}$$

$$k_3 \Big\Uparrow k_4 \qquad \qquad (9)$$

$$\text{B}^- + \text{AH} \underset{k_{-1}}{\overset{k_1}{\rightleftharpoons}} \text{A}^- + \text{BH}$$

reaction follows first-order kinetics in which the protonation–deprotonation steps, rather than the actual anionic rearrangement, are rate-limiting (21).

An interesting case consistent with this model is the rearrangement of 1-phenyl-1-trimethylsilyl-2-*tert*-butyldimethylsilylhydrazine (3) (26). Because this compound contains two different organosilyl substituents, multiple anionic rearrangements are possible. Loss of a proton from 3 produces the anion 3a, which can rearrange to 4a. The latter can become protonated to 4 or, alternatively, to 5a, which can give upon protonation 5 [Eq. (10)]. Note that two sequential rearrangements are necessary to form 5.

Compound 3 undergoes catalytic anionic rearrangements, albeit quite slowly, and the concentrations of 3, 4, and 5 can be followed

$$
\begin{array}{ll}
\textit{tert-}\text{BuMe}_2\text{Si}-\text{NH}-\underset{\underset{\text{SiMe}_3}{|}}{\text{N}}-\text{Ph} \;\rightleftharpoons\; \textit{tert-}\text{BuMe}_2\text{Si}-\underset{}{\underline{\text{N}}}-\underset{\underset{\text{SiMe}_3}{|}}{\text{N}}-\text{Ph} \\
\quad(3) \qquad\qquad\qquad\qquad\qquad (3a)
\end{array}
$$

$$
\begin{array}{ll}
\textit{tert-}\text{BuMe}_2\text{Si}-\underset{\underset{\text{Me}_3\text{Si}}{|}}{\text{N}}-\text{NH}-\text{Ph} \;\rightleftharpoons\; \textit{tert-}\text{BuMe}_2\text{Si}-\underset{\underset{\text{Me}_3\text{Si}}{|}}{\text{N}}-\underline{\text{N}}-\text{Ph} & (10)\\
\quad(4) \qquad\qquad\qquad\qquad\qquad (4a)
\end{array}
$$

$$
\begin{array}{ll}
\text{Me}_3\text{Si}-\text{NH}-\underset{\underset{\text{SiMe}_2\textit{tert-}\text{Bu}}{|}}{\text{N}}-\text{Ph} \;\rightleftharpoons\; \text{Me}_3\text{Si}-\underline{\text{N}}-\underset{\underset{\text{SiMe}_2\textit{tert-}\text{Bu}}{|}}{\text{N}}-\text{Ph} \\
\quad(5) \qquad\qquad\qquad\qquad\qquad (5a)
\end{array}
$$

by NMR spectroscopy. Results of a typical run are shown in Fig. 1. Compound 5, the double-rearrangement product, is formed *more rapidly than* 4 in the early stages of the reaction. The amount of 5 passes through a maximum after about 30 hours, then declines to the equilibrium value of about 5% as kinetic control of the product distribution is replaced by thermodynamic control. These results show that rearrangement *cannot* be rate-limiting. Protonation must be the rate-determining

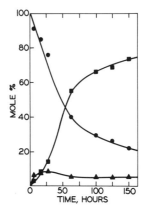

FIG. 1.   Catalytic rearrangement of compound 3 as a function of time. ●, 3; ■, 4; ▲, 5.

step in the rearrangement, and further, protonation of **5a** must be more rapid than that of **4a**.

2. *Equilibration of Silylhydrazines*

Base-catalyzed rearrangement provides a way to equilibrate isomeric silylhydrazines (*25*). Results from such equilibrations, and half-times for the equilibration reaction, are shown in Table I. Both steric and electronic effects are seen to influence the position of the equilibrium. Steric effects are important when *tert*-butyldimethylsilyl groups are present, for example 1,1-bis(*tert*-butyldimethylsilyl)hydrazine undergoes 86% rearrangement to the less hindered 1,2 isomer at equilibrium. Phenyl groups strongly stabilize isomers bearing a hydrogen on the same nitrogen   atom   as   the   aromatic   ring.   Note   that   1,1-bis(trimethylsilyl)phenylhydrazine rearranges 96% to the 1,2 isomer, and 1-trialkylsilyl-2-phenylhydrazines do not rearrange at all. The effect of phenyl groups is probably electronic, reflecting either specific stabilization of the Ph—N—H system or reduced opportunity for N→Si pi bonding in the Ph—N—Si compounds, or both.

In the absence of strong steric or electronic constraints, the bis(organosilyl)hydrazines and bis(organosilyl)methylhydrazines give *essentially equal amounts* of the N,N and N,N′ isomers at equilibrium (*25*). For these simple cases, the reaction is therefore *thermoneutral*, with no enthalpic driving force. This observation differentiates the silylhydrazine rearrangement from all other 1,2-anionic rearrangements, which are thermodynamically driven reactions proceeding essentially to completion.

TABLE I

Equilibration of Organosilylhydrazines with *tert*-Butyllithium in Benzene

Compound	Percent rearranged at equilibrium	$t_{1/2}{}^a$ (min)
$(Me_3Si)_2NNH_2$	51	1
$(EtMe_2Si)_2NNH_2$	53	1
$Me_3Si(EtMe_2Si)NNH_2$	52	1
$t\text{-}BuMe_2SiN(SiMe_3)NH_2$	73	2
$(t\text{-}BuMe_2Si)_2NNH_2$	86	4
$Me_3SiNHN(CH_3)SiMe_3$	50	3.5
$EtMe_2SiNHN(CH_3)SiEtMe_2$	47	4.5
$t\text{-}BuMe_2SiN(SiMe_3)NHMe$	68^b	80
$Me_3SiNHNHPh$	0	—
$EtMe_2SiNHNHPh$	0	—
$t\text{-}BuMe_2SiNHNHPh$	0	—
$Me_3SiNHN(Ph)SiMe_3$	96	600
$EtMe_2SiNHN(Ph)SiEtMe_2$	91	900
$t\text{-}BuMe_2SiNHN(Ph)SiMe_3$	80^b	3800^c
$t\text{-}BuMe_2SiNHNHMe$	0	—

[a] Hydrazine, 0.4 M; n-BuLi, 0.04 M.

[b] Two different products are formed (26).

[c] At 65°C.

The finding that **1** and **2**, and likewise $(Me_3SiNH)_2$ and $(Me_3Si)_2NNH_2$, have identical thermodynamic stability is quite surprising, for the silyl and methyl substituents might be expected to stabilize one isomer over the other. The simplest rationalization of the results is that the organosilyl and methyl substituents have little influence on the stability of the hydrazines. However, examination of models show that there is moderate steric interference between trimethyl- or ethyldimethylsilyl groups attached to the same nitrogen atom in the 1,1 isomers, which is relieved in the 1,2 compounds. This steric hindrance might be compensated by increased N→Si pi-bonding in the 1,1 isomers, which would stabilize the 1,1-bis(silyl) compounds electronically relative to the 1,2.

C. NMR Studies of Silylhydrazide Anions

The kinetic studies of the catalyzed hydrazine rearrangement show that the migration of silicon in the anion must be quite rapid, but give no data on the actual rate of anionic rearrangement. Information about this rate is, however, available from proton magnetic resonance spectroscopy of silylhydrazide anions (22). In these studies tris(trimethylsilyl)hydrazine (6) is a particularly useful compound, because it gives an anion with the same structure before and after rearrangement.

In hydrocarbon solvents, the lithium salt of 6 shows temperature-dependent NMR spectra illustrated in Fig. 2. A single peak is observed at room temperature, indicating that migration of silicon is rapid on the NMR time scale. This splits into two peaks of 2:1 relative intensity at $-26°C$, showing that at this temperature anionic rearrangement is slow compared to the proton relaxation time. At the coalescence temperature of $+5°C$, the half-life for exchange (0.032 second) and the free energy of activation (14.7 kcal) can be obtained from the NMR chemical shift with the use of the customary equations. The spectra are shown for hexane,

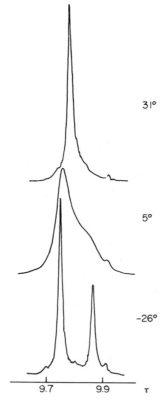

FIG. 2. Variable temperature proton nuclear magnetic resonance spectra of lithium tris(trimethylsilyl)hydrazide (6) in hexane reveals the temperature-dependent intramolecular trimethylsilyl migration. Temperatures are recorded in degrees centigrade.

but essentially identical behavior is observed in benzene and toluene. These results probably reflect exchange in the dimer, because the lithium salt of compound **6** is known to be dimeric in nonpolar solvents (*15*).

In either solution the behavior of the lithium derivative of **6** is more complex. As in hydrocarbons, a singlet is observed at room temperature, which broadens and splits to a 2:1 pattern at −20°C. However, when the solution is cooled further, a new peak appears at τ9.98, which

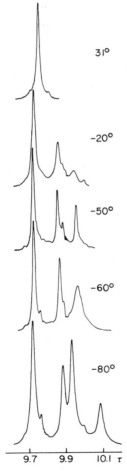

FIG. 3. Proton nuclear magnetic resonance spectra of compound **6** in diethyl ether indicate the existence of a dimer ⇌ monomer equilibrium as well as independent temperature dependence of trimethylsilyl migration in both dimer and monomer forms.

increases in intensity at the expense of the first two peaks. At still lower temperatures this peak also broadens and splits into two peaks having a 2:1 intensity ratio (Fig. 3).

These results may be interpreted in the following way (*22*): At room temperature the species exists mainly as dimer (or ion quadruplet) with rapid migration of Me₃Si groups on the NMR time scale. As the sample is cooled to −20°C, trimethylsilyl migration becomes slow so that two peaks are observed for the dimer. The related compound LiN(SiMe₃)₂ shows similar monomer–dimer equilibria in solution in THF and hydrocarbons (*12*). Below −20°C, monomer (ion pair) is observable in equilibrium with dimer. Between −20°C and −60°C silyl migration is rapid in the monomer but slow in the dimer (compared to the proton spin relaxation time). Finally at −80°C anionic rearrangement for the monomer also becomes slow, so that two pairs of 2:1 peaks are observed.

Other silylhydrazines behave similarly to **6**, although the NMR spectra are sometimes more complex. The anion of bis(*tert*-butyldimethylsilyl)methylhydrazine shows three singlets at room temperature for the

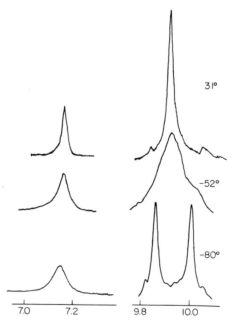

FIG. 4.   Temperature-dependent proton nuclear magnetic resonance spectrum showing the *N*-methyl (left) and dimethylsilyl (right) resonances of lithium bis(*tert*-butyldimethylsilyl)methylhydrazine in diethyl ether. Only the *N,N'*-bis(silyl) isomer is present in detectable amount.

N—CH$_3$, Si—CH$_3$, and Si—C(CH$_3$)$_3$ protons, indicating rapid exchange. As the temperature is decreased to −80°C the N—CH$_3$ resonance broadens and the other two split into clean doublets (Fig. 4). Evidently only the N,N′-bis(silyl) isomer is present in detectable amount at equilibrium. The N,N-isomer would be strongly disfavored by steric effects:

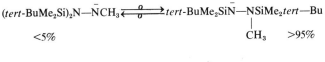

$$(\text{tert-BuMe}_2\text{Si})_2\text{N}-\overset{-}{\text{N}}\text{CH}_3 \overset{0}{\underset{0}{\rightleftharpoons}} \text{tert-BuMe}_2\text{Si}\overset{-}{\text{N}}-\text{NSiMe}_2\text{tert}-\text{Bu}$$

<5%   CH$_3$ >95%

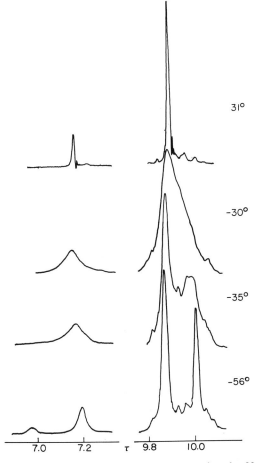

FIG. 5. Proton nuclear magnetic resonance spectra showing the N-methyl (left) and trimethylsilyl (right) resonances for lithium bis(trimethylsilyl)methylhydrazide in diethyl ether. At low temperatures a 4:1 mixture of the two isomeric anions is present, with the less hindered N,N′-bis(silyl) isomer in larger amount.

For the other bis(trimethylsilyl)methylhydrazine anions, the spectra show both species at equilibrium. Figure 5 shows the NMR of the anion of 1 (or 2); singlets for N—CH_3 and Si—CH_3 protons are again found at room temperature. As the temperature is lowered the N—CH_3 resonance is eventually split into two peaks of 4:1 intensity, and the silylmethyl proton region also shows two peaks in ratio of about 6:4; the stronger peak probably results from coincidental overlap. The spectrum is consistent with a 4:1 ratio of the ions at equilibrium [Eq. (11)]. The less hindered isomer is again

$$(Me_3Si)_2N-\overset{-}{N}CH_3 \rightleftarrows Me_3Si\overset{-}{N}-NSiMe_3 \tag{11}$$
$$\underset{CH_3}{|}$$

$$20\% \qquad\qquad 80\%$$

favored, where R = methyl or ethyl. These anions both undergo derivatization to give nearly equal amounts of 1,1 and 1,2 dimethyl and dihydrogen isomers, but the equality is now seen to be fortuitous. The 4:1 preference for the 1,2-anion at equilibrium is nearly balanced by the increased rate of reactions of the less hindered 1,1-anion, toward CH_3I and pyrrole.

The bis(trialkylsilyl)methylhydrazide anions give coalescence temper-

TABLE II

EQUILIBRIA, EXCHANGE RATES AND ACTIVATION ENERGIES FOR ORGANOSILYL HYDRAZIDE REARRANGEMENTS FROM NUCLEAR MAGNETIC RESONANCE

Anion	Solvent	Percent rearranged[a]	τ_c (sec)	T_c (°C)	$G_c\ddagger$ (kcal/mole)
$(Me_3Si)_2N\overset{-}{N}SiMe_3$	Hexane	—	0.032	+5	14.7
$(Me_3Si)_2N\overset{-}{N}SiMe_3$	Ether[b]	—	0.028	+2	14.4
$(Me_3Si)_2N\overset{-}{N}SiMe_3$	Ether[c]	—	0.025	−62	10.9
$(Me_3Si)_2N-\overset{-}{N}CH_3$	Ether	80	0.028	−33	12.5
$(Me_3Si)_2N-\overset{-}{N}CH_3$	Hexane	—	0.038	>+31	>16.2
$(EtMe_3Si)_2N-\overset{-}{N}CH_3$	Ether	80	0.028	−33	12.5
$(t\text{-}BuMe_2Si)_2N-\overset{-}{N}CH_3$	Ether	>99	0.025	−51	11.5

[a] Amount of rearrangement to N,N' isomer at equilibrium, below T_c (%).
[b] Dimer.
[c] Monomer.

atures and $G\ddagger$ values rather similar to those for $(Me_3Si)_2N$—$\bar{N}SiMe_3$ (Table II). It is interesting that the sterically hindered *tert*-butyldimethylsilyl compound shows the lowest coalescence temperature, and hence fastest rearrangement rate, in this series of compounds. This may reflect weaker ion pairing in the *tert*-butyl compound.

D. The Arylhydrazine Rearrangement

1. Phenylhydrazine Mono- and Dianions

After the discovery of the silylhydrazine rearrangement, it was of interest to search for N→N rearrangements with conventional organic migrating groups. However, all experiments designed to induce methyl migration in 1,1- and 1,2-dimethylhydrazines failed. Even after long heating at 150°C in the presence of excess base, these compounds are recovered unchanged. Likewise, solutions of the monoanions of 1,1-diphenylhydrazine and 1,1-phenylmethylhydrazine failed to rearranged after a week at 115°C (28):

$$RN—\overset{-}{N}H \xrightarrow{\quad\overset{115\,°C}{\times}\quad} R—\overset{-}{N}—NH \qquad (12)$$
$$\underset{Ph}{|} \qquad\qquad\qquad \underset{Ph}{|}$$

$$R = Ph, Me$$

Rearrangement of phenylhydrazines can be brought about, however, if more than one equivalent of base is used, so that the hydrazide *dianion* is generated. The driving force for rearrangement is very much greater for the dianion than for the monoanion. Even so, the rearrangement is slow, requiring many hours at 30°C (28) [Eq. (13)].

$$R—N—\overset{=}{N} \xrightarrow{\quad slow\quad} R—\overset{-}{N}—\overset{-}{N}—Ar \qquad (13)$$
$$\underset{Ar}{|}$$

$$R = CH_3, Ar = Ph, m\text{-}CF_3C_6H_4; R = Ar = Ph; R = Ar = p\text{-}Tol$$

2. Relative Migration Rates for Si and C

It is possible now to compare the migration rates of trimethylsilyl and phenyl groups in the hydrazine rearrangement. Because no rearrangement of the monoanion of *N,N*-diphenylhydrazine was observed at 115°C, the half-time for migration of phenyl under these conditions must be at least 10^6 seconds. The half-time for migration of R_3Si in silylhydrazide anions from the NMR results is about 3×10^{-2} second at 0 to $-60°C$ (22). The difference in temperature can be conservatively estimated to alter the rate by a factor of at least 10^4. Multiplying these factors, we conclude that in the anionic rearrangement of hydrazines, the migration rate of Me_3Si is *at least 10^{12} faster than for phenyl*. This enormous difference in mobility is a measure of the increased accessibility of the transition state when silicon is migrating. As mentioned in Section I,A, silicon can become pentacoordinate using unfilled orbitals, and so is not subject to the orbital symmetry restrictions that normally make 1,2 anionic rearrangements slow.

III

THE HYDROXYLAMINE (O→N) REARRANGEMENT

A. *Organosilyl and Germyl Hydroxylamines*

1. *Bis(organosilyl)hydroxylamines*

Molecular rearrangement of hydroxylamine anions is unknown in organic chemistry without the presence of metalloid substituents. When organosilicon groups are present, however, rapid anionic migration from oxygen to nitrogen takes place. The most thoroughly studied compounds are *N,O*-bis(organosilyl)hydroxylamines, prepared from hydroxylamine and the corresponding chlorosilane (2, 19) [Eq. (14)]. When any of

$$2RMe_2SiCl + 2Et_3N + H_2NOH \rightarrow RMe_2SiNH—OSiMe_2R + 2Et_3\overset{+}{N}HCl^-$$

(7)   R = Me
(8)   R = Et   (14)
(9)   R = Ph
(10) R = *tert*—Bu

the compounds **7–10** are deprotonated with an alkyllithium compound in THF at $-78°C$, a clear solution results. Derivatization reactions show clearly that N→O rearrangement takes place. As shown in Eq. (15) when the anions of **7–9** are treated with R_3SiCl or with CH_3I, only

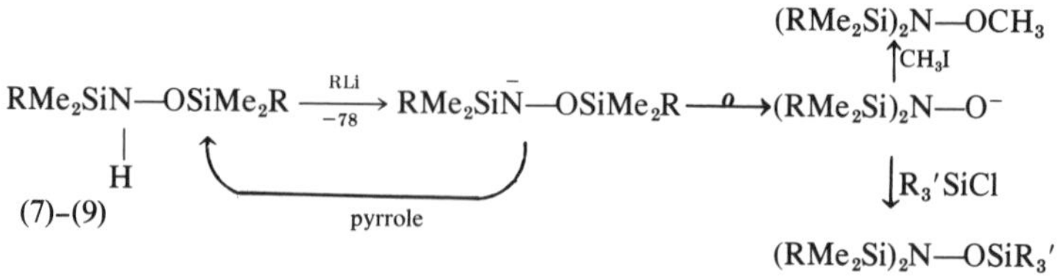

$$RMe_2SiN—OSiMe_2R \xrightarrow[-78]{RLi} RMe_2Si\bar{N}—OSiMe_2R \xrightarrow{\text{\textit{a}}} (RMe_2Si)_2N—O^-$$

$$\underset{H}{|} \qquad\qquad \underset{\text{pyrrole}}{\underbrace{}}$$

(7)–(9)

$$(RMe_2Si)_2N—OCH_3$$
$$\uparrow CH_3I$$
$$\downarrow R_3'SiCl$$
$$(RMe_2Si)_2N—OSiR_3'$$

(15)

the rearranged N,N isomers are produced (2). A crossover experiment using a mixture of **7** and **8** showed that the anionic rearrangement is intramolecular.

Reaction of the anions **7–10** with pyrrole, however, yields the *unrearranged* N,O isomers (23). The formation of the unrearranged isomer upon reprotonation is most easily explained by equilibrium between the two anion isomers and preferential reaction of the N anion with the proton source. However, protonation at oxygen followed by instantaneous dyotropic rearrangement of the N,N to the N,O compound cannot be completely ruled out. The results with *N,O*-bis(*tert*-butyldimethylsilyl)hydrazine show that in this case at least both anions are present in the solution. The products of derivatization are similar except that reaction with CH_3I gives a *1:1 mixture* of the rearranged and unrearranged products.

The silylhydroxylamine anions have also been studied by NMR spectroscopy. Upon addition of base the two proton resonances of **7** are gradually replaced by a single peak for the anion; and when 1 equivalent of base has been added, only the anion resonance is observed (Fig. 4). This anion resonance remains as a singlet down to −80°C (23). There are two possible explanations: (a) anionic migration is rapid even at low temperatures, or (b) The N,N isomer, which would show only one resonance, greatly predominates in the anionic solution.[2]

To obtain more information about the isomeric structure of the anions, their infrared spectra were studied in solution. These spectra strongly resemble those of the *N,N*-bis(silyl)-*O*-methyl derivatives, not those of the *N,O*-bis(silyl) starting materials. In particular, the anions show a strong N—Si—N asymmetric stretching absorption near 1000 cm^{-1}, present in the rearranged *O*-methyl compounds but not in the unrearranged protonated compounds (23). These results suggest that the equilibrium between anions in Eq. (15) is displaced strongly toward the rearranged isomers. If this is so, no conclusion about migration rate can be drawn from the NMR results.

However, the chemical results alone indicate that migration of silyl groups from O to N is very rapid in hydroxylamine anions. The results of quenching experiments at −78°C show that the half-life for rearrangement cannot be more than a few minutes at this temperature.

The O→N rearrangement with organosilyl migrating groups requires breaking of a strong Si—O bond and formation of a weaker Si—N bond.

---

[2] Even for the anion of X, where the N,N isomer would be subject to severe steric interference of the two *tert*-butyldimethylsilyl groups on the same nitrogen, only one SiCH$_3$ peak is found in the proton NMR.

These changes are endothermic by about 30 kcal/mole.[3] The reaction proceeds in spite of this energy deficit because of the greater stability of the O⁻ anion than the N⁻ anion.

2. Monosilylhydroxylamines

N→O anionic rearrangement has also been observed in monosilylhydroxylamines, and the results of derivatization experiments provide good evidence for equilibrium between N and O anions (2). For instance, lithiation of O-trimethylsilylhydroxylamine followed by reaction with tert-butyldimethylchlorosilane produces a 3:1 mixture of rearranged and unrearranged products (23).

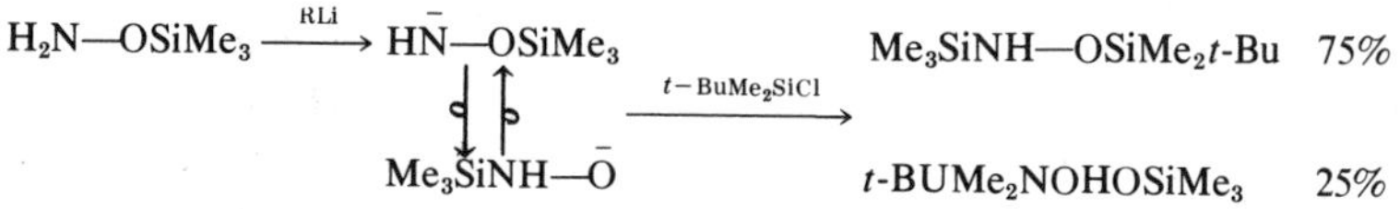

$$H_2N\!-\!OSiMe_3 \xrightarrow{\text{RLi}} H\bar{N}\!-\!OSiMe_3 \qquad\quad Me_3SiNH\!-\!OSiMe_2t\text{-Bu} \quad 75\%$$

$$\Big\updownarrow \qquad \xrightarrow{t-BuMe_2SiCl} \qquad\qquad\qquad\qquad\qquad\qquad\qquad (16)$$

$$Me_3SiNH\!-\!\bar{O} \qquad\qquad t\text{-BUMe}_2NOHOSiMe_3 \quad 25\%$$

And as shown in Eq. (17) the anion from O-trimethylsilylmethylhydroxylamine reacts with tert-BuMe₂SiCl to give the rearranged product but with CH₃I to give the unrearranged isomer:

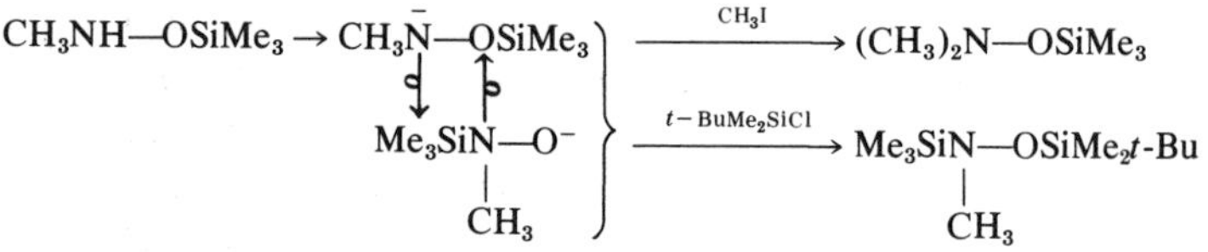

$$CH_3NH\!-\!OSiMe_3 \rightarrow CH_3\bar{N}\!-\!\underset{}{O}SiMe_3 \Bigg\rangle \xrightarrow{CH_3I} (CH_3)_2N\!-\!OSiMe_3$$

$$\Big\updownarrow$$

$$\underset{\underset{CH_3}{|}}{Me_3SiN\!-\!O^-} \Bigg\rbrace \xrightarrow{t-BuMe_2SiCl} \underset{\underset{CH_3}{|}}{Me_3SiN\!-\!OSiMe_2t\text{-Bu}} \quad (17)$$

3. Germylhydroxylamines

One example of hydroxylamine rearrangement with organogermanium migrating groups has been reported (3). The resulting anion undergoes derivatization with CH₃I or Me₃SiCl to give exclusively rearranged products:

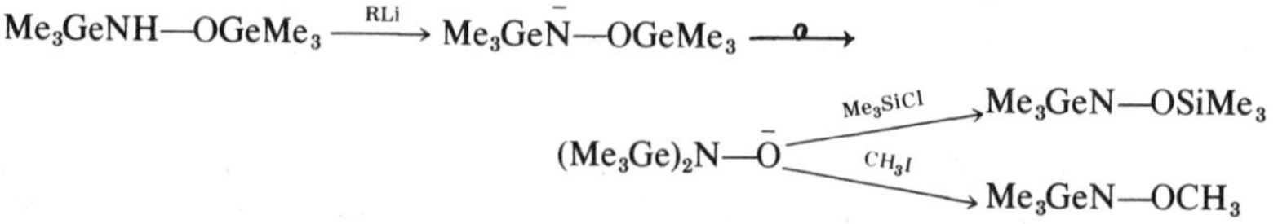

$$Me_3GeNH\!-\!OGeMe_3 \xrightarrow{\text{RLi}} Me_3Ge\bar{N}\!-\!OGeMe_3 \xrightarrow{\;\Delta\;}$$

$$(Me_3Ge)_2N\!-\!\bar{O} \overset{\xrightarrow{Me_3SiCl} Me_3GeN\!-\!OSiMe_3}{\underset{\xrightarrow{CH_3I} Me_3GeN\!-\!OCH_3}{}} \quad (18)$$

---

[3] Bond energies are about 106 for Si—O and 76 for Si—N (kcal/mole) (9).

B. *Attempts to Rearrange Arylhydroxylamine Anions*

No example of hydroxylamine rearrangement with a purely organic migrating group is yet known. Solutions of anions of N,O- and N,N-diphenylhydroxylamine undergo no rearrangement in several days at temperatures up to 80°–100°C (29). The rate for phenyl migration in hydroxylamine anions, as in hydrazides, must be many orders of magnitude slower than for organosilyl migration. Nevertheless, it seems likely that anionic rearrangement of hydroxylamines will eventually be found with suitable organic substituents as potential migrating groups.

C. *Organometallic Nitroxides from the Hydroxylamine Rearrangement*

Disubstituted nitroxides constitute a well-known class of rather stable free radicals. N,N-disubstituted anions of organometallic hydroxylamines have just one additional electron more than the corresponding nitroxide. Oxidation of the rearranged anions of silyl or germylhydroxyl amines, electrolytically or with oxygen, produces solutions of organometal nitroxides (30). Examples are shown in Eqs. (19)–(21). These nitroxides are stable in dilute solution for several days at room temperature.

$$(RMe_2Si)_2N\text{—}O^- \longrightarrow (RMe_2Si)_2N\text{—}\dot{O} \quad R{=}Me, t\text{-Bu, H, Ph, Et} \qquad (19)$$

$$
\begin{array}{c}
t\text{-BuMe}_2\text{Si} \\
\diagdown \\
N\text{—}O^- \longrightarrow \\
\diagup \\
Me_3Ge
\end{array}
\qquad
\begin{array}{c}
t\text{-BuMe}_2\text{Si} \\
\diagdown \\
N\text{—}\dot{O} \\
\diagup \\
Me_3Ge
\end{array}
\qquad (20)
$$

$$(t\text{-BuMe}_2Ge)_2N\text{—}O^- \longrightarrow (t\text{-BuMe}_2Ge)_2N\text{—}\dot{O} \qquad (21)$$

The electron spin resonance (ESR) spectra of organometallic nitroxide radicals provide important clues to chemical bonding in these species. The spectra consist of three lines due to hyperfine splitting by the nitrogen atom (Fig. 5). Organosilyl and germyl nitroxides show surprisingly low values for the ^{14}N hyperfine splitting constant (Table III), indicating that R_3Ge and especially R_3Si behave as strongly electron-withdrawing substituents in nitroxides (30). This has been explained in terms of valence bond structures (A), (B), and (C) in Eq. (22).

$$
\begin{array}{ccc}
\overset{\overline{}}{R_3M}\diagdown \overset{+}{} & R_3M\diagdown & \overset{+}{R_3M}\diagdown \\
N\text{—}\ddot{\underset{\cdot\cdot}{O}}: & N\text{—}\ddot{\underset{\cdot\cdot}{O}}: & N\text{--}\ddot{\underset{\cdot\cdot}{O}}: \\
R_3M\diagup & R_3M\diagup & R_3M\diagup
\end{array}
\qquad (22)
$$

$$(A)(B)(C)$$

TABLE III
RESULTS OF DERIVATIZATION OF ANIONS FROM BENZYLOXYTRIETHYLSILANE AND
BENZYLOXYTRIMETHYLGERMANE

Starting material	Derivatizing agent	Percent unrearranged	Percent rearranged
PhCH$_2$OSiEt$_3$	H—OAc	2	97
	Me$_3$Si—Cl	0	100
	CH$_3$—I	99	1
	CH$_3$—SO$_4$Me	40	60
PhCH$_2$OGeMe$_3$	H—OAc	0	71[a]
	CH$_3$—I	0	83[a]
	CH$_3$—SO$_4$Me	0	73
Ph$_2$CHOGeMe$_3$	H—OAc	0	70[a]
	CH$_3$—I	0	73[a]
Ph$_2$CHOSiMe$_3$	H—OAc	77	0
	CH$_3$—I	75	0

[a] Cleavage of Me$_3$Ge accounts for the remainder, to 100%.

Electron withdrawal involving donation of the lone pair on nitrogen into
d-orbitals on the metal (Form A) would localize the unpaired electron on
the oxygen atom. The lower α_N for R$_3$Si than for R$_3$Ge is consistent with
greater dative bonding from N to Si than N to Ge.

A simple molecular orbital (MO) approach can also be used to
rationalize the ESR results. The three electrons of the nitroxide bond
reside in π- and π^*-orbitals, derived from linear combinations of
nitrogen and oxygen p-orbitals. Two electrons occupy the π-orbital and
one is in the π^*-orbital (Fig. 6). When electron-withdrawing groups such
as R$_3$Si are attached to nitrogen, the effective electronegativity of
nitrogen is increased, leading to increased nitrogen character in the π
orbital but decreased nitrogen character in the π^*-orbital which contains
the unpaired electron (3).

IV

THE WITTIG (O→C) REARRANGEMENT

A. Organosilyl and Germyl Alkoxides

Although the Wittig rearrangement is the best known of conventional
organic rearrangements, the corresponding reaction with an organosilicon migrating group was not reported until 1971 (27). Like other anionic

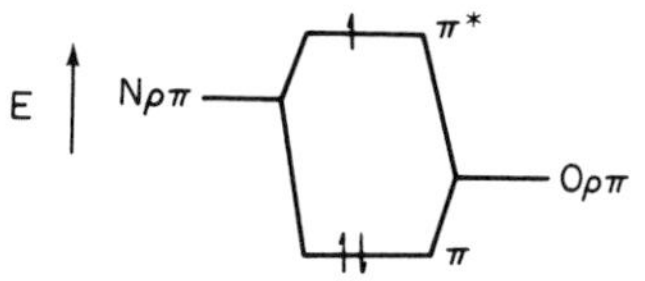

Fig. 6. Schematic molecular orbital diagram for the N—O system in nitroxide radicals.

rearrangements of silicon compounds, it is a rapid intramolecular reaction.

When a benzyloxysilane is metalated at the alpha position, the resulting carbanion rearranges immediately to give mainly the α-silylcarbinol anion (34):

$$PhCH_2OSiR_3 \xrightarrow{\text{RLi}} Ph\bar{C}HOSiR_3 \underset{O}{\overset{O}{\rightleftharpoons}} \underset{\underset{SiR_3}{|}}{PhC}{-}O^- \rightarrow \underset{\underset{SiR_3}{|}}{PhCH}{-}OH$$

$$(11) \qquad\qquad (11a) \qquad\qquad (12a) \qquad (12) \qquad R = Me, Et, Ph$$

$$(23)$$

This reaction provides a new way of generating Si—C bonds and is useful in the synthesis of silylcarbinols which are otherwise difficult to prepare.

All evidence suggests that the equilibrium in Eq. (23) lies far toward the right. The anionic solution shows only singlets in the proton NMR for the trimethylsilyl and benzyl protons, and the resonances appear consistent with the oxyanion, not the carbonion. However, a small amount of the carbanion must also be present. Derivatization with protonating agents or chlorosilanes gives exclusively rearranged products, but methyl iodide reacts to give the unrearranged C-methyl isomer, and methyl sulfate gives a mixture of both isomers. Data for a $PhCH_2OSiEt_3$, a representative example, are shown in Table III.

Benzyloxygermanium compounds also undergo the O→C anionic rearrangement (34), and in this case rearranged products are obtained exclusively even with methyl iodide as derivatizing agent (Table III). Apparently the equilibrium in Eq. (23) lies even farther to the right

$$PhCH_2{-}OGeMe_3 \xrightarrow{\text{RLi}}$$

$$Ph\bar{C}H{-}OGeMe_3 \underset{O}{\overset{O}{\rightleftharpoons}} \underset{\underset{GeMe_3}{|}}{PhCH}{-}O^- \xrightarrow{\text{RX}} \underset{\underset{GeMe_3}{|}}{PhCH}{-}OR \qquad R=H, CH_3, Me_3Si$$

$$(23)$$

than for the corresponding silicon rearrangement. In both rearrangements the oxyanion forms are favored, since the similar energies released in locating the negative charges on oxygen rather than carbon atoms outweighs those required to force the metalloid atoms onto carbon from oxygen. The latter energies are not equal, being less for germanium than silicon. Data from model compounds give $E(Ge—O)—E(Ge—C) = 17$ kcal; $E(Si—O)—E(Si—C) = 39$ kcal (34). The net energy released in the germanium anion rearrangement will be greater, and the anion equilibrium will lie farther to the right.

The difference between Ge and Si is shown even more clearly by the behavior of their benzylhydryl derivatives (α-siloxy- and α-germoxydiphenylmethane). The trimethylgermyl compound undergoes metalation to a colorless solution that gives only rearranged products upon quenching [Table III and Eq. (24)]. The anion of the trimethylsilyl compound behaves quite differently. Its solution is deep red, like diphenylmethyl anion, and gives *only unrearranged products* [Eq. (25)]. The anion must be entirely in the carbanion form—

$$Ph_2CHOGeMe_3 \xrightarrow{RLi} Ph_2\overset{-}{C}—OGeMe_3 \qquad Ph_2\underset{GeM_3}{\overset{|}{C}}—\overset{-}{O} \overset{AcOH}{\underset{CH_3I}{\Big\langle}} \begin{array}{l} Ph_2\underset{GeMe_3}{\overset{|}{C}}—OH \\[2mm] Ph_2\underset{GeMe_3}{\overset{|}{C}}—OCH_3 \end{array} \qquad (24)$$

(14)

$$Ph_2CHOSiMe_3 \xrightarrow{RLi} Ph_2\overset{-}{C}—OSiMe_3 \overset{AcOH}{\underset{CH_3I}{\Big\langle}} \begin{array}{l} Ph_2CH—OSiMe_3 \\[2mm] Ph_2\underset{CH_3}{\overset{|}{C}}—OSiMe_3 \end{array} \qquad (25)$$

no O→C rearrangement takes place for this compound (34). The additional acidity conferred on the alpha carbon atom by the second phenyl ring apparently makes the carbanion the stable form in the case of silicon, but not germanium. The bond energies mentioned above account well for this difference between silicon and germanium.

B. Relationship to the α-Silylcarbinol Rearrangement

Long before the silyl Wittig rearrangement was discovered, the base-catalyzed rearrangement of α-silylcarbinols to alkoxysilanes was known. This reaction, carefully studied by Brook and his students (4–6) takes place in exactly the opposite direction to the silyl Wittig reaction. In the α-silylcarbinol rearrangement neutral compounds are equilibrated in the presence of catalytic amounts of base (liquid NaK alloy or amines), to give the more stable alkoxysilane isomer:

$$R_3SiCHOH \xrightarrow[B^-]{\text{o}} R'CH_2OSiR_3$$

$$\underset{R'}{|}$$

(26)

(12)   (10)

R=Me; R'=Ph   R=Me; R'=Ph

The α-silylcarbinol rearrangement proceeds intramolecularly with retention of configuration at silicon (7). Kinetic studies suggest that the transition state has considerable negative charge dispersed into the aromatic ring, consistent with a transition state structure like the carbanion of the silyl Wittig rearrangement (4). And indeed the anion of benzyloxytrimethylsilane will catalyze the rearrangement of Eq. (26), R = Me (34).

How is that the rearrangement proceeds in one direction in the presence of a catalytic amount of base and in the other direction with a full equivalent of base? This can be understood in terms of bond energies. The reaction coordinate diagram of Fig. 7 shows qualitatively

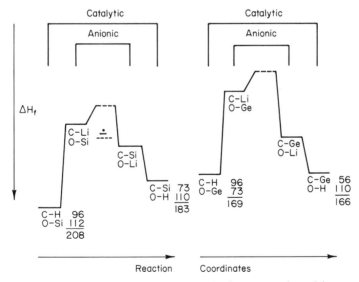

FIG. 7. Schematic energy level diagrams for the interconversion of isomers in the C—O anionic rearrangements, for silicon (left) and germanium (right) migrating groups. Relative energies are indicated, from left to right, for PhCH₂OMMe₃, PhC̄HOMMe₃, PhCH(MMe₃)O⁻, and PhCH(MMe₃)OH. Approximate thermochemical bond energies were calculated from heats of combustion for model compounds (34).

the relative heats of formation of the species involved, based on the approximate thermochemical bond energies of the bonds made and broken in the isomerizations. In the catalytic silyl rearrangement in which the neutral species **11** and **12** are being equilibrated, the sum of the C—H and Si—O bond energies in **12** exceeds the sum of the C—Si and O—H bond energies in **11**, so the equilibrium must lie toward **12**. When a full equivalent of anion is present as in the silyl Wittig rearrangement, it is apparent that the stabilization gained by having the negative charge on oxygen (in **11a**) rather than on carbon (in **12a**) outweighs the energy lost by placing silicon on carbon rather than on oxygen. Accordingly, the equilibrium between the anions lies toward **11a**.

A similar reaction coordinate diagram for germanium (Fig. 7, right) is also instructive. α-Germylcarbinols do not normally undergo catalytic rearrangement to germyloxyalkanes. The bond energies of Fig. 7, based on model compounds, suggest that there is very little energy difference between benzyloxytrimethylgermane and the corresponding germylcarbinol. Perhaps the latter compound is actually the more stable of the two, so that the catalytic rearrangement would be contrathermodynamic. However, inaccessibility of the transition state might also be responsible for the lack of catalytic rearrangement. The results of methyl iodide quenching experiments indicate that there is very little or no carbanion present in the equilibrium mixture of germyl anions in Eq. (23), consistent with a very high activation energy barrier between the oxyanion and the carbanion (Fig. 7). If the unknown catalytic rearrangement requires a transition state close to the same carbanion, the rearrangement might not be observed simply for kinetic reasons.

C. *Stereochemistry and Mechanism of the Silyl Wittig Rearrangement*

The stereochemistry at carbon in the silyl and germyl Wittig rearrangement has also been investigated with results that shed further light on the mechanism (*33*). The substrates used were *S*-(+)-benzyl-α-*d*-oxytrimethylsilane (**13**), as well as the corresponding triphenylsilane and trimethylgermane. These were prepared from *S*-(+)-benzyl-α-*d*-alcohol, obtained by yeast reduction of benzaldehyde (Scheme 1). Metalation with *tert*-butyllithium at $-78°C$ gave the deuterated anion,[4] which was

---

[4] The hydrogen/deuterium isotope ratio in this metalation, k_H/k_D had the surprising value of 24 ± 4, the highest ever reported for a metalation reaction. A study of the temperature dependence of k_H/k_D for this reaction provided evidence for quantum-mechanical tunneling in the proton transfer (*33, 35*).

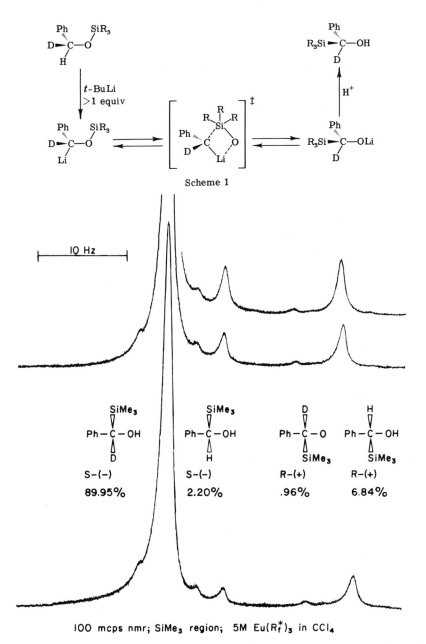

FIG. 8. Below: 100 MHz proton nuclear magnetic resonance spectrum of the trimethyl-silyl region of *S*-(+)-benzyl-α-*d*-oxytrimethylsilane (**13**) after anionic rearrangement and quenching with D$_2$O, in CCl$_4$ with chiral chemical shift reagent (*33*). Above, spectrum of the same sample with added racemic PhCH(SiMe$_3$)OH.

warmed to $-50°C$ for 2 minutes and then derivatized with glacial acetic acid at $-78°C$. Analysis of the reaction mixture was done by NMR spectroscopy in the presence of an optically active chemical shift reagent, allowing separate observation of all four possible products, the S and R isomers of both the deuterium and protium compounds (Fig. 8).

The major product was shown to have the S structure. The reaction proceeds with 99% inversion of configuration at carbon for **8** and 98% inversion for the trimethylgermyl analog **14**. Metalation of the triphenylsilyl compound required 20 hours at $-40°C$, under which conditions 93% of racemization of the anion took place, but the net stereochemistry was again inversion at carbon.

The proposed mechanism, involving a bridged transition state, is shown in Scheme 1. Inversion at carbon is also found for the inverse silylcarbinol to alkoxysilane rearrangement, and a closely similar mechanism has been suggested (4).

V
THE S→C REARRANGEMENT

The sulfur analog to the Wittig rearrangement was discovered using organosilicon and germanium migrating groups. When benzylthiotrimethylsilane or germane were metalated with lithium at the benzylic carbon, rapid rearrangement took place even at $-78°C$. PMR spectra of the anions show singlets in the $-MCH_3$ region and phenyl chemical shifts consistent with the rearranged anion. Derivatization with triethylchlorosilane, methyl iodide, methyl sulfate, or protons all gave exclusively the rearranged isomers (32):

$$PhCH_2-S-MMe_3 \xrightarrow{t-BuLi} Ph\bar{C}H-S-MMe_3 \xrightarrow{} \underset{\underset{MMe_3}{|}}{PhCH-\bar{S}} \xrightarrow{RX} \underset{\underset{MMe_3}{|}}{PhCH-S-R}$$

M = Si, Ge

$$R=CH_3, H, SiEt_3$$

$$(27)$$

The results suggest that the anionic rearrangement proceeds essentially to completion, so that very little carbanion is present at equilibrium. It is not necessary to have a phenyl group present on the carbon for it to be metalated. As seen in Eq. (28), a methylthiosilane anion also undergoes rearrangement with migration of silicon.

$$CH_3-SSiMe_2R \xrightarrow{RLi} \bar{C}H_2-SSiMe_2R \qquad RMe_2SiCH_2-\bar{S} \xrightarrow{H^+} RMe_2SiCH_2-SH$$
$$R = tert\text{-Bu}$$

$$(28)$$

Unlike silylcarbinols, silylmethylthiols do not undergo base-catalyzed rearrangement to the Si—S—C compounds. However, both silicon and germanium carbinthiols undergo reversion to sulfides under *free-radical conditions* (heating alone to 195°C or with azobisisobutynonitrile to 100°C) (*32*):

$$PhCH—SH \xrightarrow[\text{or } 195°C]{\text{AIBN, } 100°C} PhCH_2—S—MMe_3 \qquad (29)$$
$$\underset{MMe_3}{|}$$

$$M = Ge, Si$$

This unprecedented radical reaction provides a method for equilibrating the two neutral compounds. It is evident that the sulfide isomers are more stable than the α-metallobenzylmercaptans.

The opposite directions of the anionic and free-radical reactions are again explainable in terms of relative bond energies. Figure 8 shows a schematic reaction coordinate diagram for the silicon and germanium systems, which behave similarly both in the S→C (anionic) and C→S (free radical) reactions. The S⁻ anions are much more stable than the carbanions, so that anionic migration from C to S takes place effectively to completion. The unrearranged neutral isomers PhCH₂SMMe₃ are more stable than the neutral thiols, but base catalysis will not bring about reversion of the thiol to the sulfide because the anion PhC(MMe₃)—S⁻ lies too far below the anionic transition state. Free-radical catalysis however provides a lower energy pathway and allows "reverse" rearrangement to take place as in Eq. (29).

VI

THE N→C REARRANGEMENT

Recent investigations by Brook and Duff (*3a, 8a*) have shown that silicon will rearrange from carbon to nitrogen or vice versa, the direction of migration depending on the particular substituents in the substrate compound. In this the C—N compounds closely resemble the C—O systems described in Section V.

The first reaction to be described was the catalytic rearrangement of aminomethylsilanes to methylaminosilanes. This proceeds by deprotonation of the amino hydrogen and 1,2-anionic rearrangement followed by reprotonation (*3a*):

$$R_3SiCH \ R'—NHR'' \xrightarrow{n-BuLi} R_3SiC \ HR'—\bar{N}R'' \qquad (30)$$

$$\underset{SiMe_3}{\overset{|}{R'CH_2—NR''}} \xleftarrow{NH} \underset{SiR_3}{\overset{|}{\bar{C}HR'—NR''}}$$

The silyl substituents (R) were methyl or phenyl. The carbon-bonded group (R') was H or phenyl, and the groups (R'') attached to nitrogen were cyclohexyl, benzyl, or 2-propyl. Stabilization of the N^- anion by a phenyl substituent on nitrogen prevents the catalytic rearrangement (Eq. 31), but this effect may be overbalanced by placing *two* phenyl groups on carbon and so stabilizing the carbanion (Eq. 32).

$$Me_3SiCHR—NHPh \xrightarrow{n-BuLi} Me_3SiCHR—\overset{-}{N}Ph \qquad (31)$$

$$R = H, Ph \qquad\qquad (no\ rearrangement)$$

$$Me_3SiCPh_2—NHPh \xrightarrow{n-BuLi} Me_3SiCPh_2—\overset{-}{N}Ph \qquad (32)$$

$$Ph_2CH—NPh \overset{N-H}{\longleftarrow} Ph_2\overset{-}{C}—NPh$$
$$\qquad\qquad |\qquad\qquad\qquad\qquad\quad |$$
$$\qquad\qquad SiMe_3\qquad\qquad\qquad\quad SiMe_3$$

Studies of NMR spectra in hexane indicate that the position of equilibrium in the anionic rearrangement also depends on phenyl substitution at carbon. Results are given in Eqs. (33–36) (*8a*).

$$Me_3SiCH_2—\overset{-}{N}C_6H_{11} \underset{0}{\overset{0}{\rightleftharpoons}} H_2\overset{-}{C}—N(SiMe_3)C_6H_{11} \qquad (33)$$
$$\sim 100\% \qquad\qquad\qquad \sim 0\%$$

$$Me_3SiCHPh—\overset{-}{N}Me \underset{0}{\overset{0}{\rightleftharpoons}} Ph\overset{-}{C}H—N(SiMe_3)Me \qquad (34)$$
$$\sim 70\% \qquad\qquad\qquad \sim 30\%$$

$$Me_3SiCHPh—NC_6H_{11} \underset{0}{\overset{0}{\rightleftharpoons}} Ph\overset{-}{C}H—N(SiMe_3)C_6H_{11} \qquad (35)$$
$$\sim 50\% \qquad\qquad\qquad \sim 50\%$$

$$Me_3SiCPh_2—\overset{-}{N}CH_2Ph \underset{0}{\overset{0}{\rightleftharpoons}} Ph_2\overset{-}{C}—N(SiMe_3)CH_2Ph \qquad (36)$$
$$\sim 0\% \qquad\qquad\qquad \sim 100\%$$

The solvent composition also influences the position of equilibrium. Addition of large amounts of THF, or even small amounts of TMEDA, shift the equilibrium in the $Me_3SiCHPh$—NHMe system toward the carbanion. Presumably this effect results from solvation of the lithium cation, which may stabilize the Li^+—C relative to the Li^+—N^- ion pair.

The N→C anionic rearrangement is slow enough so that at least for some compounds it can be followed by NMR. A kinetic study of the rearrangement of Eq. (37) showed that the reaction is, as expected, a first-order equilibration. The energies of activation and

$$Me_3SiCHPh—\overset{-}{N}Me \underset{k_2}{\overset{k_1}{\rightleftharpoons}} Ph\overset{-}{C}H—N(SiMe_3)Me \qquad (37)$$

entropies ($\Delta S \ddagger$) are 17.0 kcal and -17.8 eu for the forward reaction and 17.4 kcal and -15.1 eu for the reverse (N→C) rearrangement.

Kinetic studies of the catalyzed C→N rearrangement have also been carried out. The results are more complex, because either the rearrangement or the protonation of the carbanion or both can be rate-determining (8a). Reversible 1,2-anionic rearrangements from N→C and C→N have also been reported recently in a purely organic system (15a). The migrating group is carboethoxy:

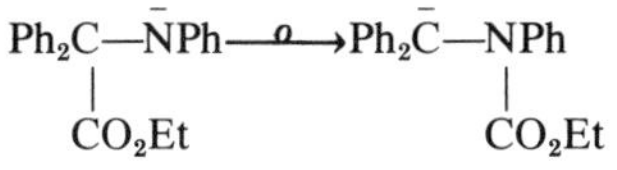

$$\begin{array}{ccc}
Ph_2C-\overset{-}{N}Ph & \overset{a}{\longrightarrow} & Ph_2\overset{-}{C}-NPh \\
\mid & & \mid \\
CO_2Et & & CO_2Et
\end{array} \tag{38}$$

VII

OTHER ANIONIC REARRANGEMENTS, KNOWN AND UNKNOWN

The preceding sections have summarized most of what is known concerning 1,2-anionic rearrangements of organometallic compounds. In this, mainly speculative, section we examine the prospects for finding other new anionic rearrangements, with the aid of organometal substituents as especially mobile migrating groups.

A. Homoatomic Rearrangements

Possible homoatomic 1,2-anionic rearrangements are classified in Figure 9. The N—N rearrangement is the topic of Section I, the C—C rearrangement with aromatic migrating groups is now well known and is treated in the chapter by Grovenstein in this volume. As expected, silicon will also undergo 1,2-anionic migration from carbon to carbon, and the experimental results show that Me_3Si migrates in preference to phenyl (10a)

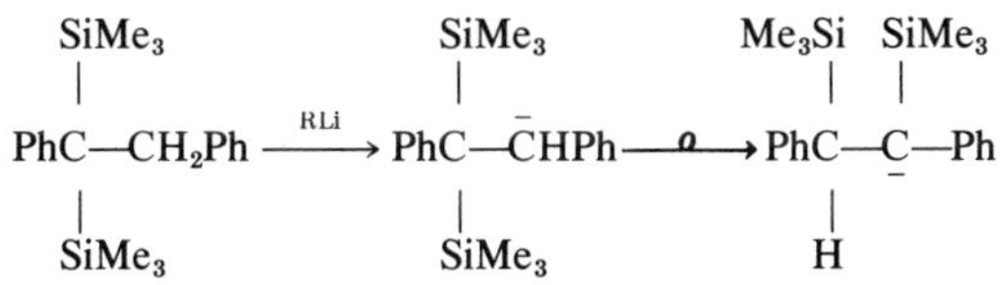

$$\begin{array}{ccccc}
SiMe_3 & & SiMe_3 & & Me_3Si\ \ SiMe_3 \\
\mid & & \mid & & \mid\ \ \ \ \mid \\
PhC-CH_2Ph & \overset{RLi}{\longrightarrow} & PhC-\overset{-}{C}HPh & \overset{a}{\longrightarrow} & PhC-C-Ph \\
\mid & & \mid & & \mid \\
SiMe_3 & & SiMe_3 & & H
\end{array} \tag{39}$$

Among the other unknown homoatomic 1,2-anionic rearrangements, several will be detectable only with great difficulty. For example, in silyl hydroperoxide anions, R_3Si—O—O$^-$, silicon may migrate from one

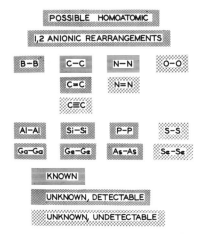

FIG. 9.   Possible homoatomic 1,2-anionic rearrangements. "Undetectable" means that observation of the rearrangement would be impossible by product analysis except through isotopic labeling.

oxygen to another but only region-specific labeling of one of the oxygens, or possibly ^{17}O NMR spectroscopy, would permit detection. Similar comments apply to the S—S, N=N and C≡C rearrangements. Of these, the most promising is the C≡C rearrangement; a study of the temperature-dependent ^{13}C NMR spectra of silylacetylide anions might be quite valuable.

Difficulties of another sort attend the study of anionic rearrangements between metalloid atoms (Si—Si, P—P, Ge—Ge, etc.). Here the chemistry is complicated by the easy cleavage of the metal–metal bonds under highly basic conditions. However, it is possible that such anionic rearrangements can be observed with careful choice of systems.

This leaves only the challenging ease of the ethylene rearrangement, C=C. No examples are known, although anionic migration in this sense, accompanied by loss of halide, is proposed as a mechanism for the Fritsch–Buttenberg–Weichell rearrangement (*12a, 13a*).

$$
\begin{array}{c}
\underset{Ar}{\overset{Ar}{\diagdown}}C=C\underset{Cl}{\overset{H}{\diagup}} \xrightarrow{\text{NaOR}} \underset{Ar}{\overset{Ar}{\diagdown}}C=C^{-}\underset{Cl}{\diagdown} \xrightarrow{\ -\ } Ar-C\equiv C-Ar + Cl^{-}
\end{array}
\tag{40}
$$

A possible 1,2-anionic rearrangement on a benzene ring has been reported, but details have not appeared (*10a*):

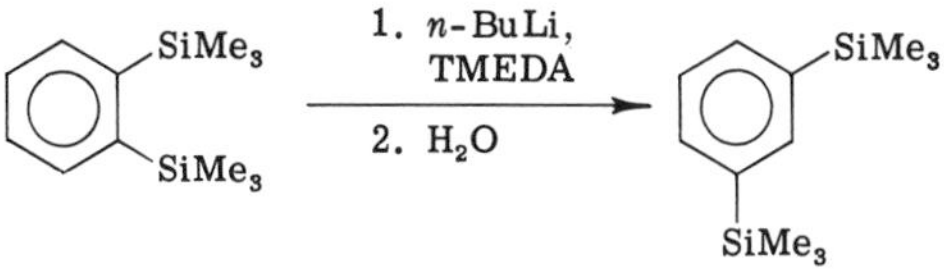

Study of anions of differentially substituted silyl olefins seems very much worthwhile.

B. *Heteroatomic Rearrangements*

Turning now to heteroatomic cases, the N—O, O—C, S—C, and N—C 1,2-anionic rearrangements have been described earlier in this chapter. Anionic rearrangements from doubly bonded nitrogen to carbon in imines, however, do not seem to have been studied. If carbanions can be generated from *N*-silylimines, N=C rearrangement seems likely:

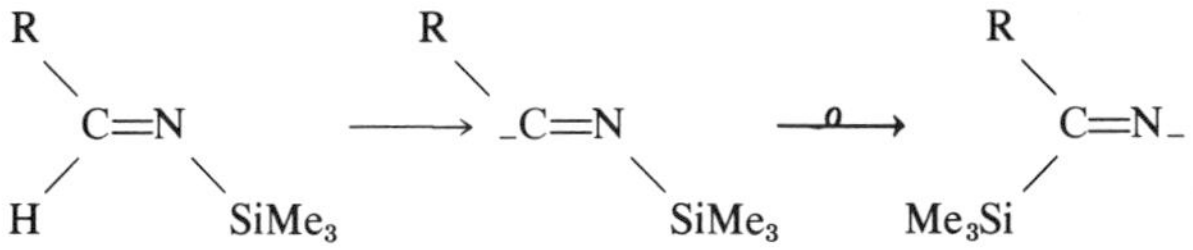

Anionic migration between sulfur and nitrogen is another possible and unknown reaction, but suitable S—N substrates may not be easy to prepare.

An interesting unknown reaction that might well repay study is the Si—C rearrangement. Metalation of hexamethyldisilane yields a carbanion in which trimethyl does not appear to rearrange from silicon to carbon, but rearrangement in the opposite direction does not appear to have been tested, and substitution of electron-withdrawing groups on the potentially migrating silicon might change the results.

$$Me_3SiSiMe_3 \xrightarrow{RLi} Me_3SiSiMe_2\bar{C}H_2 \longrightarrow \bar{S}iMe_2\!-\!CH_2SiMe_3$$

C. *Conclusion*

This review has treated only 1,2-anionic rearrangements. Anionic rearrangements to more distant sites are much less thoroughly studied, but they too can be profitably investigated using organosilicon migrating groups. A few examples of remote anionic rearrangements of organosilicon groups are covered in an earlier review (24) and in subsequent papers (10, 10b, 17).

Another direction that is almost totally unexplored is migration of metal and metalloid groups other than silicon and germanium. The compounds may be more difficult to prepare and handle, but we can expect that R_2B, R_2Al, R_3Sn, R_2As, and other metalloid substituents may similarly provide low-energy transition states for anionic migration.

As future developments take place in the study of anionic rearrangements, the exceptionally mobile organosilicon and other organometallic migrating groups seem likely to be of crucial and increasing importance.

REFERENCES

1. Bailey, R. E., and West, R., *J. Am. Chem. Soc.* **86**, 5369 (1964).
2. Boudjouk, P., and West, R., *J. Am. Chem. Soc.* **93**, 5901 (1971).
3. Boudjouk, P., and West, R., In "Intra-Science Chemistry Reports" (N. Kharasch, ed.), Vol. 7, pp. 65–82. Intra-Sci. Res. Found., Santa Monica, California, 1973.
3a. Brook, A. G., and Duff, J. M., *J. Am. Chem. Soc.* **96**, 4692 (1974).
4. Brook, A. G., LeGrow, G. E., and MacRae, D. M., *Can. J. Chem.* **45**, 239 (1967).
5. Brook, A. G., *Pure Appl. Chem.* **13**, 215 (1966).
6. Brook, A. G., *Acc. Chem. Res.* **7**, 77 (1974).
7. Brook, A. G., Warner, C. M., and Limburg, W. W., *Can. J. Chem.* **45**, 1231 (1967).
8. Dalrymple, D. L., Kruger, T. L., and White, W. N., In "The Chemistry of the Ether Linkage" (S. Patai, ed.), p. 617. Interscience, New York, 1967.
8a. Duff, J. M., and Brook, A. G., unpublished studies.
9. Ebsworth, E. A. V., In "Organometallic Chemistry of the Group IV Elements" (A. MacDiarmid, ed.), Vol. I, p. 46. Dekker, New York, 1968.
10. Eisch, J. J., Kovacs, C. A., and Rhee, S. G., *J. Organometal. Chem.* **65**, 289 (1974).
10a. Eisch, J. J., and Tsai, M.-R., *J. Am. Chem. Soc.* **95**, 4065 (1973).
10b. Gornowicz, G., and West, R., *J. Organometal. Chem.* **28**, 25 (1971).
11. Grovenstein, E., and Wentworth, G., *J. Am. Chem. Soc.* **89**, 1852 (1967).
12. Kimura, B. Y., and Brown, T. L., *J. Organometal. Chem.* **26**, 57 (1971).
12a. Köbrich, G., *Angew. Chem., Int. Ed. Engl.* **6**, 41 (1967).
13. Pine, S. H., *Org. React.* **18**, 401 (1970).
13a. Pritchard, J. G., and Bothner-By, A. A., *J. Phys. Chem.* **64**, 1271 (1960).
14. Schöllkopf, U., *Angew. Chem., Int. Ed. Engl.* **9**, 763 (1970).
15. Seppelt, K., and Sundermeyer, W., *Chem. Ber.* **103**, 3939 (1970).
15a. Smith, J. G., and Sheepy, J. M., *J. Chem. Soc., Chem. Commun.* 339 (1976).
16. Stevens, T. J., *Prog. Org. Chem.* **7**, 48 (1968).
17. Stewart, H. F., Koepsell, D. G., and West, R., *J. Am. Chem. Soc.* **92**, 1846 (1970).
18. Wannagat, U., and Liehr, W., *Z. Anorg. Allg. Chem.* **299**, 341 (1959).
19. Wannagat, U., and Smrekar, O., *Monatsh. Chem.* **100**, 750 (1969).
20. West, R., Ishikawa, M., and Bailey, R. E., *J. Am. Chem. Soc.* **88**, 4648 (1966).
21. West, R., and Ishikawa, M., *J. Am. Chem. Soc.* **89**, 4072 (1967).
22. West, R., and Bichlmeir, B., *J. Am. Chem. Soc.* **94**, 1649 (1972).
23. West, R., and Boudjouk, P., *J. Am. Chem. Soc.* **95**, 3987 (1973).
24. West, R., *Pure Appl. Chem.* **19**, 291 (1969).
25. West, R., Ishikawa, M., and Bailey, R. E., *J. Am. Chem. Soc.* **89**, 4068 (1967).

26. West, R., and Ishikawa, M., *J. Am. Chem. Soc.* **89,** 4981 (1967).
27. West, R., Lowe, R., Stewart, M. F., and Wright, A, *J. Am. Chem. Soc.* **93,** 282 (1971).
28. West, R., and Stewart, H. F., *J. Am. Chem. Soc.* **92,** 853 (1970).
29. West, R., and Kuo, C., unpublished results.
30. West, R., and Boudjouk, P., *J. Am. Chem. Soc.* **95,** 3983 (1973).
31. Woodward, R. B., and Hoffmann, R., "The Conservation of Orbital Symmetry." Academic Press, New York, 1970.
32. Wright, A., and West, R., *J. Am. Chem. Soc.* **96,** 3222 (1974).
33. Wright, A., and West, R., *J. Am. Chem. Soc.* **96,** 3227 (1974).
34. Wright, A., and West, R., *J. Am. Chem. Soc.* **96,** 3214 (1974).
35. Wright, A., Ph.D. Thesis, University of Wisconsin-Madison (1973).
36. Zimmerman, H. E., *In* "Molecular Rearrangements" (P. deMayo, ed.), Vol. I, p. 278. Interscience, New York, 1965.

Dyotropic Rearrangements and Related σ–σ Exchange Processes

M. T. REETZ

Fachbereich Chemie der
Universität, Lahnberge
Marburg, West Germany

I

INTRODUCTION

Dyotropic rearrangements were defined as a new class of pericyclic valence isomerizations in which two σ-bonds (i.e., groups) migrate intramolecularly (*1*, *2*). Reactions in which the two migrating groups interchange their positions were designated as type I. Those of type II involve migration to new bonding sites in a manner that avoids positional exchange. A theoretical treatment of the stereochemistry of these processes based on simple orbital symmetry arguments, qualitative perturbation theory, and self-consistent field (SCF) calculations has been described (*3*). In the following a brief summary is presented, intending to show that the search for these intriguing transformations is best conducted in the field of organometallic chemistry.

Type I rearrangements involve two migrating groups and two stationary substrate atoms. Equation (1) depicts a reaction in which the stationary atoms a and b are directly bonded to one another. The assumed anti conformation forces the groups to migrate on opposite sides of the a–b framework, each suprafacially, necessarily leading to inversion of configuration at both a and b. Migration with retention of configuration at both R groups is a thermally forbidden, photochemically allowed $[\sigma_a^2 + \sigma_a^2]$ process.[1] Allowing one group to migrate with inversion implies a thermally allowed $[\sigma_s^2 + \sigma_a^2]$ reaction, but leads to a sterically unfavorable transition state.

$$\tag{1}$$

Energetically favorable processes can nevertheless be constructed (*3*). Employing ethane as a model, the forbiddenness of a hypothetical double hydrogen migration is easily traced to the highest occupied molecular orbital of the transition state. It has the antibonding form (**1**) with a nodal plane through the migrating hydrogen atoms. This suggests the utilization of migrating groups that have additional empty orbitals of proper symmetry. Candidates include groups with low-lying *d*- or *f*-orbitals, such as silyl functions (**2**) or transition metals. A different form of stabilization may be envisioned by introducing substituents at a or b

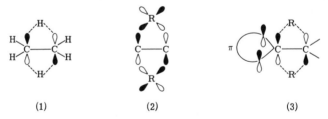

(1)   (2)   (3)

---

[1] The sterically unfavorable syn migration is a thermally forbidden $[\sigma_s^2 + \sigma_s^2]$ process. The latter designation was erroneously applied to the anti migration (*3*); this does not affect the general conclusions, however.

with symmetric lowest unoccupied orbitals (3), which also diminish the electron density in the antibonding region of the transition state (1, 2, 4).

An orbital symmetry allowed valence isomerization also results if one of the migrating groups contains additional π-systems, e.g., Eq. (2). These models are to be considered as working hypotheses, the actual bonding situation being more complicated if a or b are heteroatoms with lone electron pairs (3).

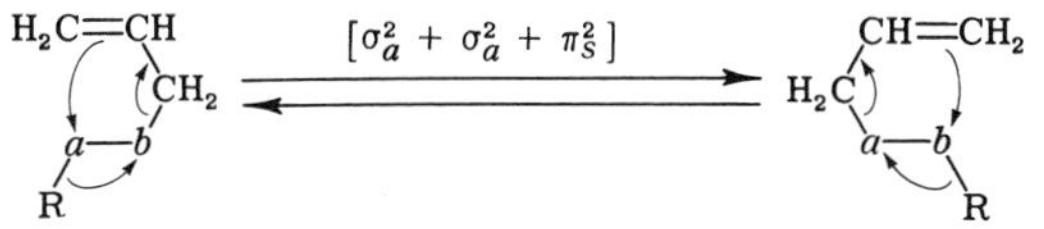

$$H_2C\!\!=\!\!CH \qquad [\sigma_a^2 + \sigma_a^2 + \pi_s^2] \qquad CH\!\!=\!\!CH_2 \qquad (2)$$

Since the simultaneous cleavage of two bonds—although not rare—is generally considered to be a high-energy process, dyotropic reactions are not expected to occur as generally as sigmatropic shifts (5). Also, multistep processes must be considered in each case. Reactions are most likely to occur if the R groups are known to have a high migratory aptitude in sigmatropic rearrangements. Thus, since silyl and germyl groups undergo extremely rapid metallotropic shifts (6–8), it is reasonable to begin investigations using these functions. The purpose of this chapter is not only to describe the present author's own systematic studies involving organosilicon compounds, but also to motivate others in investigating dyotropic processes in general.

II

THERMAL REARRANGEMENT OF 1,2-BIS(SILYL) DERIVATIVES

In this section the thermal behavior of bis(silyl) compounds with respect to reversible degenerate valence isomerization 4 $\rightleftarrows$ 5 is described.

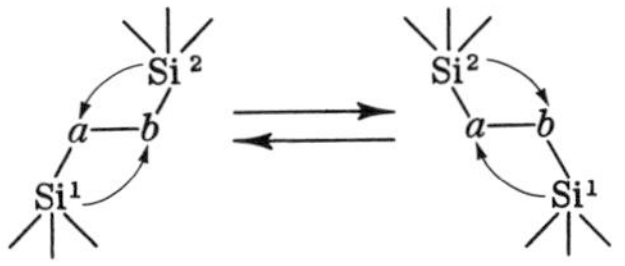

(4)   (5)

a: a = carbon, b = oxygen
b: a = carbon, b = sulfur
c: $a = b$ = carbon

All three classes of compounds fulfill the above-mentioned requirements concerning empty d-orbitals in both migrating groups. However, whereas the ethers and thioethers contain atoms with lone electron pairs, which may interact with silicon d-orbitals, no such process is possible in the 1,2-bis(silyl)ethane derivatives.

A. Thermal Rearrangement of Silyl Silylmethyl Ethers

1. Structural Effects on Migratory Aptitude

The observation of a degenerate dyotropic exchange in silyl silylmethyl ethers [Eq. (3)] requires labeled silyl groups. Thus, syntheses were designed with Si^1 = trimethylsilyl and Si^2 = D_9-trimethylsilyl as well as Si^1 = trimethylsilyl and Si^2 = dimethylphenylsilyl (9, 10).

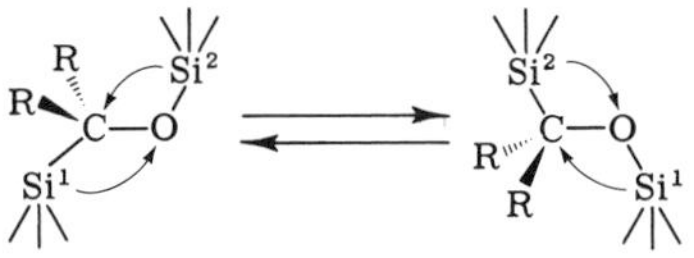

$$\begin{array}{c} \text{(3)} \end{array}$$

In order to avoid undesired side reactions, such as silanol elimination, compounds were chosen with R = aryl. Such substituents were also expected to facilitate the rearrangement process, partly owing to electron withdrawing effects as discussed in Section I, and partly owing to the increased reactivity of benzyl bonds in general (11). The synthesis of **6, 8, 10, 12,** and **14** required several different methods and has been described in detail (10).

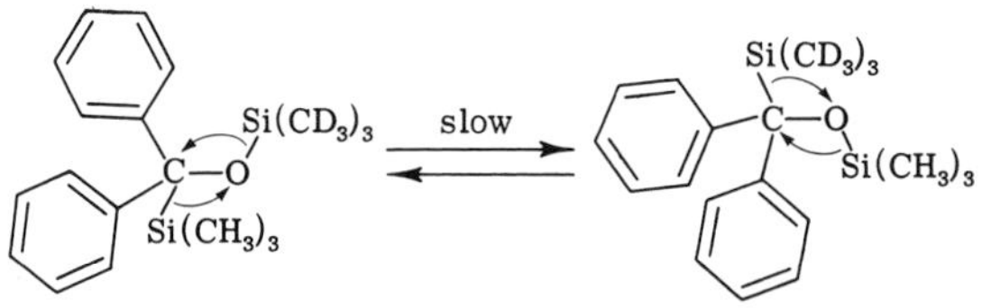

(6) slow (7)

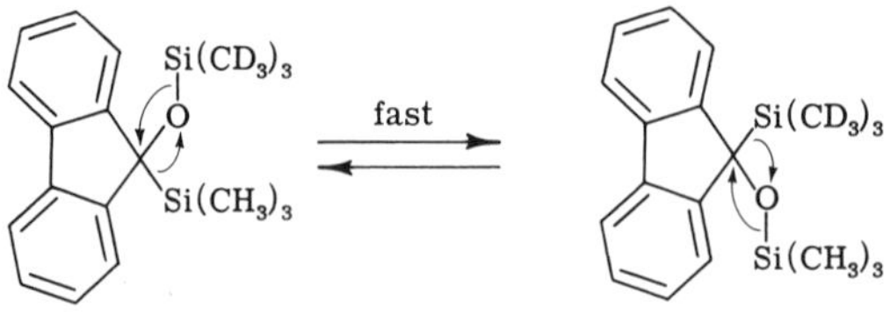

(8) fast (9)

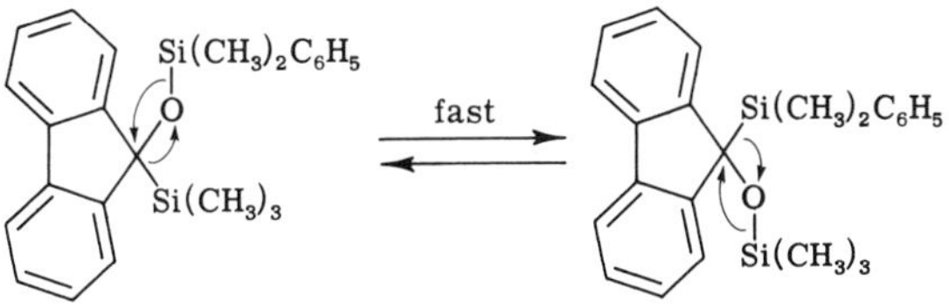

(10)   (11)

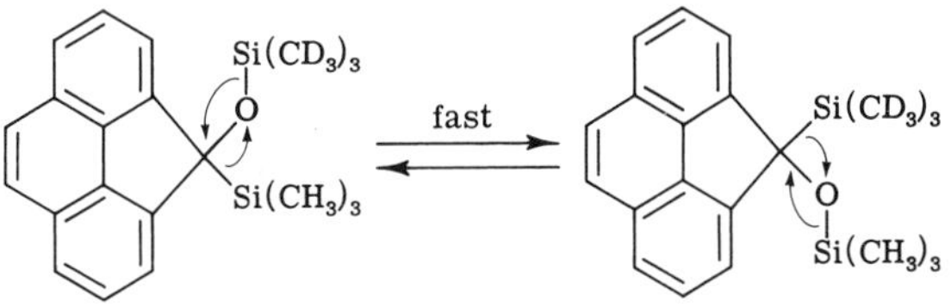

(12)   (13)

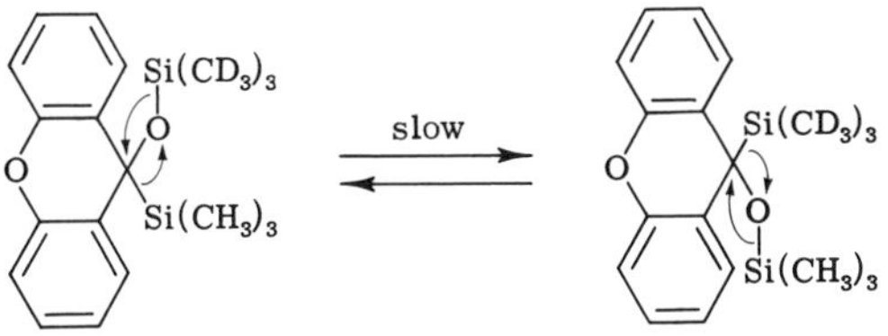

(14)   (15)

The thermolysis of **8, 10,** and **12** at 140°–175°C in inert solvents (e.g., benzene) does indeed afford the rearrangement products **9, 11,** and **13,** respectively, the reactions leading to equilibria that are easily detected by H—NMR spectroscopy (*10*). Equilibrium is reached rapidly, e.g., within 2 hours at 175°C. Heating an independently synthesized sample of **11** leads to the same equilibrium as **10.** All reactions are clean, no traces of side products being observed. The fluxional behavior of **8** and **12** is strictly thermoneutral, i.e., the equilibrium constants have the value $K = 1.0$ within experimental error. The valence isomerization **10** $\rightleftarrows$ **11** is not quite thermoneutral (Table I).

In sharp contrast to **8, 10,** and **12,** the remaining compounds, **6** and **14,** rearrange very sluggishly. At 140°–175°C no reaction whatsoever is observed, even after several days. Forced conditions (more than 10 hours at 230°–235°C) induce slow rearrangements leading to at most 10–15% of **7** and **15,** respectively. Under such drastic conditions side products of unknown structure are formed, a process that prevents equilibrium from being established. The dramatic and unexpected difference in migratory aptitude within the above series is an important factor in deriving a mechanism (Section II,A,7).

TABLE I

Thermodynamic Data of $10 \underset{k_{-1}}{\overset{k_1}{\rightleftharpoons}} 11$ in Benzene

Temperature (°C, ±0.1)	$K = k_1/k_{-1}$
143.8	0.850
153.7	0.878
164.1	0.890
173.8	0.895

2. Kinetics

Activation parameters were obtained by monitoring H—NMR spectroscopically the equilibration of benzene solutions of **8** and **10** in the range 140°–175°C (Table II). The negative activation entropy indicates some loss of degrees of freedom in reaching the transition state (12). No accurate activation parameters could be derived for **6** or **14**. Instead, approximate relative rates are listed in Table III. The results point to rate differences of several orders of magnitude.

TABLE II
Activation Parameters

Rearrangement	E_a (Arrhenius) (kcal/mole)	$\Delta H\ddagger$ (kcal/mol)	$\Delta S\ddagger$ (e.u.)
8 → 9	30.8 ± 0.6	30.0 ± 0.6	−8.8 ± 0.8
10 → 11	31.1 ± 0.6	30.3 ± 0.6	−8.6 ± 0.8

TABLE III
Relative Rates

Rearrangement	k_{rel} (155°C)	k_{rel} (230°C)
6 → 7	—	$\approx 1 \times 10^{-4}$
8 → 9	1.5	1.5
10 → 11	1.0	1.0
12 → 13	1.2	—
14 → 15	—	$\approx 6 \times 10^{-5}$

3. *Crossover Experiments*

Because dyotropic rearrangements involve migration of two groups, at least two crossover experiments are required to prove intramolecularity. A series of such experiments utilizing **8, 10,** and **12** as well as appropriate derivates showed that the rearrangements are 100% intramolecular, even after prolonged reaction times (*12*). This means that the two silyl groups not only "remain together" during the reaction, they also do not depart as a unit from the substrate.

4. *Solvent Effects*

The possible influence of solvent polarity was studied by measuring relative rates of rearrangement of **10** in different solvents (*12*). The k_{rel} values in Table IV show that the rate is not influenced to any appreciable extent. However, this cannot be viewed as proof of a nonpolar transition state, since $\Delta S\ddagger$ may become larger (more negative) and $\Delta H\ddagger$ smaller as the solvent polarity increases. Such a cancellation could lead to similar rates, as has been shown in other systems (*13*). Therefore, the activation parameters for **10** → **11** were measured in acetonitrile. The results (Table IV) lend support to the initial conclusion regarding the absence of an ionic or zwitterionic transition state (*9*).

TABLE IV
SOLVENT EFFECTS (X) → (XI)

Solvent	E_t (*14*)	k_{rel}	E_a (kcal/mole)	$\Delta H\ddagger$ (kcal/mole)	$\Delta S\ddagger$ (e.u.)
Benzene	35	1.0	31.1 ± 0.6	30.3 ± 0.6	−8.6 ± 0.8
o-Dichlorobenzene	—	1.8	—	—	—
Acetonitrile	46	2.8	30.4 ± 0.6	29.7 ± 0.6	−8.1 ± 0.8

5. *Hammett Study*

In order to gain information concerning the electronics at the stationary carbon atom in the transition state, a Hammett study utilizing fluorenyl substrates **8** substituted in the para position was carried out (*12*). Because no derivatives with electron-withdrawing substituents (e.g., nitro) could be synthesized (*15*), the investigation was limited to

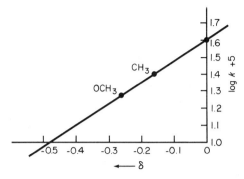

FIG. 1. Hammett plot of the rates of 3-substituted fluorenyl ethers (**8**).

hydrogen, methyl, and methoxy. The results, shown in Fig. 1, give rise to a rho (ρ) value of $+1.2$, which proves that some negative charge accumulates at the stationary carbon atom in the transition state. This is in line with theory (Section I). However, the effect is rather small, unsubstituted **8** rearranging only about twice as fast as the methoxy derivative. The intermediacy of ionic species is unlikely.

6. *Stereochemistry at the Migrating Silyl Groups*

The expectation that the rearrangement of a silyl group from carbon to oxygen proceeds with retention of configuration was fully substantiated

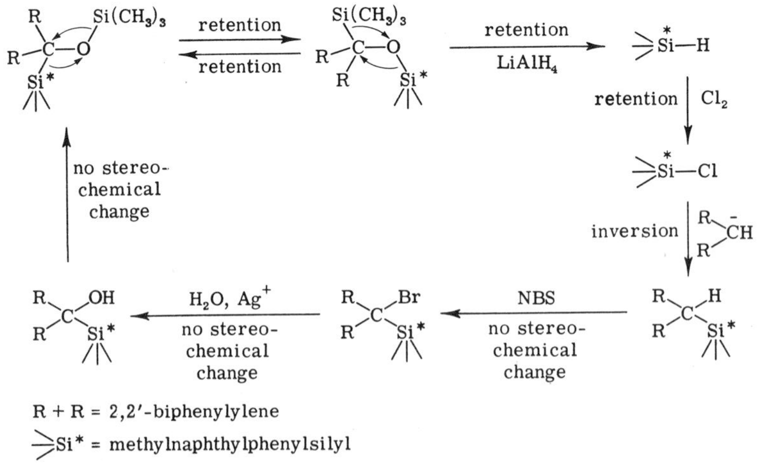

R + R = 2,2'-biphenylylene

$\geqslant$Si* = methylnaphthylphenylsilyl

SCHEME 1.   Stereochemistry of dyotropic silyl group migration.

using appropriate optically active compounds (*12*). The complete Walden cycle, which bears some resemblance to the stereochemical studies of Brook concerning base-catalyzed 1,2 shifts from carbon to oxygen in α-silyl alcohols (*16*), is depicted in Scheme 1. The stereochemistry at the chiral silyl group in the dyotropic rearrangement is retained for at least 18 half-lives. This proves retention of configuration not only for the migration from carbon to oxygen, but also for the reverse process, i.e., the shift from oxygen to carbon.

7. Stereochemistry at the Stationary Carbon Atom

Theory predicts that every rearrangement event is accompanied by inversion of configuration at the stationary carbon atom (*3*). Consequently, the rate of racemization (k_{rac}) must equal the rate of rearrangement (k_{rearr}) [Eqs. (4) and (5)]. Unfortunately, all attempts to synthesize appropriate compounds for such intriguing studies have failed so far (*17*).

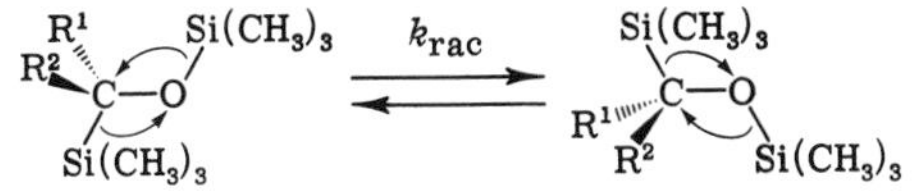

$$(4)$$

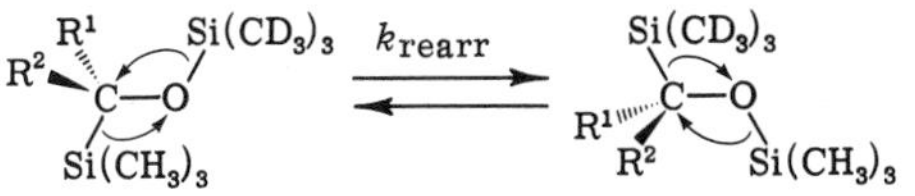

$$(5)$$

8. Mechanism

In discussing the mechanism, the following facts must be considered: (a) The rearrangement is 100% intramolecular; (b) the silyl groups migrate with retention of configuration; (c) the rate is nearly independent of solvent polarity; (d) substituent effects are small ($\rho = +1.2$); the activation parameters amount to $E_a \approx 32$ kcal/mole and $\Delta S\ddagger \approx -9$ e.u. Besides a dyotropic shift as previously defined, several alternative multistep mechanisms are possible, as depicted in Scheme 2. Paths A and B involve the intermediacy of silyl radicals. Such processes are not in line with the crossover and stereochemical experiments, particularly in view of the reversibility of the rearrangement and the prolonged reaction time during which intramolecularity and stereochemical integrity is maintained. Even in case of an efficient cage effect—which is unlikely at such high temperatures—the radicals should have a definite chance to escape.

Path C begins with rate-determining α-elimination followed by readdi-

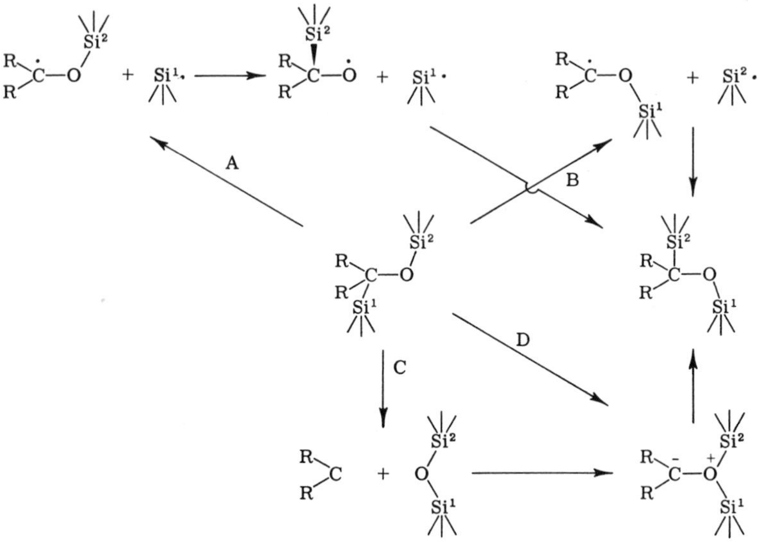

SCHEME 2. Alternative multistep mechanisms.

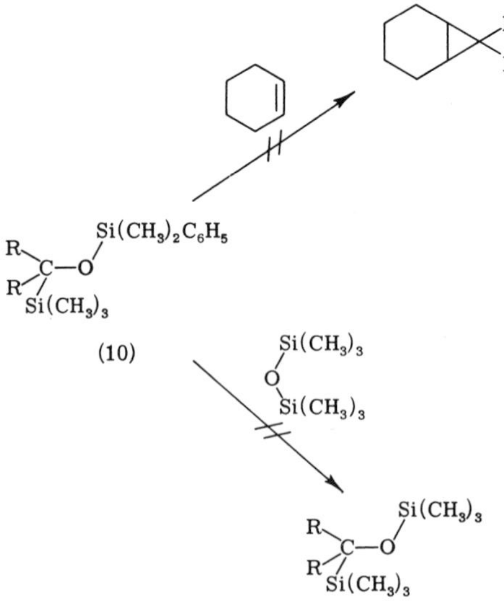

tion to form an oxygen ylid, which rearranges either to a product or a starting compound. Related eliminations have been reported (*18*). However, they involve the generation of stabilized nucleophilic carbenes that could be trapped. In the present case, all attempts to trap possible carbenes failed. For example, heating **10** in cyclohexene for prolonged reaction times does not result in carbene addition products. Similarly, the thermolysis of **10** in hexamethyldisiloxane fails to induce incorporation of two trimethylsilyl groups.

Arguments against path D are less straightforward. It can be viewed as a "two-step concerted" process (*19*). Rate-determining formation of an oxygen ylid is followed by rapid rearrangement, orbital symmetry being conserved throughout. Although oxygen ylids have not been isolated or spectroscopically observed to date, their existence as short lived, high energy species is generally accepted (*20, 21*). However, no evidence for ylid formation via 1,2-shifts is presently known. It may be argued that the small solvent effects and the small ρ value speak against such a mechanism. Path D is the limiting case of a dyotropic mechanism. Ylid formation is characterized primarily by the motion of *one* silyl group, whereby the second group has begun its journey only to the extent that the silicon–oxygen bond is likely to be loosened (stretched) in the transition state. It is generally difficult to distinguish between one-step and two-step processes. In this connection the above-mentioned stereochemical study at the stationary carbon atom would be most useful; in case of ylid formation $k_{\text{rac}} > k_{\text{rearr}}$.

At present all experimental results are easily incorporated within a dyotropic mechanism. Do both silyl groups migrate equally fast in such a case? Or does one group have a "head start" leading to an asymmetric transition state? Present theories predict asymmetric transition states for *all* pericyclic reactions (*22, 3*). Furthermore, a number of orbital symmetry-controlled reactions proceed via local minima along the reaction coordinate (*23, 24*). In the silyl–silyl exchange, it is appealing to invoke a short-lived intermediate that represents such a minimum on the hyperfine potential energy surface (*12*). Rate-determining formation of a three-membered ring (**16**) based on d–p bonding is plausible and consistent with all experimental observations including the negative activation entropy. Rapid motion of the second silyl group toward the backside of the carbon atom stabilizes the developing p-orbital and avoids formation of high-energy zwitterions (in contrast to path D). d–p interaction between Si^2 and the stationary carbon atom in **16** cannot be rigorously excluded, but is probably small (*12*).

(16)   (17)

This picture is clearly more complicated than the qualitative models

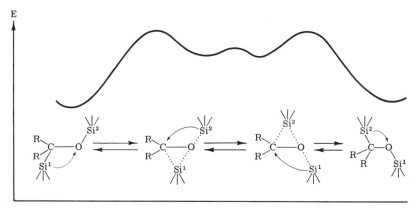

FIG. 2.   Possible reaction profile of the dyotropic silyl–silyl exchange.

mentioned in Section I. Thus, besides *d*-orbital participation as originally discussed, interaction according to **17** is likely to be operating. One does not exclude the other (*12*). Interestingly, the assumption of **16** means that the complete reaction coordinate must involve *two* local minima, as shown in Fig. 2. Because the valence isomerization is degenerate, the principle of microscopic reversibility applies.

The above dyotropic mechanism (Fig. 2) can be invoked to explain the pronounced differences in rate within the series of compounds studied. Fluorenyl and structurally similar substrates are not only electronically well suited for dyotropic shifts owing to their electron-withdrawing ability (*10*), they are also flat, reducing steric hindrance at the backside of the carbon atom to a minimum **18**. In contrast, two phenyl groups sterically prevent smooth backside attack (**19**). The xanthenyl system is sterically less demanding, but electronically unfavorable owing to the

(18)   (19)

electron donating property of the oxygen atom in the ring. Although coordination **16** is likely in these systems, the motion of the second silyl group is no longer rapid, the potential energy surface no longer resembling Fig. 2.

These comments underline the stereoelectronic requirements for dyotropic valence isomerizations of silyl silylmethyl ethers. A final, and rather subtle, argument supporting the above mechanism evolves from observations concerning related rearrangements of allyl silylmethyl ethers and will be discussed in Section III,J.

B. *Thermal Rearrangement of Silyl Silylmethyl Thioethers*

In order to gain information concerning the role of the stationary substrate atoms, the thioether **20** was synthesized (*17*).

Surprisingly, **20** is stable under reaction conditions that cause complete equilibration of the oxygen analog **8**. Under more drastic conditions (>10 hours at 230°C) about 10% of the rearrangement product (**21**) is observed (*25*). Equilibrium is not established owing to a competing irreversible desulfurization leading to **22**. Kinetically it was found that **20** rearranges several orders of magnitude more slowly than the oxygen analog (**8**), the relative rates being $k_{sulfur}/k_{oxygen} \cong 10^{-4}$.

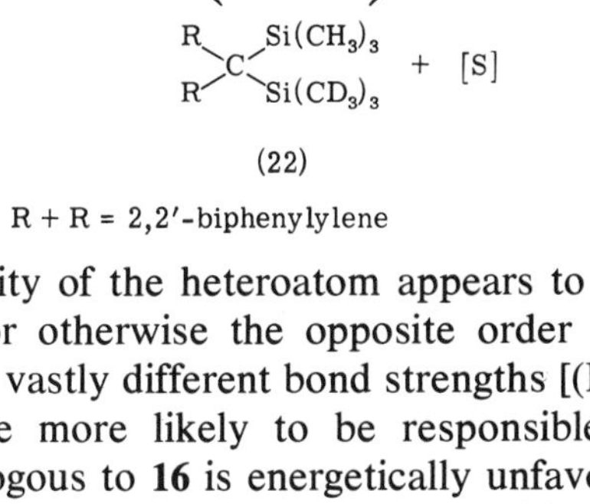

(20) (21)

(22)

R + R = 2,2'-biphenylylene

The nucleophilicity of the heteroatom appears to play no role in the rearrangements, for otherwise the opposite order of reactivity would have to apply. The vastly different bond strengths [$(E_{Si-O} - E_{Si-S}) \approx 30$ kcal/mole] (*26*) are more likely to be responsible. Rate-determining complexation analogous to **16** is energetically unfavorable for the sulfur compound owing to the lower strength of the silicon–sulfur bond. The longer carbon–sulfur bond (*27*) may also contribute to the lower rate.

C. *Thermolysis of 1,2-Bis(silyl) Ethane Derivatives*

Substituting oxygen by carbon in the above silyl ethers results in 1,2-bis(silyl)alkanes lacking heteroatoms. Thus, coordination between silicon and lone electron pairs is no longer possible. However, dimetalloal-kanes, e.g., 1,2-bis(silyl)- and 1,2-bis(stannyl)alkanes, show strong σ–σ interaction between the two metal–carbon bonds in the *ground* state, as demonstrated by Traylor's interesting photoelectron studies (*28*). Intui-tively, such σ–σ conjugation may enhance vibrational coupling neces-sary for concerted dyotropic shifts (*3*). Thus, **23** was synthesized (*29*).

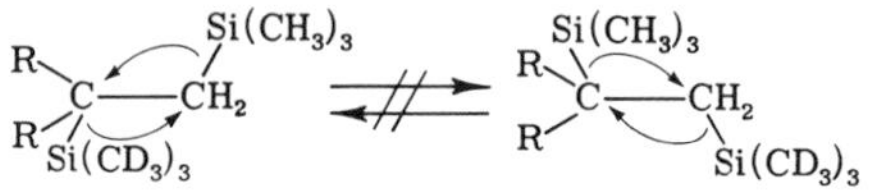

(23)   (24)

R + R = 2,2'-biphenylylene

Heating **23** at 180°C for 24 hours fails to induce rearrangement to **24**. Even at 240°C no rearrangement or decomposition occurs (*29*). These results show that *at least one* of the stationary atoms must have lone electron pairs.

D. *The Question of 1,4 Double Migration*

Since 1,4 *sigmatropic* shifts [e.g., Eq. (6)] of silyl groups have been known for some time (*30*), it was tempting to test the feasibility of dyotropic 1,4 processes.

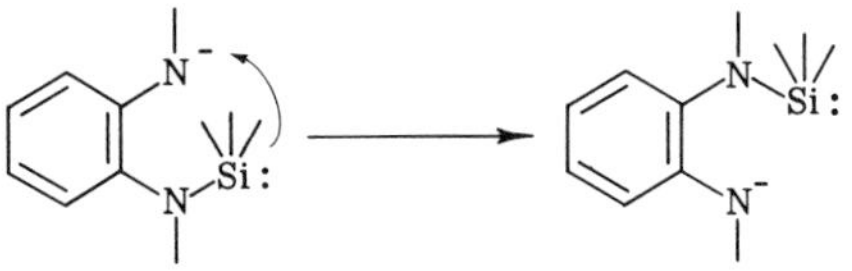

(6)

Separate thermolysis of **25** and **26** do *not* lead to interconversion, even at 230°C (>24 hours), the compounds being stable (*29*). In view of the facile 1,4 sigmatropic shift [Eq. (6)], this may appear surprising. How-ever, closer inspection of the geometry of an assumed complex (**27**)

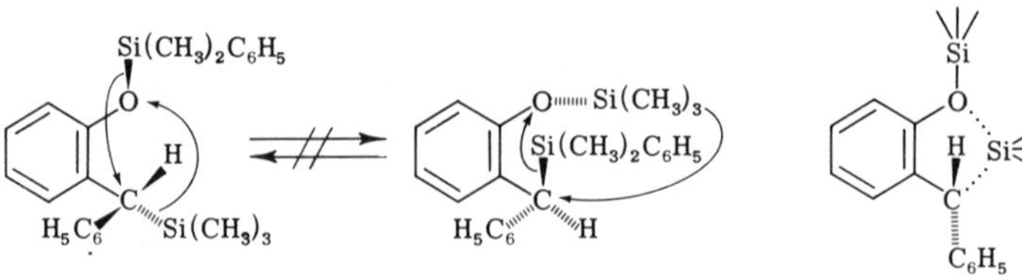

(25)   (26)   (27)

between the attacking silyl group and the oxygen atom reveals that the second silyl group points away from the stationary carbon atom. Unlike the close proximity and correct geometry of the second silyl group in the rearrangements described in Section II, the necessary stereoelectronic requirements appear not to be met in the present compounds.

III

THERMAL REARRANGEMENT OF ALLYL SILYLMETHYL ETHERS

The working hypothesis that incorporation of the π-bond of allyl groups in the rearrangement process leads to a symmetry-allowed [$\sigma_a^2 + \sigma_a^2 + \sigma_s^2$] reaction was delineated in Section I. Allyl ethers of the type **28** were chosen, in the expectation that the exchange of allyl and silyl should afford the thermodynamically more stable silyl ethers **29** (31, 32).

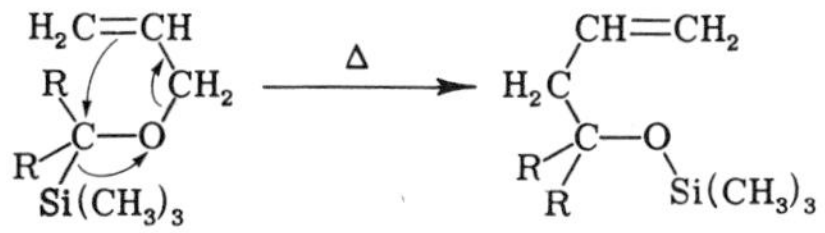

(28)   (29)

a: R = R = phenyl
b: R + R = 2,2'-biphenylylene
c: R = phenyl; R = 1-naphthyl

As before, R = aryl, providing activation and preventing side reactions such as retro-ene and silanol eliminations.

A. Structural Effects on Migratory Aptitude

Compounds **28a–c** were found to rearrange smoothly at 150°–190° to **29a–c**, respectively. The reactions are extremely clean, the yields being greater than 96% in all cases (31, 32). Kinetic data for the rearrangements are summarized in Table V. In striking contrast to the silyl–silyl

TABLE V
RATE DATA FOR ALLYL–SILYL EXCHANGE

Compound	k_{rel}	E_a (kcal/mole)	$\Delta H\ddagger$ (kcal/mole)	$\Delta S\ddagger$ (e.u.)
28a	0.5	32.8 ± 0.6	31.9 ± 0.6	−7.7 ± 0.8
28b	1.0	32.3 ± 0.6	31.4 ± 0.6	−8.9 ± 0.8
28c	1.8	32.7 ± 0.8	31.8 ± 0.8	−7.0 ± 1.0

exchange, no appreciable structural effects on the rates are observed within the series. Furthermore, the activation parameters are quite similar to those for the rapidly reacting silyl silylmethyl ethers (Section II,A,2). These unexpected observations are valuable in deriving a mechanism.

B. *Crossover Experiments*

Utilizing appropriate compounds in the fluorenyl series it was shown that the silyl groups migrate 100% intramolecularly (*33*). On the other hand, allyl migration was *not* found to be strictly intramolecular, *inter*molecularity amounting to about 28%.[2] This means that at least the latter portion does not adhere to a strict dyotropic mechanism.

C. *Stereochemistry of Allyl Group Migration*

In order to test whether the allyl groups migrate with allyl inversion, the α,α-dideuterio compounds **30a–c** were synthesized (*33*). Thermolysis of **30a** proceeds with a high degree of stereospecificity, only 20% of the "wrong" isomer (**32a**) being formed. In contrast, **30b** and **30c** afford statistical amounts of α,α and γ,γ products.

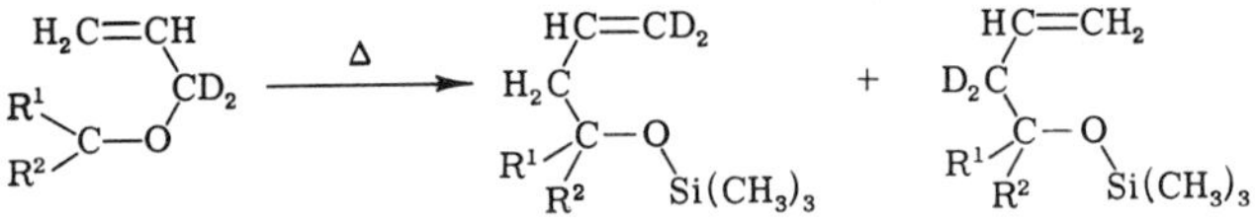

	(31)	(32)
a: $R^1 + R^2$ = 2,2'-biphenylylene	80%	20%
b: $R^1 = R^2$ = phenyl	50%	50%
c: R^1 = phenyl; R^2 = 1-naphthyl	50%	50%

The results show that the rearrangements do not proceed via a single pathway. They suggest that in the case of **30a** 60% migrates concertedly with allyl inversion, the remaining 40% reacting by a multistep mechanism, such as allyl radical formation.

D. *Radical Trapping Experiments*

The thermolysis of **28a** in the presence of *p*-benzoquinone as a radical trapping agent yields **29a** in 60% yield (normally 98%). Apparently allyl

[2] This is an approximate value, the experimental error being large because the intensities of the parent peaks in the mass spectra are very weak (*33*).

radicals are trapped, thereby preventing recombination. The same effect is observed in the thermolysis of **30a**. Significantly, *the rearrangement product (60% yield) contains >97% of the γ,γ isomer* **31a** *(33)*. This substantiates the view that about 60% of the rearrangement adheres to a concerted dyotropic shift with clean allyl inversion. The trapping agent has no effect on the rate of disappearance of **28a**.

E. *Secondary Deuterium Kinetic Isotope Effects*

Insight into the bonding at the C^α and C^γ atoms of the allyl group in the transition state was gained by secondary deuterium kinetic isotope effects *(33)*. If the hybridization changes from sp^3 to sp^2, a normal isotope effect ($k_H/k_D > 1$) is to be expected. An $sp^2 \rightarrow sp^3$ change results in an inverse effect ($k_H/k_D < 1$). For example, in the Cope rearrangement values of 1.19 and 0.94 per two deuterium atoms were found *(34)*. In the allyl–silyl exchange, values of 1.05 and 1.00 were observed:

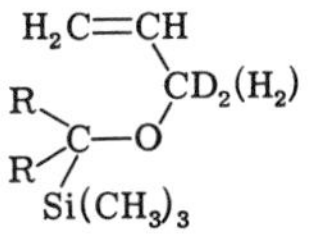

$$H_2C\!\!=\!\!CH$$
$$R\underset{R}{\overset{}{\diagup}}\!\!\overset{CD_2(H_2)}{\underset{}{\diagdown}}$$
$$R\diagup\!\!C\!\!-\!\!O$$
$$Si(CH_3)_3$$

$$(H_2)D_2C\!\!=\!\!CH$$
$$R\underset{R}{\overset{}{\diagup}}\!\!\overset{CH_2}{\underset{}{\diagdown}}$$
$$R\diagup\!\!C\!\!-\!\!O$$
$$Si(CH_3)_3$$

$k_H/k_D = 1.05 \pm 0.03$ 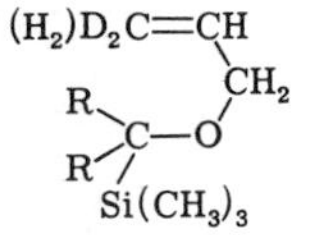 $k_H/k_D = 1.00 \pm 0.03$

R + R = 2,2'-biphenylylene

The results show that the oxygen–carbon bond is stretched to a small extent in the transition state, meaning that silyl migration is far more advanced than allyl migration. Furthermore, the transition state does not involve bonding between the C^γ atom of the allyl group and the C—9 atom of the fluorenyl substrate.

F. *Solvent Effects*

The influence of solvent polarity on rearrangement rate is rather small (Table VI), thereby excluding ions or zwitterions as intermediates *(31, 33)*.

TABLE VI

SOLVENT EFFECTS IN **28b** → **29b**

Solvent	Decalin	Benzene	o-Dichlorobenzene	Propylenecarbonate
k_{rel}	1.0	1.2	2.8	5.1

G. Stereochemistry of Silyl Group Migration

In determining the stereochemistry of silyl migration, a Walden cycle involving **33** and **34** was completed (*34*), similar to previous investigations (Section II,A,5). Here, as before, the silyl groups migrate with >96% retention of configuration (*33, 35*). Thus, all presently known cases of sigmatropic (*7*), dyotropic, and fragmentation processes (*18*) involving migration of silicon to oxygen proceed with retention of configuration.

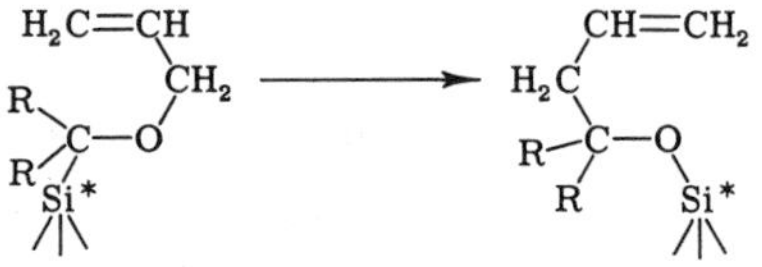

(33)   (34)

Si* = methylphenyl-1-naphthylsilyl
R + R = 2,2'-biphenylylene

H. Control Experiments

Several control experiments with radical initiators and inhibitors clearly exclude a radical chain mechanism (*33*). Also, various elimination–addition mechanisms are unlikely on the basis of appropriate control experiments (*33*). For example, Eq. (7) does not apply, since fluorenone fails to react with allyltrimethylsilane.

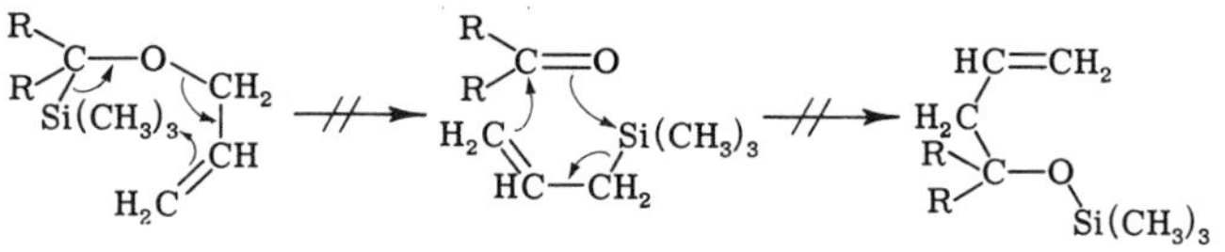

(7)

J. Discussion of the Mechanism

The allyl-silyl exchange reactions are clearly more complicated than the 100% intramolecular rearrangements of silyl silylmethyl ethers. Nevertheless, the present author believes that they are closely related. It is rather striking that both processes (in the fluorenyl series) have nearly identical activation parameters. Solvent effects are small in both cases. Thus, it appears reasonable to assume similar transition states, namely rate-determining coordination between silicon and oxygen. For the fluorenyl substrate a mechanistic dichotomy evolves (Scheme 3). The

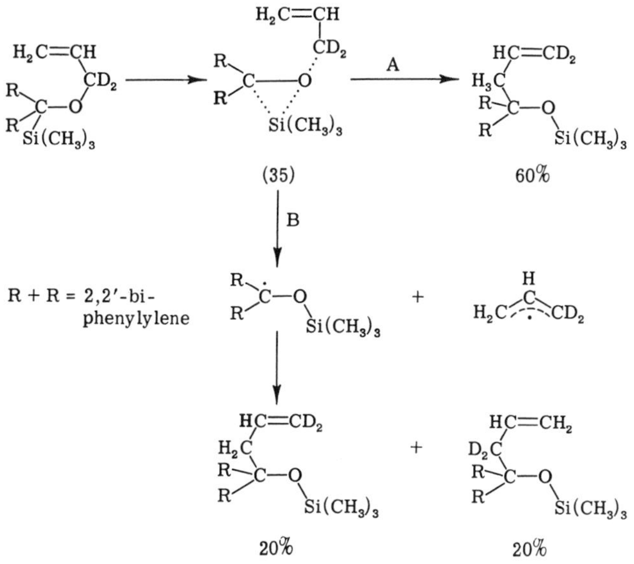

$$H_2C=CH$$

$$H_2C=CH$$
$$CD_2$$
$$R-C-O$$
$$R-Si(CH_3)_3$$

$$\longrightarrow$$

$$H_2C=CH$$
$$CD_2$$
$$R-C-O \quad A$$
$$R-Si(CH_3)_3$$

$$(35)$$

$$\xrightarrow{A}$$

$$HC=CD_2$$
$$H_3C$$
$$R-C-O$$
$$R-Si(CH_3)_3$$

$$60\%$$

$$\downarrow B$$

R + R = 2,2'-bi-
phenylylene

$$R-C-O$$
$$R-Si(CH_3)_3$$

$$+$$

$$H$$
$$C$$
$$H_2C-CD_2$$

$$\downarrow$$

$$HC=CD_2$$
$$H_2C$$
$$R-C-O$$
$$R-Si(CH_3)_3$$

$$+$$

$$HC=CH_2$$
$$D_2C$$
$$R-C-O$$
$$R-Si(CH_3)_3$$

$$20\% \qquad\qquad 20\%$$

SCHEME 3. Mechanism of the rearrangement of **30a**.

high-energy intermediate **35** rearranges either concertedly with all inversion (60%) (path A), or collapses into radicals which recombine statistically (40%) (path B). In this manner an apparent 80% allyl inversion results. Competing radical and concerted mechanisms are not unusual in rearrangements (*36*).

The diphenyl and phenyl-1-naphthyl systems react solely via path B, since the concerted attack of the allyl group at the backside of the carbon atom is sterically inhibited. This induces 100% allyl radical formation, in line with the observed deuterium scrambling in the allyl group. Such a process represents a unique case of anchimerically accelerated unimolecular bond homolysis and will be treated in detail in Section IV.

A detailed discussion concerning alternative mechanisms similar to those in Section II will not be presented here (*33*). However, the possibility of carbene formation or 1,2 silyl shifts generating oxygen ylides (analogous to those in Scheme 2) deserves attention. Further arguments against such mechanisms can now be construed. Significantly, the silyl–silyl dyotropic exchange occurs smoothly only in flat fluorenyl systems, whereas the allyl–silyl rearrangement is independent of the nature of the aryl substituents. Such pronounced differences are not to be expected if carbene or ylid formation operates. In contrast, the hypothesis of rate-determining coordination leads to a unified mechanistic picture. The allyl–silyl exchange in the diphenyl and phenyl-1-naphthyl systems occurs rapidly because the allyl groups—although incapable of concerted intramolecular flip due to steric inhibition—have a "way out." They split off as resonance stabilized allyl radicals. In case of the silyl–silyl rearrangement of **6**, no such low-energy alternative pathway is available; silyl radicals are generally not easily formed by unimolecular bond homolysis (*37*).

IV

THERMAL REARRANGEMENT OF BENZYL SILYLMETHYL ETHERS: A CASE FOR ANCHIMERICALLY ACCELERATED UNIMOLECULAR BOND HOMOLYSIS

The present study was motivated by the fact that neighboring group participation in radical reactions is exceedingly rare (*38*). Anchimeric assistance in unimolecular bond homolysis has been demonstrated only in certain cases of thermal perester decomposition (*38*). Despite numerous attempts, no examples in which the assisting group *migrates* to the radical center had previously been registered. Evidence for such a process was first reported in the study of allyl silylmethyl ethers (*31*) (Section III). In contrast to those derivatives, benzyl silylmethyl ethers should thermally rearrange solely via a radical mechanism in which the silyl groups participate anchimerically.

A. *Structural Effects on Migratory Aptitude*

A series of alkyl silylmethyl ethers **36** were synthesized (*17, 39, 40*) in the expectation that thermolysis should afford benzyl–silyl exchange products **38** as shown in Eq. (8).

$$
\begin{array}{ccc}
\underset{\text{Si(CH}_3)_3}{\overset{\underset{\displaystyle R^1 \quad R^3}{\diagdown \diagup}}{R^2 - C - O}} & \xrightarrow{\Delta} & \underset{\text{Si(CH}_3)_3}{\overset{\underset{\displaystyle R^1}{\diagdown}}{R^2 - \dot{C} - O}} \quad + \quad \dot{R}^3 \longrightarrow \underset{R^2 \quad \text{Si(CH}_3)_3}{\overset{R^3}{R^1 - C - O}}
\end{array}
\tag{8}
$$

(36)   (37)   (38)

a: $R^1 = R^2 = $ phenyl; $R^3 = $ benzyl
b: $R^1 = $ phenyl; $R^2 = $ 1-naphthyl; $R^3 = $ benzyl
c: $R^1 + R^2 = $ 2,2′-biphenylylene; $R^3 = $ benzyl
d: $R^1 + R^2 = $ 2,2′-biphenylylene; $R^3 = p$-methylbenzyl
e: $R^1 + R^2 = $ 2,2′-biphenylylene; $R^3 = p$-methoxybenzyl
f: $R^1 + R^2 = $ 2,2′-biphenylylene; $R^3 = p$-nitrobenzyl
g: $R^1 + R^2 = $ 2,2′-biphenylylene; $R^3 = $ 2-furanylmethyl
h: $R^1 + R^2 = $ 2,2′-biphenylylene; $R^3 = $ methyl

TABLE VII
RATE DATA FOR BENZYL–SILYL EXCHANGE REACTIONS

Compound	k_{rel} (185°C)	E_a (kcal/mole)	$\Delta H\ddagger$ (kcal/mole)	$\Delta S\ddagger$ (e.u.)
36a	0.3	—	—	—
36b	0.9	—	—	—
36c	1.0	32.6 ± 0.6	31.7 ± 0.6	−8.6 ± 0.8
36d	1.5	—	—	—
36e	2.2	—	—	—
36f	0.2	—	—	—
36g	0.3	—	—	—

Mechanistically significant rate differences are observed (39, 41). Whereas 36a–36g all rearrange smoothly at 160°–190°C to form 37a and 37b (>96% yields), 36h is stable under the reaction conditions. For example, thermolysis of 36c at 180°C for 8 hours induces complete conversion to 37c, whereas 36h shows no signs of rearrangement or decomposition for at least 24 hours at this temperature. This reflects the relative leaving group ability, methyl radical not being stabilized by aromatic or heteroaromatic substituents. Rate data are summarized in Table VII. The relative rates do not vary significantly. The negative entropy of activation indicates a tight transition state, in line with neighboring group participation (39).

Equally revealing is the observation that the *t*-butyl analog 39 of 36c is stable at 190°C for at least 5 days (39). Thus, steric acceleration cannot be the driving force in the benzyl–silyl rearrangements.

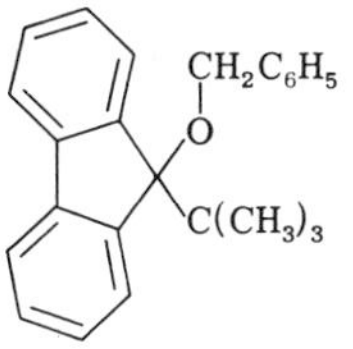

(39)

B. *Electron Spin Resonance Evidence*

Proof of the generation of ketyl radicals was gained electron spin resonance (ESR) spectroscopically (41). Thermolysis of a neat sample of 36c in an ESR spectrometer at 180°C gives rise to a well resolved spectrum due to 37c having a^H = 3.48, 0.83, 3.78, and 0.83 G for the proton couplings at the 1, 2, 3, and 4 positions of the fluorenyl ring.[3] The values correspond to those reported by Neumann, who recently prepared **37c** by an independent route (*42*). Under identical conditions the rearrangement product (**38c**) does not give rise to any ESR signals, proving the absence of possible redissociation of the benzyl group following rearrangement.

C. Further Mechanistic Studies

A number of additional observations, the details of which will be omitted here (*39, 40, 41*), include the following: (a) benzyl radicals can be trapped by appropriate trapping agents; (b) the silyl groups migrate 100% intramolecularly; (c) solvent polarity has no significant effect on the rearrangement rate; (d) radical initiators or inhibitors have no influence on the rate; (e) the corresponding *germyl* ethers decompose several orders of magnitude more slowly (*43*).

D. Discussion of the Mechanism

An extensive discussion will not be presented here (*39, 40, 41*). Suffice it to say that all present observations are in line with a mechanism according to Eq. (8), similar to the previously discussed pathway of the radical portion of allyl–silyl exchange (Section III). This involves silicon–oxygen coordination, which weakens the oxygen–carbon bond. Collapse of the high-energy, three-membered cyclic species leads to stabilized ketyl and benzyl (or 2-furanylmethyl) radicals and the formation of the strong silicon–oxygen bond (110–130 kcal/mole). These may be necessary conditions for anchimerically accelerated unimolecular bond homolysis in general (*41*). It should be noted that the degree of anchimeric assistance by the silyl groups is rather pronounced. For example, the unimolecular bond homolysis in simple ethers leading to alkoxy and allyl radicals requires activation energies of about 50 kcal/mole (*44, 45*). The slow rate of reaction of the germyl ethers is reasonable, the germanium–oxygen bond being weak (≈ 85 kcal/mole). Parenthetically, detailed electron impact studies including direct analysis of daughter ions (DADI) spectra of the above and related compounds reveal novel anchimeric assistance by silyl and germyl groups (*46, 47*).

---

[3] The 2- and 4- proton couplings are coincidentally identical (*42*).

V

THERMAL REARRANGEMENT OF *N,N*-BIS(TRIMETHYLSILYL)-*O*-ALLYLHYDROXYLAMINE

The title compound was prepared in order to test the effect of replacing the stationary carbon atom in allyl silylmethyl ethers (Section III) by nitrogen. Thermolysis at 170°C leads within several hours to nearly quantitative formation of 1,5-hexadiene and hexamethyldisiloxane (*48*). Although no mechanistic studies have been initiated, allyl–silyl exchange followed by α-elimination according to Scheme 4 is one of several possibilities.

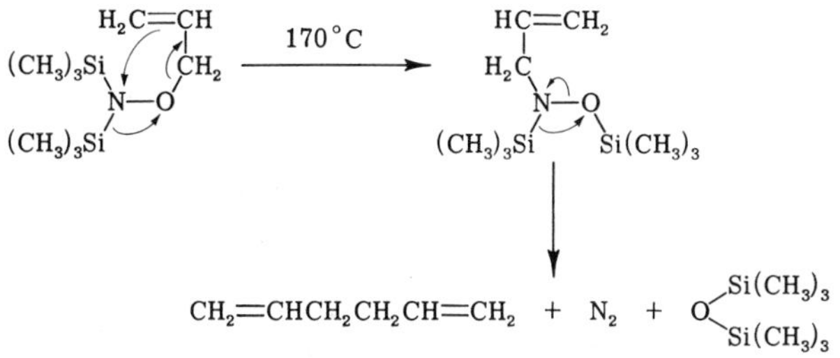

SCHEME 4. Possible decomposition mechanism involving allyl–silyl exchange.

VI

THERMAL REARRANGEMENT OF BIS- AND TRIS(SILYL)HYDROXYLAMINES

Several examples of silyl–silyl valence isomerization in hydroxylamines are presently known. Frainnet and co-workers have described such a process involving trimethyl and triethylsilyl groups [Eq. (9)] (*49*). No symmetric products are formed, showing that the silyl moieties remain together during the rearrangement. Further crossover experiments with respect to the substrate were not reported. The authors postulate a four-center intermediate (**40**) based on *d–p* interaction between silicon and lone pairs on oxygen and nitrogen.

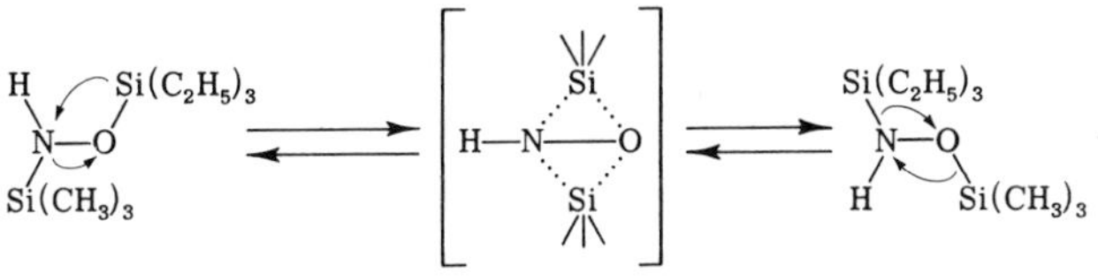

$$(9)$$

(40)

Systematic studies of related rearrangements in tris(silyl) hydroxylamines have been reported by West and co-workers (*50, 51*), the results of which are described below.

A. Structural Effects on Migratory Aptitude

A series of hydroxylamines **41a–d** and **42a–d** having variously substituted silyl groups were synthesized (*51*).

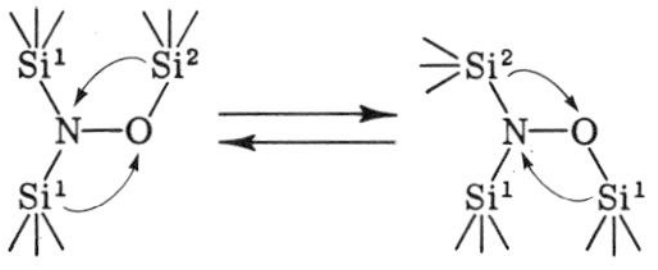

$$(10)$$

(41)   (42)

a: Si^1 = trimethylsilyl; Si^2 = dimethylsilyl
b: Si^1 = trimethylsilyl; Si^2 = dimethylphenylsilyl
c: Si^1 = trimethylsilyl; Si^2 = triethylsilyl
d: Si^1 = trimethylsilyl; Si^2 = dimethyl-t-butylsilyl
e: Si^1 = trimethylsilyl; Si^2 = trimethylsilyl

Thermolysis reveals that reversible rearrangement [Eq. (10)] is general, but that it is subject to severe steric hindrance imposed by bulky substituents at silicon (*51*). Compounds **41a** and **41b** rearrange cleanly at temperatures below 165°C. Thermolysis of independently synthesized **42a** and **42b** allows entry into the same equilibria. Compounds **41c** and **42c** rearrange similarly at 170°C, but also undergo a competing irreversible rearrangement to silylaminosiloxanes (*52*). In contrast, **41d** and **42d** fail to exhibit silyl–silyl exchange, the compounds undergoing alternative irreversible transformations at higher temperatures. Apparently,

TABLE VIII

RATE DATA FOR DYOTROPIC SILYL–SILYL EXCHANGE IN HYDROXYLAMINES

Rearrangement	$k \times 10^{-4}$ (sec^{-1})	Temperature (°C)	K	E_a (kcal/mole)	$\Delta H\ddagger$ (kcal/mole)	$\Delta S\ddagger$ (e.u.)
41a → 42a	2.44	139				
	6.78	150	0.24	30.6 ± 0.2	29.8 ± 0.2	−3.3 ± 0.8
	17.5	162				
41b → 42b	0.431	142				
	0.820	148	0.49	38.6 ± 0.2	37.8 ± 0.2	−11.2 ± 0.8
	2.98	160				

increased steric bulk at silicon decreases the ability to reach optimal geometry necessary for double migration. Rate data are shown in Table VIII. First-order rate constants were obtained. The rates are unaffected by changes in concentration or solvent polarity.

B. Crossover Experiments

Thermolyses of **41a** or **41b** do not afford crossover products (*51*). Thus **41e** is not formed, proving 100% intramolecularity. In contrast to Eq. (9), this single experiment shows that the two exchanging silyl groups stay together *and* do not depart from the substrate.

C. Discussion of the Mechanism

Since the activation parameters are quite similar to those of the silyl–silyl exchange in silyl silylmethyl ethers (Section II), the authors suggest that the rearrangements in the hydroxylamine series may also occur via a dyotropic mechanism (*51*). Furthermore, insensitivity to solvent polarity appears to rule out the intermediacy of ions or zwitterions. The possibility of an asymmetric transition state was discussed (*51*).

Although the above evidence points to a dyotropic exchange, it might be mentioned that the thermolysis of N-phenyl-N,O-bis(trimethylsilyl)hydroxylamine (**43**) at 100°C (16 hours) results in phenylnitrene formation [Eq. (11)], as shown by trapping experiments (*53*). This is reminiscent of Scheme 4. Possible silyl–silyl exchange in

$$
\underset{(43)}{\underset{\text{Si(CH}_3)_3}{\overset{\text{C}_6\text{H}_5}{\underset{|}{\text{N}}}}\text{O}^{\text{Si(CH}_3)_3}} \longrightarrow [\text{C}_6\text{H}_5\!-\!N] \quad + \quad \text{O}\overset{\text{Si(CH}_3)_3}{\underset{\text{Si(CH}_3)_3}{\diagdown}} \tag{11}
$$

this system does not occur fast enough to be observed by dynamic NMR (*54*). However, this does not preclude such processes. It would be interesting to determine whether rearrangement and α-elimination are independent reactions, or whether silyl equilibration comes about by nitrene elimination–insertion. Similar questions apply to the decomposition of allyl hydroxylamines (Section V).

Finally, silyl–silyl valence isomerization has been invoked by Wannagat to explain certain transformations in silylated hydrazines (*55*) in which both stationary atoms are nitrogen [Eq. (12)]. No rate data or mechanistic studies have been reported to date, although d–p interaction appears to be involved (56).

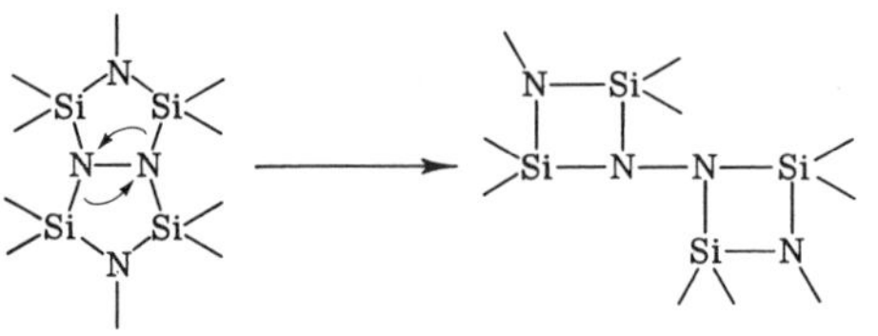

(12)

VII

THERMAL REARRANGEMENT OF SQUARIC ACID BIS(TRIMETHYLSILYL) ESTER

Type II dyotropic rearrangements involve double migration but no positional exchange. One such possibility is the squaric acid system in which thermally induced migration of the two ester functions toward the two carbonyl oxygen atoms is an orbital symmetry-allowed [$\sigma_s^2 + \sigma_s^2 + \pi_s^6$] process. Since sigmatropic silyl migration between two oxygen atoms is well known as exemplified by rapid rearrangements of acetylacetone and tropolone trialkylsilyl ethers (57, 58), the bis(trimethylsilyl) derivative (43) was synthesized[4] in hope of observing a degenerate rearrangement according to Eq. (13) (59).

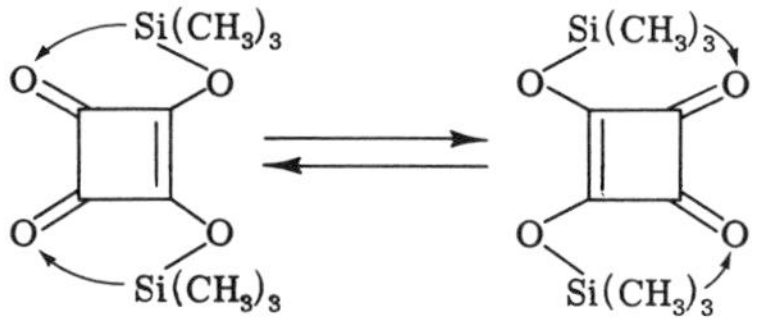

(13)

Because labeling experiments are likely to be tedious, the method of choice is dynamic ^{13}C—NMR utilizing Fourier transform. Indeed, at room temperature two sharp singlets at $\delta = 187.9$ and $\delta = 194.6$ (TMS internal standard, D_5-nitrobenzene) are observed that broaden on raising the temperature (59). At 83°C coalescence is observed, and at temperatures above 150°C a single signal at $\delta = 191.2$ appears. Upon cooling of the original, two singlets reappear, thereby ruling out decomposition. Utilizing $k_{coal} = \pi \delta \nu / \sqrt{2}$, the rate constant at coalescence was calculated to be $2.8 \ 10^{-4}$ sec^{-1}. Application of the Eyring equation results in an activation energy of $\Delta G\ddagger = 16.9$ kcal/mole.

---

[4] Compound **43** was independently synthesized by West (60, 61).

In order to test intramolecularity, crossover experiments were performed using **43** and the D_{18}-derivative **44** (*59*). Upon standing for less than 30 minutes at room temperature, a 1:1 molar mixture of **43** and **44** results in a statistical distribution of **43, 44,** and **45**, as demonstrated by mass spectroscopy.

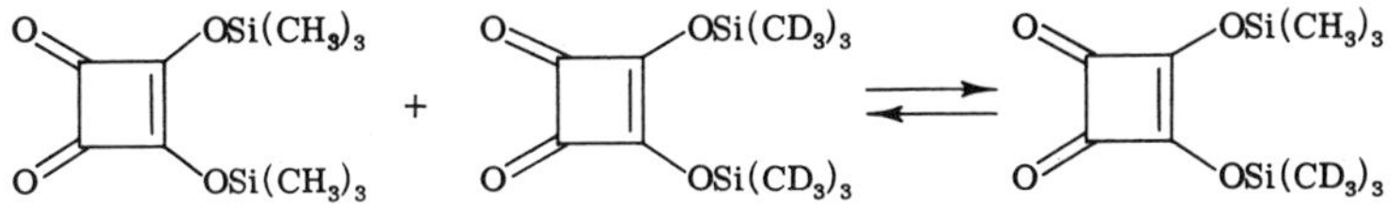

1	:	**1**	:	**2**
(**43**)		(**44**)		(**45**)

Clearly, the equilibration of the carbon atoms is due to rapid *inter*molecular silyl transfer. Of course, these results do not rule out a simultaneously occurring slow dyotropic process. The nature of the intermolecular mechanism is uncertain. Compound **43** is highly hygroscopic. Trace impurities such as moisture may cause rapid desilylation, resilylation setting in at different oxygen atom. However, addition of catalytic amounts of water or squaric acid itself has no influence on the reaction rate (*62*). Alternatively, desilylation may proceed according to Scheme 5. Unfortunately, studies of rate dependency on concentration are not possible for dynamic ^{13}C—NMR at present.

Relevant to the mechanism of Scheme 5 is a report of an *inter*molecu-

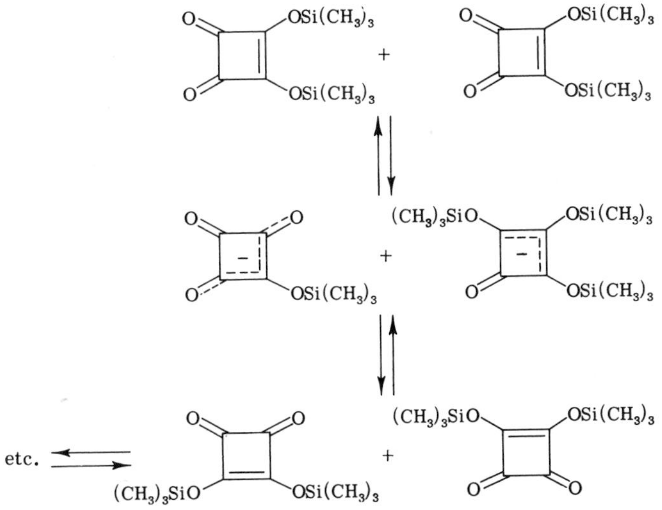

SCHEME 5. Possible mechanism for carbon atom equilibration (**43**).

lar component in the silyl migration of certain acetonylacetone silyl ethers (63), in contrast to the generally accepted intramolecularity due to the cis form (57). The former process may be related to the rapid rearrangement of **43** and needs further clarification.

Compounds **46, 47,** and **48** were also synthesized, since similar intermolecular silyl migrations should likewise induce complete equilibration of the relevant carbon atoms (59). However, the ^{13}C—NMR spectra do not reveal dynamic behavior up to 130°C. Nevertheless,

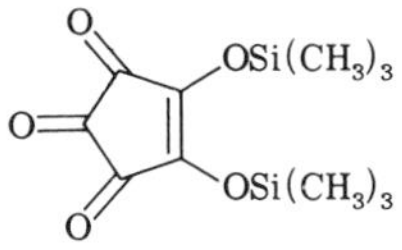

(46) 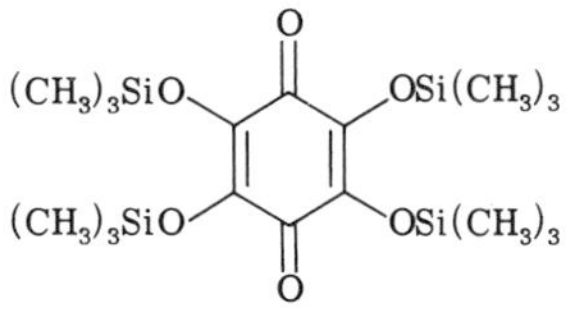　　　　　　　　　(47) 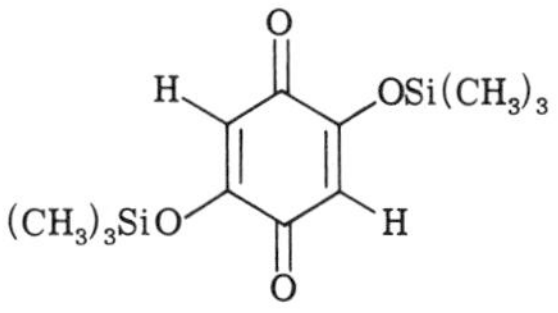　　　　　　　　　(48)

recent crossover experiments utilizing **46** and **48** in conjunction with their respective per-deutero analogs prove the occurrence of intermolecular silyl migration similar to those of **48** (43). Thus, carbon atom equilibration appears to be a general phenomenon, occurring more rapidly in the squaric acid system than in the other substrates.

VIII

MISCELLANEOUS σ–σ EXCHANGE REACTIONS

A number of formal σ–σ exchange reactions involving organometallic compounds have been reported, some of which are briefly mentioned here. Unfortunately, in most cases mechanistic studies of sufficient detail are lacking. An interesting halogen–halogen exchange has been described by Haszeldine and co-workers [Eq. (14)] (64). The rearrangements have first-order kinetics; intramolecularity, activation parameters, and solvent effects are presently unknown.

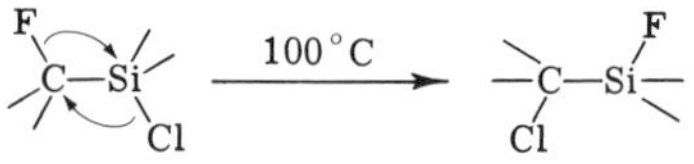

$$\text{(14)}$$

Similar processes involving phenyl groups and such nucleofugal moieties as fluorine, chlorine, and acetate have been reported by Brook (65). Closely related is the thermally induced quantitative exchange of acetate and methyl groups as shown in Eq. (15) (41). Initial mechanistic studies include first-order kinetics with E_a = 33.6 kcal/mole and $\Delta S\ddagger$ = −5.8 e.u., as well as solvent effects; the relative rates in benzene and acetonitrile are 1.0:2.1, demonstrating insensitivity to solvent polarity. Further studies such as crossover experiments and migratory aptitude of acetate versus trifluoroacetate have not been carried out, so that no definite conclusion regarding the mechanism can be drawn at present.

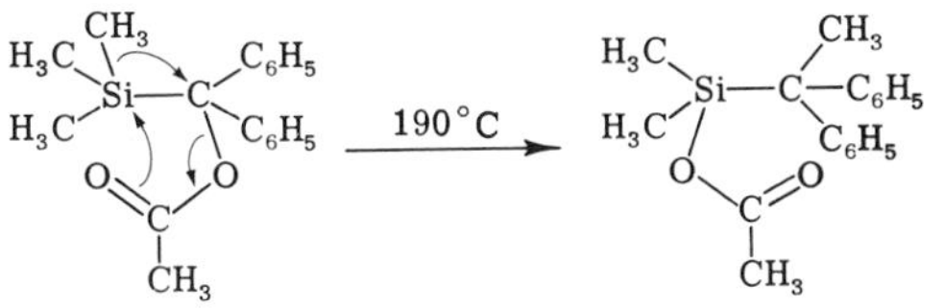

(15)

More drastic conditions are required for phenyl–hydrogen exchange according to Eq. (16) (66). In this case silyl radicals may well be involved (67).

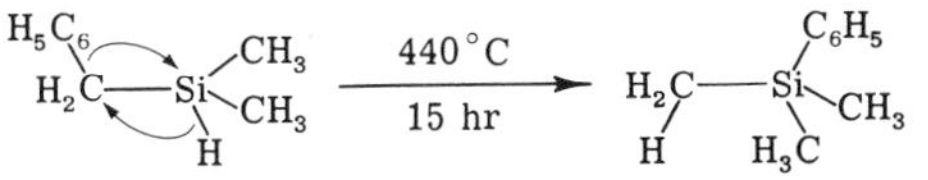

(16)

The well studied Wittig rearrangement is generally classified as a *sigmatropic* shift in which an alkyl group migrates from oxygen to the neighboring lithiated carbon atom (68). Little is known concerning the role of the gegenion. If the latter is included, a dyotropic mechanis evolves. Indeed, the empty *p*-orbital of lithium may interact according to Eq. (17), transforming the symmetry-forbidden 1,2 shift into an allowed

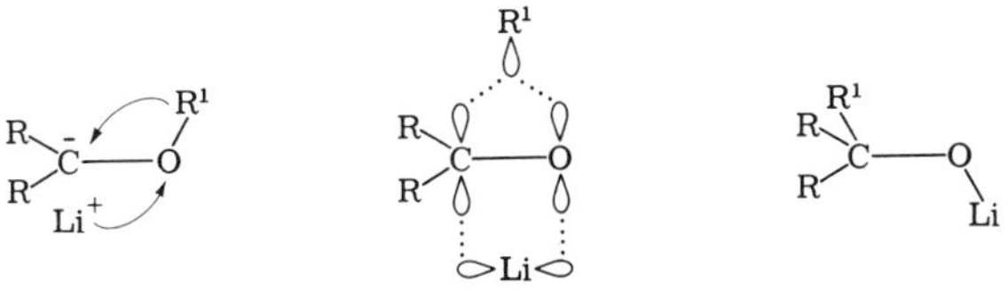

(17)

$[\sigma_s^2 + \sigma_a^2]$ process (5) in which R migrates with retention and lithium with "inversion." However, this is not a requirement, since the fluoride ion catalyzed rearrangement of allyl and benzyl (silylmethyl) ethers (**28, 36**; Sections III and IV) occurs rapidly at room temperature via the same type of alkoxy anions that have no metal gegenions, but rather ammonium ions (41).

Concerted and multistep radical mechanisms have been proposed (68, 69). Lithium complexation appears to be involved in the aza–Wittig rearrangement (70).

Exchange processes are well known in the field of transition metal chemistry, elimination addition mechanisms often prevailing, e.g., Eq. (18) (71).

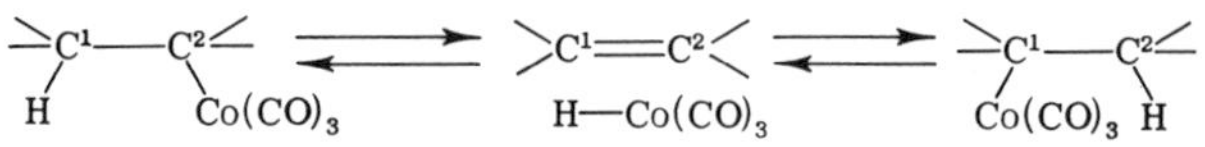

$$\tag{18}$$

Formal dyotropic processes are also well known in a number of coenzyme B_{12}-mediated isomerizations involving cobalt [Eq. 19] (72).

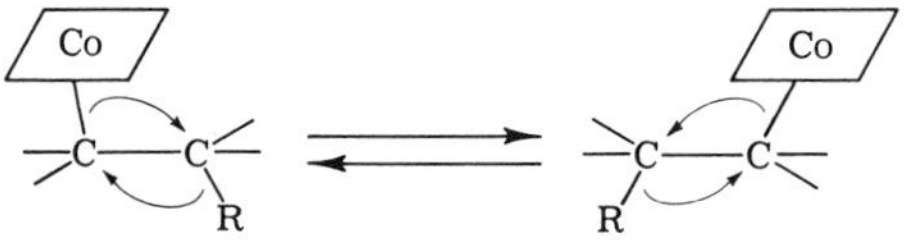

$$\tag{19}$$

An unusual iron–iron exchange according to Eq. (20) has been described (73).

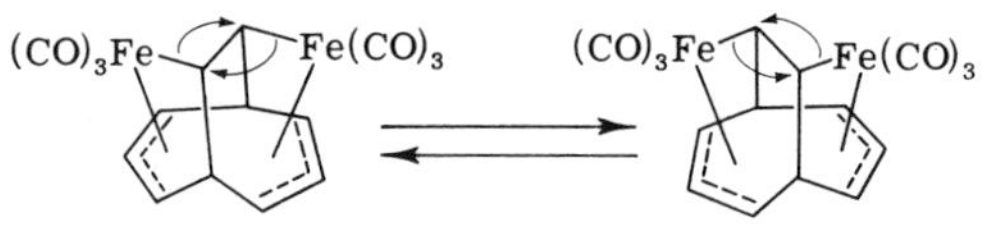

$$\tag{20}$$

Finally, several cases in which the dyotropic *transition* state is actually the *ground* state having a doubly bridged geometry are known, e.g., Eq. (21) (74) and Eq. (22) (75).

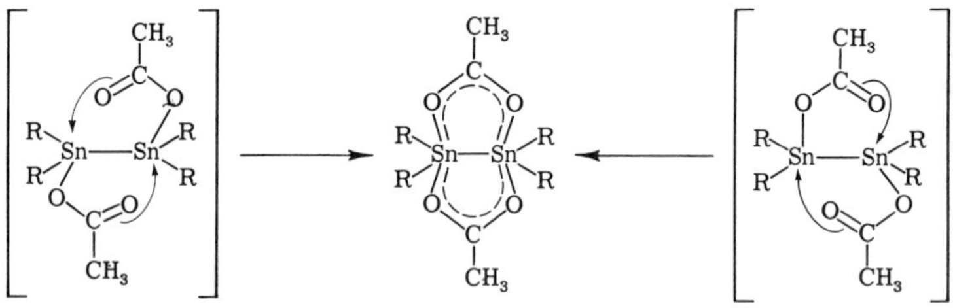

$$\tag{21}$$

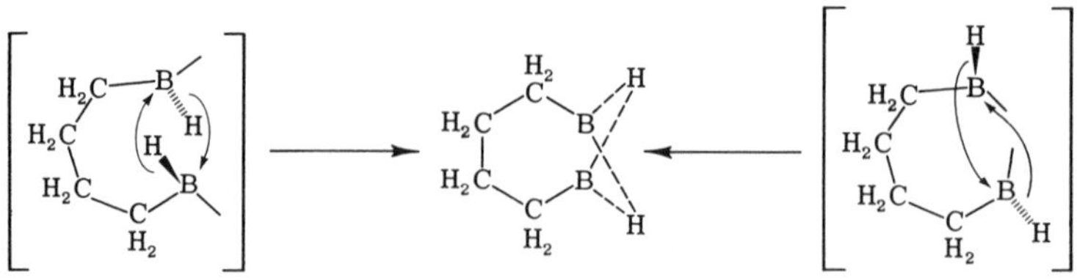

$$\tag{22}$$

IX

OUTLOOK

This chapter provides a summary of the mechanistic methods applicable to dyotropic rearrangements. To date the most detailed studies have been carried out with organosilicon compounds. The presence of heteroatoms appears to be necessary. However, the present author believes that dyotropic processes are not restricted to these elements. σ-σ exchange reactions—irrespective of the mechanism—involve far-reaching molecular transformations and are thus of potential synthetic interest (76).

ACKNOWLEDGMENT

The author would like to thank Professor Dr. R. W. Hoffmann for numerous discussions. Support by the Deutsche Forschungsgemeinschaft and the Fonds der Chemischen Industrie is gratefully acknowledged.

REFERENCES

1. Reetz, M. T., *Angew. Chem., Int. Ed. Engl.* **11**, 129 (1972).
2. Reetz, M. T., *Angew. Chem., Int. Ed. Engl.* **11**, 131 (1972).
3. Reetz, M. T., *Tetrahedron* **29**, 2189 (1973).
4. Hoffmann, R., and Williams, J. E., *Helv. Chim. Acta* **55**, 67 (1972).
5. Woodward, R. B., and Hoffmann, R., *Angew. Chem., Int. Ed. Engl.* **8**, 781 (1969).
6. Larrabee, R. B., *J. Organomet. Chem.* **74**, 313 (1974).
7. Brook, A. G., *Acc. Chem. Res.* **7**, 77 (1974).
8. West, R., and Boudjouk, P., *J. Am. Chem. Soc.* **95**, 3987 (1973).
9. Reetz, M. T., Kliment, M., and Plachky, M., *Angew. Chem., Int. Ed. Engl.* **13**, 813 (1974).
10. Reetz, M. T., Kliment, M., and Plachky, M., *Chem. Ber.* **108**, 2716 (1976).
11. Golden, D. M., and Benson, S. W., *Chem. Rev.* **69**, 125 (1969).
12. Reetz, M. T., Kliment, M., Plachky, M., and Greif, N., *Chem. Ber.* **108**, 2728 (1976).
13. Yankee, E. W., Bodea, F. D., Howe, N. E., and Cram, D. J., *J. Am. Chem. Soc.* **95**, 4210 (1973).
14. Dimroth, K., and Reichardt, C., *Fortschr. Chem. Forsch.* **11**, 1 (1968).
15. Greif, N., Diplomarbeit, Universität Marburg, 1975.
16. Brook, A. G., Warner, C. M., and Limburg, W. W., *Can. J. Chem.* **45**, 1231 (1967).
17. Kliment, M., Dissertation, Universität Marburg, 1976.
18. Brook, A. G., and Dillon, P. J., *Can. J. Chem.* **47**, 4347 (1969).
19. Baldwin, J. E., and Fleming, R. H., *Fortschr. Chem. Forsch.* **15**, 281 (1970).
20. Kirmse, W., and Kapps, M., *Chem. Ber.* **101**, 994 (1968).
21. Ando, W., Konishi, K., Hagiwara, T., and Migita, T., *J. Am. Chem. Soc.* **96**, 1601 (1974).
22. McIver, J. W., *Acc. Chem. Res.* **7**, 72 (1974).
23. Dewar, M. J. S., and Wade, L. E., *J. Am. Chem. Soc.* **95**, 290 (1973).

24. Andrist, A. H., *J. Org. Chem.* **38**, 1772 (1973).
25. Reetz, M. T., and Kliment, M., *Tetrahedron Lett.* 2909 (1975).
26. Haas, A., *Angew. Chem., Int. Ed. Engl.* **4**, 1014 (1965).
27. Ebsworth, E. A. V., *in* "The Bond to Carbon" (A. G. MacDiarmid, ed.), Vol. I, p. 52. Dekker, New York, 1968.
28. Hosomi, A., and Traylor, T. G., *J. Am. Chem. Soc.* **97**, 3682 (1975).
29. Reetz, M. T., and Plachky, M., unpublished results, 1973.
30. Stewart, H. F., and West, R., *J. Am. Chem. Soc.* **92**, 846 (1970).
31. Reetz, M. T., *Angew. Chem., Int. Ed. Engl.* **13**, 416 (1974).
32. Reetz, M. T., *Chem. Ber.* **110**, 954 (1977).
33. Reetz, M. T., *Chem. Ber.* **110**, 965 (1977).
34. Humski, K., Malojcic, R., Borcic, S., and Sunko, D. E., *J. Am. Chem. Soc.* **92**, 6534 (1970).
35. Reetz, M. T., *Tetrahedron Lett.* 817 (1976).
36. Baldwin, J. E., and Patrick, J. E., *J. Am. Chem. Soc.* **93**, 3556 (1971).
37. Davidson, J. M. T., *Q. Rev., Chem. Soc.* **25**, 111 (1971).
38. Martin, J. C., *in* "Free Radicals" (J. K. Kochi, ed.), Vol. II, p. 502. Wiley, New York, 1973.
39. Reetz, M. T., and Kliment, M., *Tetrahedron Lett.* 797 (1975).
40. Reetz, M. T., Kliment, M., and Greif, N., *Chem. Ber.*, submitted.
41. Reetz, M. T., and Greif, N., in preparation.
42. Neumann, W. P., Schroeder, B., and Ziebarth, M., *Liebigs Ann. Chem.* 2279 (1975).
43. Reetz, M. T., and Greif, N., unpublished results, 1976.
44. Mobra, M. J., and Ariza, E., *An. R. Soc. Esp. Fis. Quim., Ser. B* **56**, 851 (1960).
45. Kwart, K., Sarner, S. F., and Slutsky, J., *J. Am. Chem. Soc.* **96**, 5234 (1973).
46. Schwarz, H., Kliment, M., Reetz, M. T., and Holzmann, G., *Org. Mass Spectrom.* **11**, 989 (1976).
47. Schwarz, H., and Reetz, M. T., *Angew. Chem., Int. Ed. Engl.,* **15**, 705 (1976).
48. Reetz, M. T., Kliment, M., and Plachky, M., unpublished results, 1973.
49. Frainnet, E., Duboudin, F., Dabescat, F., and Vincon, G., *C. R. Acad. Sci., Ser. C* **276**, 1469 (1973).
50. West, R., Nowakowski, P., and Boudjouk, P., *Int. Symp. Organosilicon Chem., 3rd, 1972*.
51. Nowakowski, P., and West, R., *J. Am. Chem. Soc.* **98**, 5616 (1976).
52. West, R., Nowakowski, P., and Boudjouk, P., *J. Am. Chem. Soc.* **98**, 5620 (1976).
53. Tsui, F. P., Vogel, T. M., and Zon, G., *J. Am. Chem. Soc.* **96**, 7144 (1974).
54. Zon, G., Private communication (1975).
55. Wannagat, U., *Organomet. Chem. IUPAC Symp.* **1**, 274 (1966).
56. Wannagat, U., Private communication (1973).
57. Pinnavaia, T. J., Collins, W. T., and Howe, J. J., *J. Am. Chem. Soc.* **92**, 4544 (1970).
58. Reich, H. J., and Murcia, D. A., *J. Am. Chem. Soc.* **95**, 3418 (1973).
59. Reetz, M. T., Neumeier, G., and Kaschube, M., *Tetrahedron Lett.* 1295 (1975).
60. Eggerding, D., and West, R., *J. Am. Chem. Soc.* **97**, 207 (1975).
61. Eggerding, D., and West, R., *J. Am. Chem. Soc.,* **98**, 3641 (1976).
62. Neumeier, G., Diplomarbeit, Universität Marburg, 1975.
63. Kusnezows, I, K., Rühlmann, K., and Gründemann, E., *J. Organomet. Chem.* **47**, 53 (1973).
64. Bevan, W. I., Haszeldine, R. N., Middleton, J., and Tipping, A. E., *J. Organomet. Chem.* **23**, C 17 (1970).
65. Brook, A. G., and Jones, P. F., *Chem. Commun.* 1324 (1969).

66. Sakurai, H., Hosomi, A., and Kumada, M., *Chem. Commun.* 521 (1969).
67. Sakurai, H., and Hosomi, A., *J. Am. Chem. Soc.* **90**, 7507 (1970).
68. Schöllkopf, U., *Angew. Chem., Int. Ed. Engl.* **9**, 763 (1970).
69. Garst, J. F., and Smith, C. D., *J. Am. Chem. Soc.* **95**, 6870 (1973).
70. Reetz, M. T., and Schinzer, D., *Tetrahedron Lett.* 3485 (1975).
71. Coates, G. E., Green, M. L., and Wade, K., "Organometallic Compounds," Vol. II, p. 334. Methuen, London, 1968.
72. Retey, J., and Zagalak, B., *Angew. Chem., Int. Ed. Engl.* **12**, 671 (1973).
73. Aumann, R., *Angew. Chem., Int. Ed. Engl.* **10**, 189 (1971).
74. Plazzogna, G., Peruzzo, V., and Tagliavini, G., *J. Organomet. Chem.* **66**, 57 (1974).
75. Brown, H. C., Negishi, E., and Burke, P. L., *J. Am. Chem. Soc.* **92**, 6649 (1970).
76. Ager, D. J., and Fleming, I., *J. Chem. Res. S* **1**, 6 (1977).

Rearrangements of Unsaturated Organoboron and Organoaluminum Compounds

JOHN J. EISCH

Department of Chemistry
State University of New York at Binghamton
Binghamton, New York

I

INTRODUCTION

A. Historical Background

The first syntheses of boron and aluminum alkyls were accomplished over one hundred years ago, barely a decade after Edward Frankland had laid the foundations for the development of organometallic chemistry. In the period 1849–1855 Frankland and his co-workers were led to prepare alkyls of zinc and other elements as part of studies aimed at understanding the combining capacity or, in modern parlance, valence of an element (50–52). Indeed, Frankland and Duppa reported the first synthesis of boron alkyls from zinc alkyls in 1860 (53). Then in 1865 Buckton and Odling were able to prepare aluminum alkyls by the displacement reaction between aluminum metal and mercury alkyls, a method that was to prove to be broadly applicable to active metals (9).

Despite this long and prominent history, only within the last 20 years has much attention been given to the tendency of boron and aluminum organometallics to undergo rearrangement. Several examples of the redistribution reaction with mixed boron alkyls were discovered in the 1930s; if the dimeric character of alkylboron hydrides is noted, such a redistribution may be considered as an isomerizing rearrangement (84):

$$(R_2BH)_2 \rightleftharpoons (R_3B \cdot BH_3) \rightleftharpoons R_3B + BH_3 \tag{1}$$

Although no isomerizing rearrangements had been reported for aluminum in the 1930s, derivatizing rearrangements (cf. Section I,D) were recognized (57):

$$(C_6H_5CH_2)_3Al \xrightarrow[\text{2. } H_2O]{\text{1. } H_2CO} C_6H_5CH_2CH_2OH \quad + \quad \text{CH}_3\text{-C}_6\text{H}_4\text{-CH}_2\text{OH} \tag{2}$$

Not until the mid 1950s did the ease and variety of isomerizing rearrangements with these metal alkyls become apparent. In attempting to synthesize pure tri-t-butylborane, Hennion and co-workers found that the resulting boron alkyl contained both t-butyl and isobutyl groups (61). Subsequent studies by Köster (69) and Brown (5) provided abundant evidence that such isomerizations are general and ensue by a dehydroboration–hydroboration sequence [Eq. (3)].

$$-\overset{|}{\underset{B}{C}}-\overset{H}{\underset{H}{C}}-H \rightleftharpoons C=C \overset{H}{\underset{H}{}} + H-B \rightleftharpoons -\overset{|}{\underset{H}{C}}-\overset{H}{\underset{H}{C}}-B \tag{3}$$

By thermal treatment of suitable acyclic or cyclic boron alkyls, the most stable isomer or mixture of isomers will be formed via the following isomerizations: (a) movement of the boron along a given carbon chain (tertiary → secondary → primary); (b) interconversion of cyclic boranes and acyclic boron hydrides; (c) change in the ring size of the cyclic borane (six-membered ring favored); and (d) isomerization of ring substituents between cis and trans positions. Not surprisingly, similar isomerizations were encountered in contemporaneous, parallel studies of aluminum alkyls and of alkene–aluminum hydride interactions (*7, 72, 73, 96, 97*).

The remarkable impact that the studies of aluminum alkyls by Karl Ziegler (*98*) and of hydroboration by Herbert Brown (*5*) have had on organic synthesis has intensified interest in the fundamental chemistry of boron and aluminum organometallics during the past decade. This impetus, coupled with significant advances in scientific instrumentation, has led to extensive work on the rearrangement tendencies of these metal alkyls. Besides being interesting chemical processes in themselves, such rearrangements may constitute useful *intramolecular* model systems for the understanding of significant intermolecular reactions (*14, 15*). Thus, allylic isomerizations of group III alkyls [Eq. (4)] can be viewed as an intramolecular interaction of the vacant metal np-orbital with the filled C$=$C π-orbital.

$$\overset{\text{O}}{\underset{\text{O}}{M}}-CH_2-CH\overset{*}{=}CH_2 \rightleftharpoons M\overset{CH_2}{\underset{CH_2}{\overset{*}{\diagdown}}}CH \rightleftharpoons M-CH_2-C\overset{*}{=}CH_2 \qquad (4)$$

An evaluation of the electronic, steric, and stereochemical factors involved in the kinetics and thermodynamics of this kind of process may be most helpful in elucidating the course of the Ziegler carbometallation process [Eq. (5)], central to the industrial production of long-chain alcohols and olefins from ethylene and to the stereoregular polymerization of α-olefins (*18, 21*).

$$\overset{\text{O}}{\underset{\text{O}}{M}}-R + CH_2=CHR' \longrightarrow \underset{R}{M}\overset{CH_2}{\underset{CHR}{\overset{\vdots}{\vdots}}} \longrightarrow M-CH_2-CHRR' \qquad (5)$$

B. *Previous Surveys*

Several aspects of organoboron and organoaluminum rearrangements have been reviewed recently. The thermal rearrangement of saturated organoboranes [Eq. (3)] has been discussed in connection with the synthetic utility of hydroboration, boron heterocycle formation and the basic chemistry of organoboranes (*69, 82*). Allylic or benzylic derivatives of boron (*78*) and aluminum (*11, 64*) have been surveyed individually and in the context of the general structural and chemical properties of such rearrangement-prone systems [e.g., Eq. (2)]. Finally, the significance of either isolable or assumed four-coordinate organoboron intermediates and their 1,2-migrations has been recognized for many boron–carbon bond insertions (82). Depending upon the reaction conditions and the stability of the rearranged organoborane [i.e., 1 and 2 in Eqs. (6) and (7)], the migration may be followed by deboration.

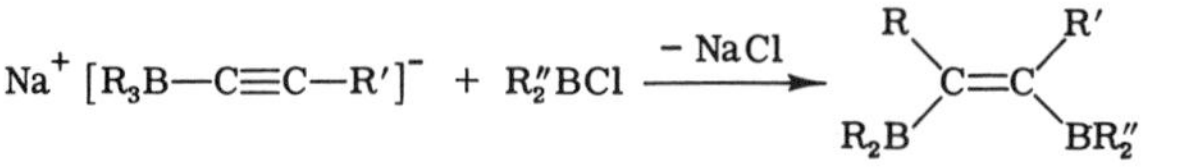

$$
\text{Na}^+ [R_3B—C\equiv C—R']^- + R''_2BCl \xrightarrow{\;-\text{NaCl}\;} \underset{R_2B}{\overset{R}{}}C=C\underset{BR''_2}{\overset{R'}{}} \qquad \text{ref. (3)} \qquad (6)
$$
$$
(1)
$$

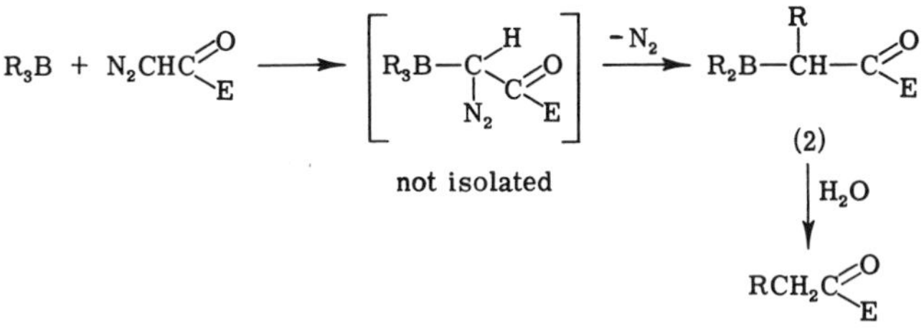

$$
R_3B + N_2CHC{\overset{O}{\underset{E}{\diagdown}}} \longrightarrow \left[R_3B—\underset{N_2}{\overset{H}{\underset{|}{C}}}\underset{E}{\overset{O}{\diagdown C\diagdown}} \right] \xrightarrow{\;-N_2\;} R_2B—\overset{R}{\underset{|}{C}}H—C{\overset{O}{\underset{E}{\diagdown}}} \qquad \text{ref. (63)}
$$

$$
\text{not isolated} \qquad\qquad (2) \qquad \Big\downarrow H_2O \qquad (7)
$$

$$
RCH_2C{\overset{O}{\underset{E}{\diagdown}}}
$$

C. Scope of the Present Treatment

Since considerable attention has already been given to the rearrangements of saturated organic groups attached to boron (69) and aluminum (7, 72, 96, 97), discussion here will be limited to the rearrangements of unsaturated groups. The multitude of known migrations involving tetracoordinate boron [Eqs. (6) and (7)] having been ably surveyed (82), such rearrangements will be discussed here only to the extent that mechanistic insights can be gained. Thus, three principal aspects of molecular reorganization will be considered: first, the occurrence of sigmatropic rearrangements, of which allylic isomerization is a special case; second, the organic group migrations of the [1,n] type occurring by way of isolable tetracoordinate organoboron or organoaluminum complexes; and third, the relevance of such rearrangement tendencies to the intermolecular reactions of these metal alkyls with organic substrates [cf. Eqs. (3) and (4)].

D. Classification of Molecular Rearrangements

Since the classification of chemical reactions by means of current perceptions of their mechanisms is one of man's most ephemeral activities, it is deemed more prudent here to classify rearrangements on the basis of net structural change. With the recognition of ionic and both σ- and π-covalent structures among organometallic compounds, a useful distinction could be made between isomerizing and derivatizing rearrangements (16). In the former, a constant aggregrate of atoms $[(RM)_n]$ might undergo various structural permutations [cf. Eqs. (1) and (3)]; in the latter, the interaction of an organometallic reagent (RM) with a substrate (A———B) would yield a derivative having an isomeric structure for the original R group or possibly a different chirality at M. A further subdivision of either isomerizations or derivatizations can be made as to the kind of structural change: (a) the concatenation of atoms in R and hence a *skeletal* change could ensue; (b) the skeleton could be maintained, while the *skin* of hydrogen, halogen, or metal groups might migrate to new sites of attachment; or (c) the three-dimensional *orientation* (stereochemistry or chirality) of R or M might change without altering the skeletal or skin rearrangement in R (16). The phenomenological classification of rearrangements into skeletal, skin, and orientational types will be used in this chapter.

E. *Pericyclic Processes and the Role of Intermediate Complexes*

Clearly, the isomerizing and derivatizing rearrangements of group III should be viewed in light of the Woodward–Hoffmann principle of the conservation of orbital symmetry (55, 95). But the assurance that a suggested reaction pathway is consistent with this principle does not remove the obligation for critical experimentation. Various physical organic techniques must be used to learn whether the reaction is actually concerted; here, often only nonconcertedness can be firmly established—detection of intermediates, catalytic acceleration or retardation, loss of stereochemistry, observation of crossover products, among other signs. Those reactions adhering to this principle, so-called pericyclic processes, undergo first-order changes in their bonding relationships in concert on a closed curve. The isomerizations of triallylborane (78, 80) and but-1-en-3-yl(dimethylamino)ethylborane (50) appear to fit the criteria for a pericyclic process [Eq. (4)].

$$\tag{8}$$

Because organoboranes and organoalanes form relatively stable intermediate complexes with various substrates as a prelude to final product
formation, it seems permissible to try to extend the scope of the
Woodward-Hoffmann principle to the reorganization pathways of such
complexes. Thus, it would be useful, for example, to consider whether
the chemical behavior of an allylic aluminum system complexed with a
ketone (3) might resemble the thermally allowed [3, 3] sigmatropic
rearrangement (4). The value of viewing the collapse of such complexes
as potential pericyclic processes will become evident in Section IV,C,
where the interplay of kinetic versus thermodynamic control on ketone
insertions into carbon–metal bonds is discussed.

II

SIGMATROPIC REARRANGEMENTS OF ORGANOBORANES

A. Allylic Rearrangements

An extensive array of derivatizing rearrangements undergone by
allylic boranes has been uncovered by B. M. Mikhailov and the value of
such reactions in organic and organometallic synthesis has been recently
reviewed (78). Illustrative of the scope of such carboborations is the
preparation of the novel 1- and 2-boraadamantane skeletons (79). The
strain in the former system is revealed in the unusual formation of a
stable etherate; apparently, the change to sp^3-hybridization at boron
upon complexation is strongly favored.

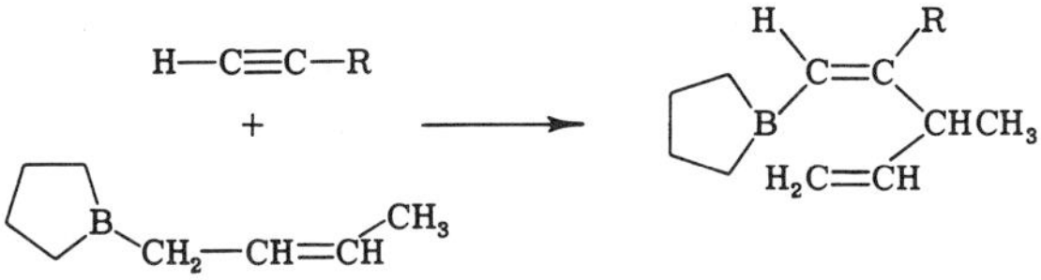

$$H—C\equiv C—R$$

$$\text{ref. } (8) \qquad (9)$$

Also, a prominent feature of such allylic boranes is that reaction is
accompanied by an allylic skin rearrangement (78) [Eq. (9)]. Such a
reaction may be favored over the direct insertion of the alkyne into the
2-butenyl-boron bond because of the possible pericyclic character of the
rearrangement. If no intermediate is involved, the reaction in Eq. (9) can
be viewed as an allowed ene-reaction (55), a $[_\sigma 2_s + _\sigma 2_s + _\pi 2_s]$ process; if
an intermediate collapse is involved, its collapse to the product can be
viewed as a converted process, analogous to that shown in Eq. (8).

The isomerizing rearrangements of such allylic boranes vary markedly in their ease, depending upon substituents attached to boron: triallylborane isomerizes with an activation of ca. 11 kcal mol^{-1} (78, 80); but-1-en-3-yl(dimethylamino)ethylborane isomerizes less readily with E_a = 24.6 kcal mol^{-1} (59); and but-1-en-3-yl(dimethylamino)methoxyborane shows no rearrangement up to 150°C (60). If the vacant 2p-orbital on boron is necessary for the sigmatropic shift [cf. Eq. (4)], the presence of unshared electron pairs on the ligands would be expected to retard the rearrangement, and this is in accord with the facts.

B. Cyclic Allylic Boranes

1. Synthesis from Boroles

Recent research on the chemistry of the boracyclopentadiene or borole system has brought to light further manifestations of such allylic shifts and some novel consequences. Since previous reports on the properties of boroles have proved to be erroneous (4), some comments on their successful synthesis and unusual properties are in order (36). Treatment of 1,4-dilithiotetraphenylbutadiene with phenylboron dihalide in ethyl ether solution will lead to pentaphenylborole as a pale yellow ethyl etherate (5) [Eq. (10)]; the ether can be removed under reduced pressure to yield the deep blue-green[1] uncomplexed borole [7 in Eq.

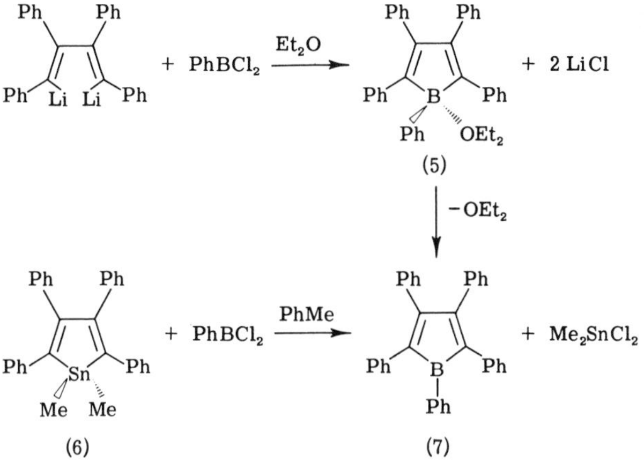

$$\tag{10}$$

$$\tag{11}$$

[1] Solid samples of pentaphenylborole can appear to have deep blue-green or deep-blue colors; the greenish cast may be the iridescence of reflected light or the tint due to traces of residual complexed dimethyltin dichloride. Microcrystalline samples of repeatedly washed 7 appeared deep blue (32).

(11)] (32). A more convenient general approach to such uncomplexed boroles is the interaction of the air-stable stannoles, (e.g., **6**), with a boron halide in a hydrocarbon or halocarbon solvent [Eq. (11)]. The long-wavelength spectral maxima exhibited by the various boroles synthesized in this manner are given in Table I (32). The resulting boroles are extraordinarily air-sensitive: brief exposure to air is sufficient to discharge the deep-blue or deep-violet colors of their solutions. This sensitivity may explain why previous workers erroneously concluded that **7**, a sample that was isolated in air and recrystallized from ethanol, is a stable, pale yellow solid (4). When a sample of genuine **7** is exposed to air and then worked up by protolysis, tetraphenylfuran (**8**) and 3-benzylidene-1,2-diphenylindene (**9**) can be isolated. Hot glacial

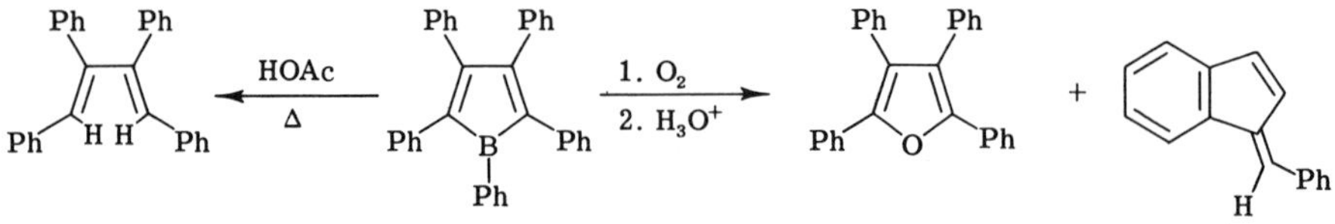

(12)

acetic acid slowly deborates **7** to (E,E)-1,2,3,4-tetraphenylbutadiene (**10**).

Treatment of these boroles with alkynes leads to a prompt discharge of the color, even with the ordinarily less reactive diphenylacetylene,

TABLE I

LONG-WAVELENGTH MAXIMA IN THE VISIBLE
SPECTRA OF SUBSTITUTED BOROLES[a]

Boroles	λ_{max} (nm)
2,3,4,5-Tetraphenylboroles	
1-Methyl-	542
1-Chloro-	560
1-Phenyl-	567[b]
2,3,4,5-Tetra-p-tolylboroles	
1-Chloro-	580
1-Phenyl-	602

[a] Spectral maxima measured on solutions of the boroles, which resulted from the interaction of the stannole and E—BX$_2$.

[b] Spectral measurement on a sample of the purified borole.

and the formation of the Diels–Alder adducts (*30*). The resulting 7-borabicyclo[2.2.1]heptadiene system in **11** has proved to be most valuable, both for the study of several novel rearrangements and for the probing of electronic interactions between boron and olefinic bonds. These possibilities stem from the recognition that **11** is a bisallylic borane,

$$\text{(13)}$$

as well as a bishomoboracyclopropene (*33*). This allylic character might lead to boron–carbon bond-switching on the central C_6-skeleton; the homoboracyclopropene relationship might be revealed in delocalization of electron density onto boron.

2. Degenerate Allylic Rearrangements

That systems such as **11** may be undergoing a net, dual allylic rearrangement near room temperature became evident when the dienophile reacting with **7** bore groups other than phenyl. Thus, the reaction of **7** with di-*p*-tolylacetylene at 20°–25°C and subsequent purification at or below 25°C gave a colorless adduct that by mass spectral (ms) and nuclear magnetic resonance (NMR) spectral analyses exhibited the molecular ion and aromatic : methyl proton ratio, respectively, expected for adduct **12** (*31*). Incongruous with structure **12**, however, was the presence of two methyl signals in the NMR spectrum of the purified Diels–Alder adduct. The ratio of the methyl signals varied from 3.2 at 47°C to 2.0 at 100°C, where overlapping began to occur. Heating up to 115°C caused merging, but not quite coalescence. Although such spectral changes proved to be reversible up to about 100°C, complete coalescence could not be achieved, for heating such systems at 110°C slowly causes transformation to the fluorescent green borepin system (see below). No such color development was noticed during variable-temperature NMR measurements conducted up to 100°C.

Since a net, dual allylic rearrangement would interconvert **12** and **13**, it is reasonable to ascribe the more intense, downfield methyl signal in the product of Eq. (14) as arising from *p*-tolyl groups in **12** and **13** being attached to the olefinic linkage, and the higher field methyl signal to the *p*-tolyl group at the bridgehead in **13**. With the assumptions that the proportion of **12** and **13** at equilibrium is governed only by probability factors and that the methyl signals of p-tolyl groups attached to olefinic carbon would coincide and be deshielded, a 2:1 signal ratio would be expected. The ratio observed in the spectrum measured at 100°C approaches this value rather satisfactorily. That the methyl ratio at lower temperatures is higher can be ascribed to an electronic preference for a tolyl group to be attached to the olefinic linkage (i.e., an increase in the proportion of **12**). Corroboration that boroles interact with di-p-tolylacetylene to give mixtures of 7-borabicyclo[2.2.1]heptadienes having both bridgehead and nonbridgehead tolyl groups will be offered later when nondegenerate rearrangements of borate salts are discussed.

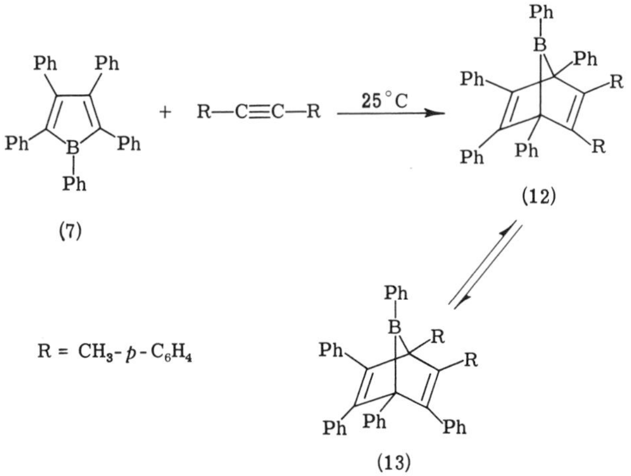

3. Nondegenerate Allylic Rearrangements

The net, dual allylic rearrangement depicted in Eq. (14) should not be construed as necessarily being a simultaneous, pericyclic process; in fact, the occurrence of another, slower rearrangement at temperatures above 100°C makes the importance of sequential [1,3] boron–carbon shifts seem likely for both rearrangements. Thus, a single 1,3-shift would lead from **11** to **14** (k_1); although strained, **14** could undergo a further 1,3-shift to restore the 2.2.1-bicyclic system (k_2) (**11**). At a slower rate ($k_3 < k_1 < k_2$), however, **14** could also undergo an electrocyclic ring opening

---

[2] The 3-benzoborepin system has been synthesized, but by the more conventional route of an exchange between the corresponding stannepin and boron halides (75).

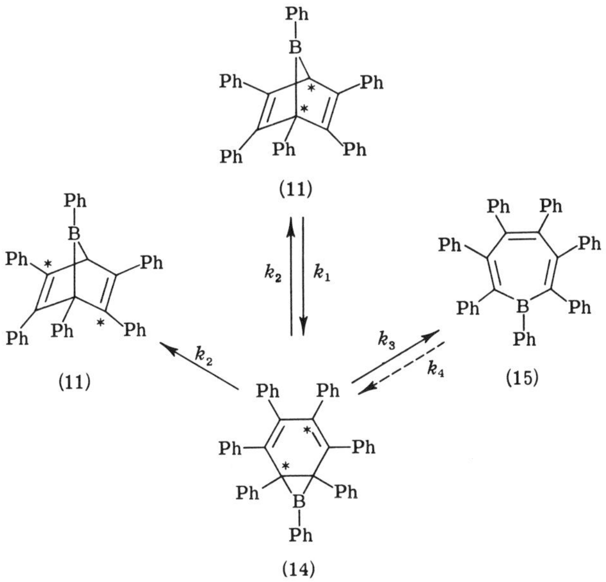

Scheme 1

of its 1,3-cyclohexadiene system to yield the borepin ring **15**[2] (Scheme 1) (*30*).

As to the feasibility of the facile 1,3-shift leading to **14**, the ground-state properties of **11**, especially its electronic spectrum, are worthy of comment. The ultraviolet spectrum of **11** in ethyl ether (with which it does not complex) displays a maximum at 318 nm; its pyridine adduct **16**, however, shows only a shoulder at 273 nm as its longest wavelength absorption. These data are consistent with the view that the available 2*p*-orbital of boron in **11** interacts with the *cis*-stilbene-like double bonds causing π-electron delocalization and hence a bathochromic shift (**17**) (*31*). Whether the boron orbital interacts selectively with one double bond, leading to a bending of the boron bridge to one side (**18**), remains to be investigated by crystallographic measurements[3] (*44*). But the spectrum of the pyridine adduct **16** does support the conclusion that the vacant boron orbital is the cause of the bathochromic shift; the UV absorption of **16** is that expected of an isolated *cis*-stilbene chromophore (*cis*-stilbene absorbs at 280 nm). Thus, it seems likely that such ground-

---

[3] The isoelectronic character of the 7-borabicyclo[2.2.1]heptadiene with the 7-norborna-dienyl cation (*93*) means that similar problems of structure and of the nature of electronic interactions must arise (*76*).

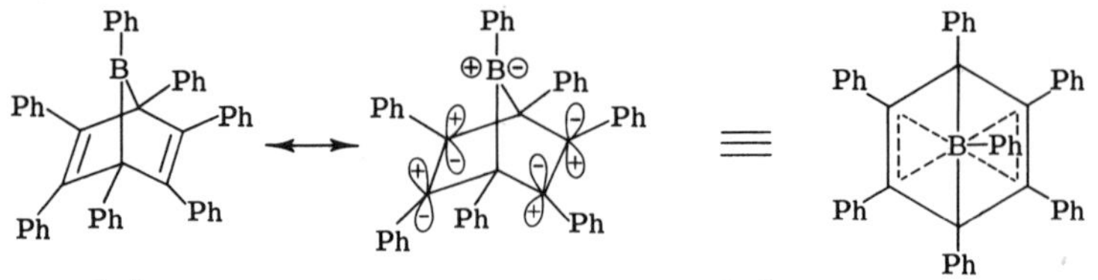

(11)   (17)

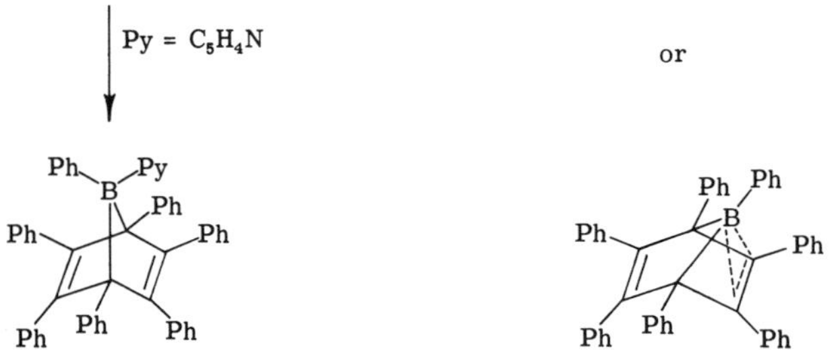

or

Py = C₅H₄N

(16)   (18)

state interactions as those suggested in **17** and **18** facilitate the 1,3-shift proposed in Scheme 1. Of course, the available orbital on boron permits the 1,3-sigmatropic rearrangement to occur in suprafacial manner.

C. *Electrocyclic Rearrangements*

Few electrocyclic reactions leading to boron heterocycles have been reported as such; some examples can be identified, however. Since it is unlikely that the formation of boroles from stannoles occurs by two simultaneous boron-tin exchanges (*36*) [Eq. (11)], the presence of an open-chain boron–tin intermediate can be inferred (**19**). A conrotatory ring-closure of the 1,3-alkadienylborane (analogous to the pentadienyl cation), followed by elimination of Me_2SnCl_2, would yield the borole.

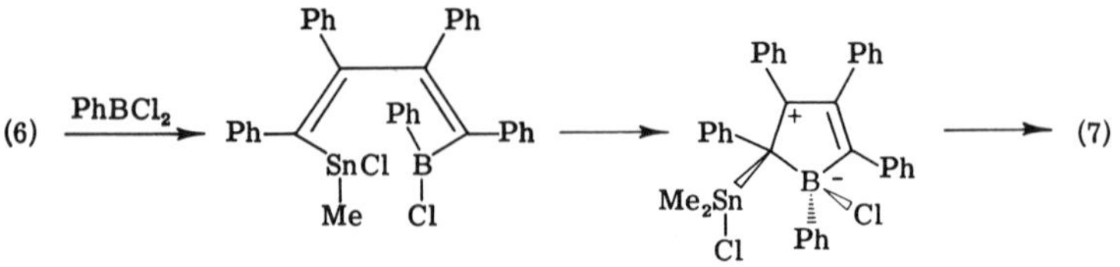

$$(6) \xrightarrow{PhBCl_2} \quad (19) \longrightarrow (20) \longrightarrow (7) \quad (15)$$

A similar interpretation would seem appropriate for the formation of **21** from sodium triethyl-1-propynylborate and acetyl chloride [Eq. (16)]

and of **23** from the irradiation of dicyclohexyl(E-3-methylbuta-1,3-dien-1-yl)borane [Eq. (17)]. The reasonable intermediates, **22** and **24**, can be viewed as undergoing electrocyclic ring closure to yield tetracoordinate boron intermediates (**25** and **26**), which can subsequently undergo a 1,2-alkyl shift from boron to carbon, a migration that can be either thermally or photochemically induced (cf. Sections III,A and III,B).

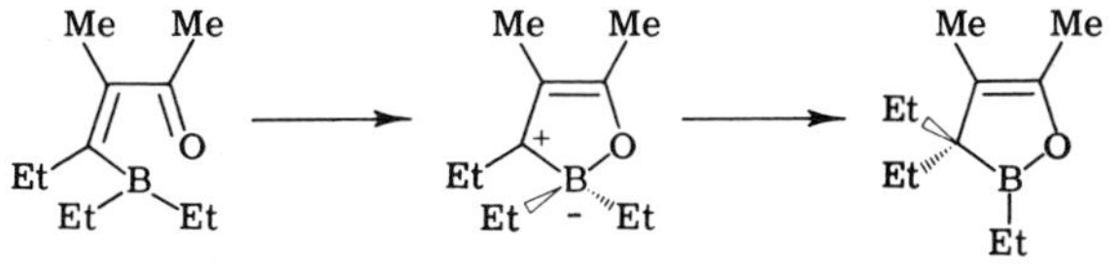

$$\text{(22)} \longrightarrow \text{(25)} \longrightarrow \text{(21)} \qquad \text{ref. } (2) \qquad (16)$$

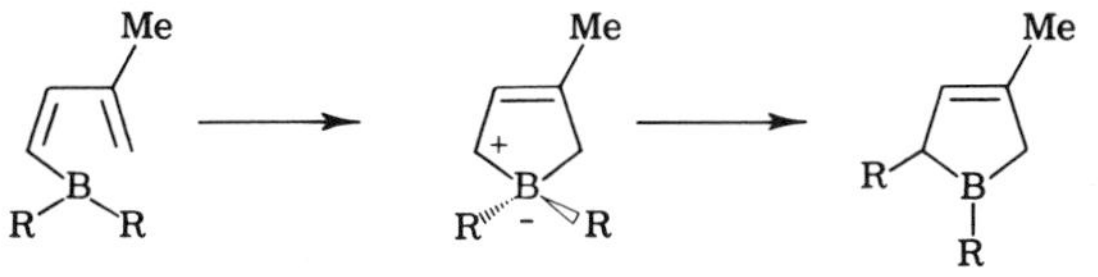

$$\text{(24)} \longrightarrow \text{(26)} \longrightarrow \text{(23)} \qquad \text{ref. } (10) \qquad (17)$$

In none of these examples, however, are the stereochemical strictures on electrocyclic processes observable. The smooth generation of the borepin nucleus from the 7-borabicyclo[2.2.1]heptadiene, via the assumed 7-borabicyclo[4.1.0]heptadiene (Scheme 1), does seem to be the first unambiguous example among organoboranes of a thermally stereochemically permitted electrocyclic ring opening (30). The question of the detectable reversibility of this reaction is still unresolved [i.e., the relative magnitude of k_3 and k_4 in Eq. (18)], but some evidence for the possible, simultaneous presence of **27** and **28** may be offered. When 1-phenyl-2,3,4,5-tetra-p-tolylborole was allowed to react with di-p-tolylacetylene and the resulting adduct **29** was heated to yield the borepin (**30**, R = p-MeC_6H_4), this borepin displayed six methyl peaks of approximately the same intensity. Although coalescence to a pattern of four peaks could be achieved at 113°C, further reduction in the number of peaks was not feasible. These results suggest that the expected borepin **30a** is in equilibrium with its bicyclic valence isomer **30b** or some other nonplanar structure. Clearly, chemical degradation evidence for the coexistence of **27** and **28** (R=C_6H_5) is open to many objections, but it is noteworthy that various protolytic degradations of borepin lead to

both the expected hexaphenylhexatrienes and *cis*-1,2-dihydrohexaphen-
ylbenzene, the protodeboronation products of **27** and **28**, respectively.

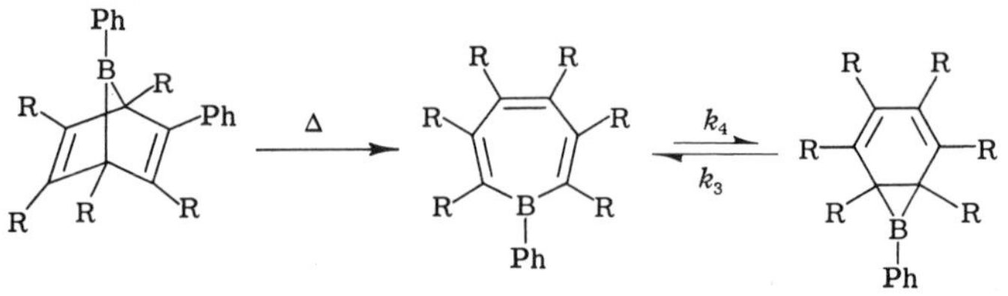

D. *Cycloadditions and Cycloreversions*

Although cycloadditions or their reversal are not rearrangements,
skeletal reorganization often accompanies such reactions ($7 \rightarrow 11 \rightarrow 15$).
The remarkable ease with which pentaphenylborole reacts with alkynes
(*30*) [4 + 2 cycloaddition] should be contrasted with the failure of
heptaphenylborepin to undergo either a [4 + 2] cycloaddition with
diphenylacetylene at 150°C or a [6 + 2] cycloaddition when irradiated in
the presence of the same alkyne (*32*). The difference in behavior
between the five- and seven-membered boron rings may be adduced as
evidence for the antiaromatic character of the former 4-π-electron ring
and for the aromatic character of the 6-π-electron borepin ring, although
effects of planarity and ring strain cannot readily be eliminated as
causes.

The hope might be entertained that the 7-borabicyclo[2.2.1]heptadiene

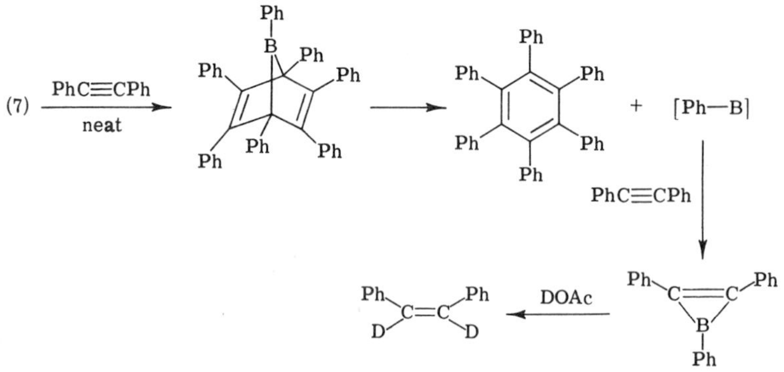

might undergo the elimination of phenylboron(I) due to the aromatization driving force. Apparently the possible rearrangements of this
system (Scheme 1) compete successfully with the cheletropic elimination
of PhB. In attempts to trap the latter subvalent borane, pentaphenylborole was heated at 200°–250°C with four equivalents of diphenylacetylene, and the resulting products were heated with CH_3COOD. Up to
45% of *cis*-stilbene was isolated that by MS and NMR analyses was
shown to consist of 75% α-deuterio, 8% α,α-dideuterio, and 17% protio-
cis-stilbene (*32*). Although a portion of the alkyne may have captured
some subvalent boron fragment [Eq. (19)], clearly the dominant pathway
must involve the generation of a boron hydride during the pyrolysis. In
fact, prolonged heating of various phenylboranes with alkynes leads,
upon acetolysis, to *cis*-alkenes, as well as to the products shown in
Scheme 2 (*34*). The generation of the boron hydride (**31**) accounts for the
formation of a monodeuterated *cis*-alkene in the foregoing reaction; the
hydrocarbons **32** and **33** can be viewed as arising from carboboration and
electrocyclic processes:

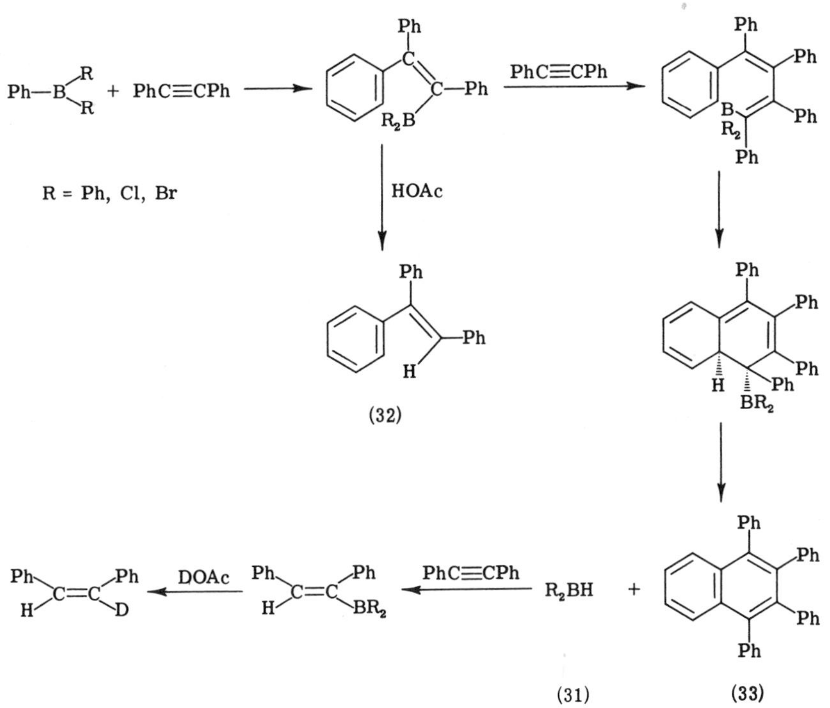

Scheme 2

III

MIGRATIONS IN TETRACOORDINATE ORGANOBORANES

A. Thermally or Chemically Induced Migrations with Retention of Boron

Much of the chemistry of such coordinated organoboranes has been recently and ably surveyed (82). Characteristic thermal and chemically induced migrations have already been cited in foregoing discussions [Eqs. (6), (7), (16), and (17)]. Of interest here are electronic and steric factors determining these migrations. Generally considered, the interaction of a carbanion R^- with an α,β-unsaturated borane 34 could lead to three modes of attachment (77) [Eq. (20), 35–37]. Coordination at boron

$$\text{$>$B--C}\overset{...}{\equiv}\text{C--R'} + \text{R}^- \longrightarrow \text{R--}\overset{|}{\underset{|}{\text{B}}}\text{--C}\overset{...}{\equiv}\text{C--R'} + \text{$>$B--}\overset{\overset{\text{R}}{|}}{\text{C}}\overset{...}{\text{--}}\bar{\text{C}}\text{--R'} + \text{$>$B--}\bar{\text{C}}\overset{...}{\text{---}}\text{C}\overset{R'}{\underset{R}{<}}$$

$$\qquad(34)\qquad\qquad\qquad\qquad(35)\qquad\qquad(36)\qquad\qquad(37)$$

$$(20)$$

is highly favored unless the groups on boron are bulky; β-attachment of R^- is then favored by the delocalization possible for a carbanion α- to boron. Thus, t-butyllithium can add to the β-position of 1-t-butyl-1-boracycloheptadiene-2,5 (41):

$$(21)$$

If the R group in 35 can stabilize negative charge, attack at the α-carbon may result. Such attack may occur via the borate complex 35, which may then rearrange thermally and with high stereoselectivity. The E- and Z-isomers of β-chloro-α-methylstyryllithium, for example, coordinate with triphenylborane and then undergo the rearrangements shown in Eqs. (22) and (23) (68). The observed stereoselectivity requires some synchronous loss of chloride and 1,2-migration of phenyl; in a

$$\underset{\text{Me}}{\overset{\text{Ph}}{>}}\text{C}=\text{C}\overset{\overset{-}{\text{BPh}_3}}{\underset{\text{Cl}}{<}} \longrightarrow \underset{\text{Me}}{\overset{\text{Ph}}{>}}\text{C}=\text{C}\overset{\text{Ph}}{\underset{\text{BPh}_2}{<}} \qquad (22)$$

$$\underset{\text{Me}}{\overset{\text{Ph}}{>}}\text{C}=\text{C}\overset{\text{Cl}}{\underset{\overset{-}{\text{BPh}_3}}{<}} \longrightarrow \underset{\text{Me}}{\overset{\text{Ph}}{>}}\text{C}=\text{C}\overset{\text{BPh}_2}{\underset{\text{Ph}}{<}} \qquad (23)$$

similar way to Eq. (6), attack by an electrophile unleashes a 1,2-migration:

$$
\begin{array}{ccc}
& R'' & & R'' \\
R & \overset{|}{\underset{}{C}}\overset{-}{B}R_2'' & & R & \overset{|}{\underset{}{C}}\overset{-}{B}R_2'' \\
\underset{R'}{\overset{}{C}}=\underset{\delta+\ Cl\ \delta^-}{C} & & vs. & \underset{R_3'B\ \delta^-}{\overset{}{C}}=\underset{\delta+}{C} \\
\end{array}
$$

The unusual reversal of these 1,2-shifts can be observed where **35** is more stable. Thus, certain boronate complexes of the 7-borabicyclo[2.2.1]heptadiene system will cleave, if certain steric factors permit (*31*). Treatment of 1-halo- or 1-methyl-2,3,4,5-tetraphenylborole with diphenylacetylene, reaction of the Diels–Alder adduct with excess methyllithium and heating leads, upon protolytic workup, to >80% yields of pentaphenylbenzene. Other labeling techniques showed that the phenyl group had migrated from carbon to boron (Scheme 3).

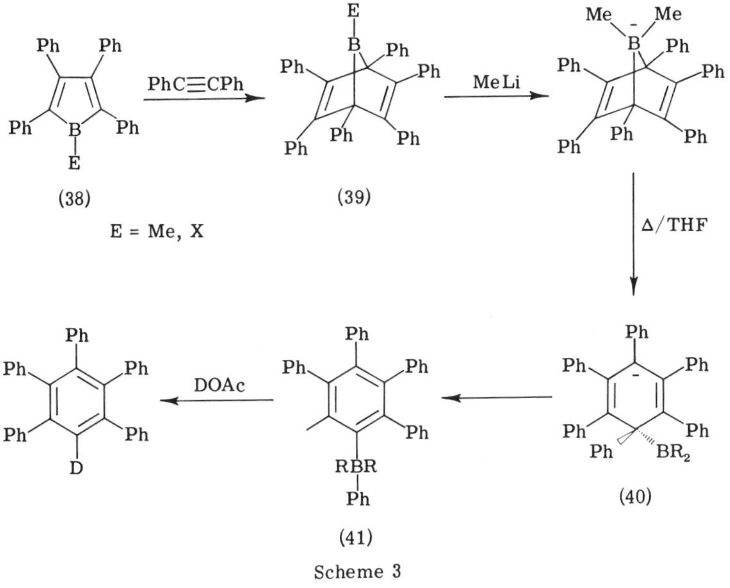

Scheme 3

However, if E in **38** is phenyl or if E is methyl and phenyllithium is used, no rearrangement occurs. Possibly the attachment of one phenyl group to boron in **40** causes too much congestion in the transition state (**42**, where the B-phenyl and *ortho*-phenyls would impede overlap of the boron $2p_z$-orbital and the π-cloud).

As corroboration that the reaction of pentaphenylborole with di-*p*-tolylacetylene leads to the mixture of 7-borabicyclo[2.2.1]heptadienes

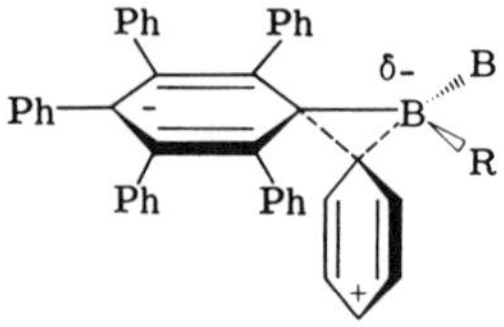

(42)

shown in Eq. (14) (**12** and **13**), it should be noted that an analogous reaction between **38** (E=Me) and di-*p*-tolylacetylene, followed by the reaction sequence in Scheme 3, led to a mixture of tetraphenyl-*p*-tolyl- and triphenyldi-*p*-tolylbenzenes. Such hydrocarbons could only have arisen from intermediates (**12** and **13**) having either phenyl or *p*-tolyl groups α to boron (cf. **39**, **40**, and **42**).

B. *Photochemical Migration with Retention of Boron*

The extensive and informative studies of the photochemistry of tetracoordinate boron have usually been done in the presence of oxygen or protolyzing solvents (*90, 91*), where the organoborane products did not survive. By a study of mesitylborate salts, however, the interesting rearrangement leading to **43** was observed (*58*). Labeling of the mesityl

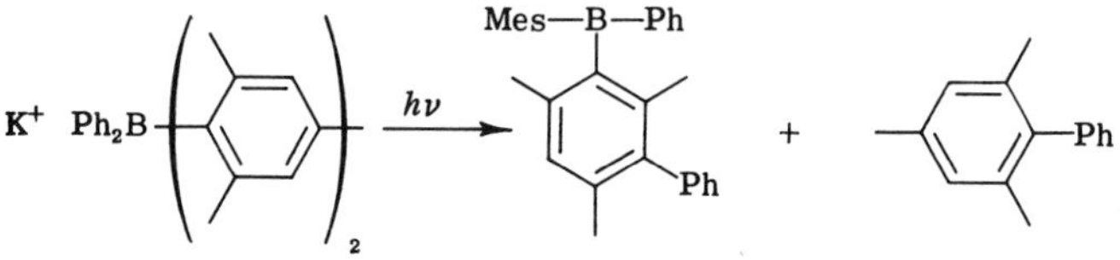

(43)

revealed that the boron altered its site of attachment in forming **43**, presumably through borabicyclic intermediates similar to **11** or **14** (Scheme 4).

Photorearrangements of boron aryls seem to require the presence of donor solvents (*92*); one reason for this may be that 1,2-migrations from boron to carbon involve bridging transition states (cf. **42**). Clearly in trigonal tricoordinate boron, groups that must interact are farther apart than in tetrahedral boron complexes:

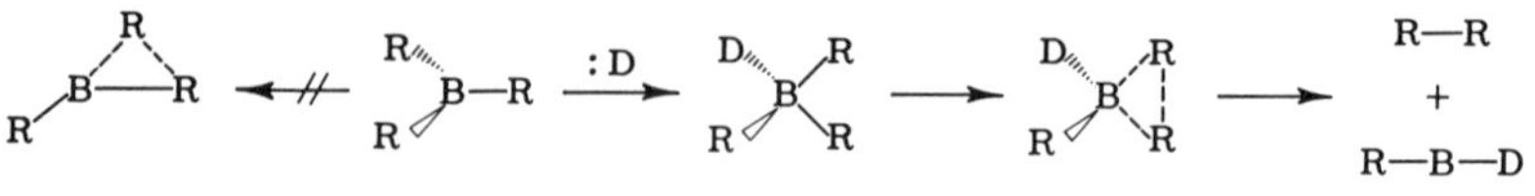

$$(25)$$

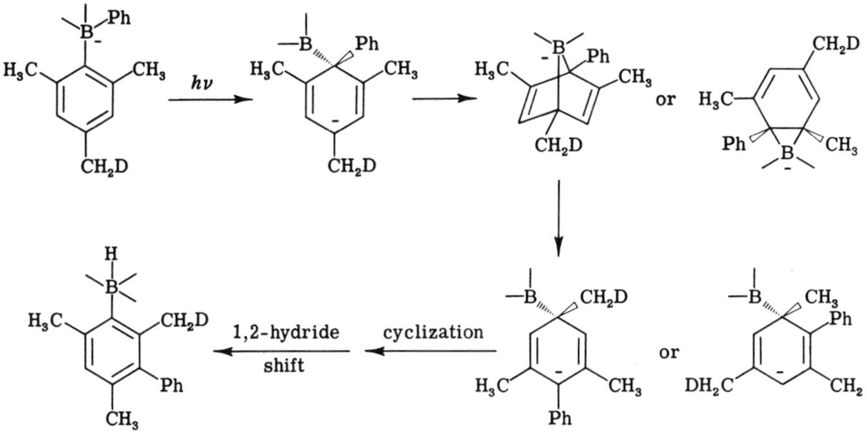

Scheme 4

In addition, if any boron(I) fragment is to be generated, then solvation by a Lewis base D would also be advantageous. Although the irradiation at 254 nm of the triphenylborane pyridine complex under nitrogen in dry tetrahydrofuran (THF) does yield 8% of biphenyl after 24 hours, inclusion of 25% (v/v) of methanol in the THF raises the yield of biphenyl to 20% (45). Possibly this protic medium aids the trapping of labile intermediates.

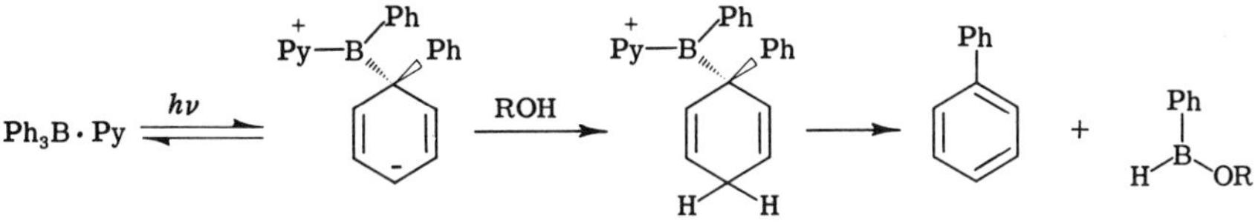

$$(26)$$

The irradiation of sodium tetraphenylborate(III) at 254 nm in the absence of oxygen and protic solvents (in dry THF under nitrogen) provides evidence supporting the ideas of Scheme 4 and Eqs. (25) and (26) (46). The work-up of the irradiation solution with o-deuterioacetic acid (DOAc) led to the isolation of 40–70% of biphenyl and 16% of m- and p-terphenyls; the biphenyl was about 66%, the m-terphenyl 50%, and the p-terphenyl about 70% monodeuterated. In addition, treatment with DOAc caused the evolution of hydrogen gas (80% yield, based on the moles of NaBPh$_4$), which was composed of 75% D$_2$, 20% DH, and 5% H$_2$. A reaction pathway consistent with these findings and those of Williams and co-workers (58, 90, 91) is presented in Scheme 5. The m-terphenyl products could arise from alternative isomerizations of intermediates like 44 (cf. Scheme 4), and hence the position of deuteration in

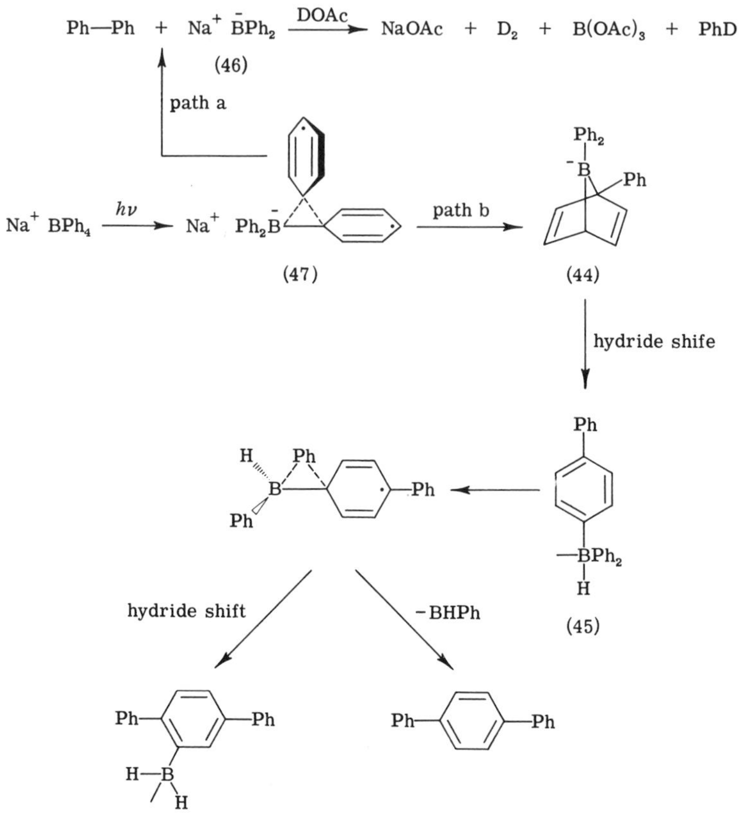

Scheme 5

biphenyl could vary with the isomers of **45** that are formed. The hydride shift in the aromatization of **44** and its analogs is very similar to the phenyl shift discussed in Section III,A (Scheme 3). From the relative amounts of D_2 and undeuterated hydrocarbons, path a is at least 2–3 times more important than path b.

The inability of such photolyzate solutions to hydroborate 1-octyne rules out the presence of neutral boron hydrides. However, the detection of HD, upon the work-up with DOAc, and the formation of undeuterated toluene, upon treatment of the photolyzate with benzyl chloride and deuterolytic work-up, clearly support the presence of borohydrides, such as **45** (6). Finally, evidence supporting the generation of sodium diphenylborate(I) (**46**) or a similar product was obtained by conducting the photolysis in the presence of diphenylacetylene. Since monomeric **46** is formally isoelectronic with a carbene, adducts like **48** might be expected. Work-up with DOAc did yield principally *cis*-stilbene that was 20% α,α'-dideuterated, a finding consistent with the presence of **48**.

$$\text{Ph—C≡C—Ph} \; + \; \text{Na}^+ \text{BPh}_2 \longrightarrow \underset{\substack{\displaystyle | \\ \text{Ph}_2}}{\overset{\text{Ph}}{\underset{\text{B}^-}{\diagdown}}}\text{C}{=}\text{C}\overset{\text{Ph}}{\diagup} \xrightarrow{\text{DOAc}} \underset{\text{D}}{\overset{\text{Ph}}{\diagdown}}\text{C}{=}\text{C}\underset{\text{D}}{\overset{\text{Ph}}{\diagup}} \quad (27)$$

$$\begin{array}{cc} (46) & \\ & (48) \end{array}$$

The behavior of mixed and substituted tetraarylborate salts upon irradiation gives further delineation to the mechanism of these rearrangements (*46*). Photolyses of sodium phenyl(tri-*m*-tolyl)borate (**49**) and sodium phenyl(tri-*p*-tolyl)borate (**50**) in THF, followed by acetolytic work-up, yielded mixtures of *m*-methylbiphenyl and *m,m'*-bitolyl (1.1:1.0) and of *p*-methylbiphenyl and *p,p'*-bitolyl (1.1:2.7), respectively. The absence of any isomerized bitolyls, and especially of biphenyl itself, rules out interionic coupling and any type of biaryl coupling other than at carbon–boron bonds. Finally, the photoinertness of sodium bis(*o,o'*-biphenylene) (**51**) borate upon prolonged irradiation at 254 nm in THF, exactly those conditions under which sodium tetraarylborates readily react, provides telling support for the importance of bridging transition states, such as **47**, in such 1,2-photomigrations. Since the spiro structure in **51**, where the biphenylene rings are perpendicular, could not attain such bridging (**52**) without great strain, the photomigration would be unattainable [Eq. (28)].

$$\begin{array}{ccc} (51) & & (52) \end{array}$$

Photoisomerizations of tetracoordinate 1-alkynylboranes offer some synthetic prospects of unusual significance, namely, an entree into the boracycloprene system (*45*). Thus, irradiation of lithium triphenyl(phenylethynyl)borate in THF and acetolytic work-up gave, in high conversion, a 1:5.6 ratio of biphenyl and *cis*-stilbene. The low yield of biphenyl (15%) and hydrogen gas (13%) shows that the cleavage of the path a type (Scheme 5) is minor here. When the photolyzate was treated with DOAc, the isolated *cis*-stilbene was 45% α,α'-dideuterated, 44% α-deuterated, and 11% undeuterated. The principal course of this reaction is, therefore, that depicted in Scheme 6. There is no evidence that **54** persists as a monomer, but there cannot be a large proportion of **53** present; such arylvinyllithium compounds are known to undergo rapid cis,trans-isomerization in THF (*12*).

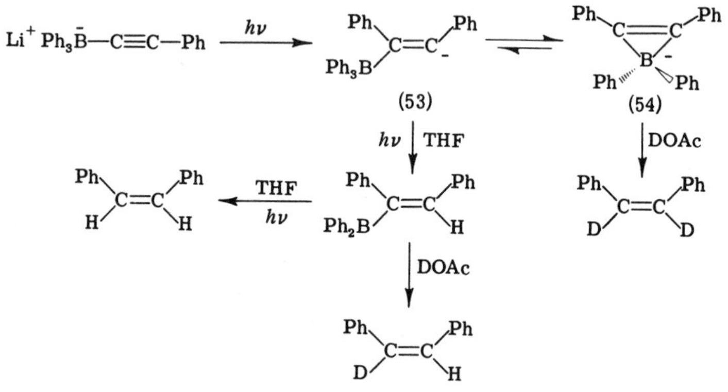

Scheme 6

With the aspiration of preparing the pyridine complex of the potentially aromatic triphenylboracyclopropene (**56**), a similar photolysis has been conducted with diphenyl(phenylethynyl)borane pyridinate (**55**) (*45*). Again, however, the photolysis in THF and work-up with DOAc provided almost exclusively *cis*-stilbene (only 6% of the *trans*-isomer), which was 49% α,α'-dideuterated, 41% α-deuterated, and 10% undeuterated; in this case, no biphenyl was detected. Accordingly, a photopathway similar to that in Scheme 6 seems to exist. By use of lower temperatures and shorter irradiation times, attempts are being made to optimize and preserve the concentration of the long-sought initial product **56**.

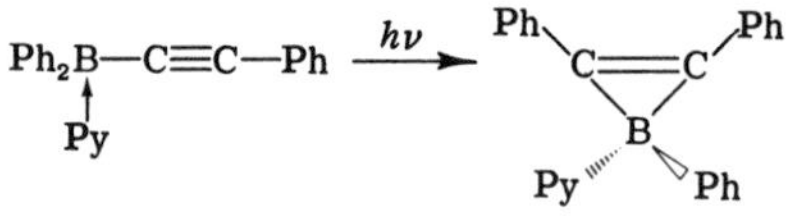

(29)

C. Migration with Deboration

Of related interest to the foregoing photomigrations is the production of biaryls by the oxidation of tetraarylborate salts [Eq. (30)]. For this purpose, electrooxidation (*54*), halogen (*81*), ceric (*54*), ferric (*83*), hexachloroiridate(IV) (*1*), and singlet oxygen (*13*) have been employed.

$$M^+BAr_4 \xrightarrow{-2e^-} Ar\!-\!Ar + M^+ + Ar_2B^+ \qquad (30)$$

By the use of mixtures of tetraphenylborate and its perdeuterated derivative, the coupling reaction with ceric ion or electrochemical means has been shown to be intraionic (*54*). A further examination of the response of the potassium salts of the phenyl(tri-*m*-tolyl)borate (**57**) and the phenyl(tri-*p*-tolyl)borate (**58**) ions to various oxidizing agents has revealed varying preferences to the mode of biaryl coupling (*47*). These differences in the composition of the resulting biaryls point to a change in reaction mechanism (Table II) as the oxidant is varied from a transition metal ion (Ce^{4+} and Fe^{3+}) or organic oxidant (DDQ)[4] to a halogen source (I, Br, and NBS). Ceric ion oxidation having already been shown to occur via an electron-transfer process, one might expect that other oxidants would lead to similar ratios of methylbiphenyl to bitolyl, if such oxidations also occurred by electron transfer. Although the oxidation of the meta anion (**57**) uniformly favors the formation of the bitolyl (as is partly explicable on statistical grounds), oxidation of the para anion (**58**) with halogen oxidants favors, instead, the dominance of the methylbiphenyl. A similar contrast is seen in the behavior of potassium bis(*o,o'*-biphenylene)borate (**59**) toward the known electron-

TABLE II
OXIDATIVE COUPLING OF POTASSIUM
PHENYL(TRI-*m*-TOLYL)BORATE AND POTASSIUM
PHENYL(TRI-*p*-TOLYL)BORATE

	Ratios of the bitolyl and methylbiphenyl products	
Oxidant	*m*-Tolyl salt (**57**)	*p*-Tolyl salt (**58**)
I_2	$3.4:1.0^a$	$1.0:3.3^b$
Br_2	$1.8:1.0$	$1.0:2.4$
NBS^c	$1.7:1.0$	$1.0:2.0$
DDQ^c	$1.9:1.0$	$2.0:1.0$
Fe^{3+}	$1.4:1.0$	$2.2:1.0$
Ce^{4+}	$1.4:1.0$	$2.5:1.0$

^{a} Ratio is that of *m,m'*-bitolyl to *m*-methylbiphenyl.
^{b} Ratio is that of *p,p'*-bitolyl to *p*-methylbiphenyl.
^{c} NBS = *N*-bromosuccinimide; DDQ = 2,3-dichloro-5,6-dicyanobenzoquinone.

[4] DDQ = 2,3-dichloro-5,6-dicyanobenzoquinone; NBS = *N*-bromosuccinimide.

transfer oxidant Ce^{4+}, as compared with the halogen oxidant I_2. The anion **59** is readily attacked by ceric ion and, upon protodeborative work-up with HOAc, a 75% yield of o,o'-quaterphenyl is obtained. Iodine oxidation and the usual protodeborative work-up on the other hand, gives a >90% yield of o-iodobiphenyl and biphenyl; no quaterphenyl is detectable.

Because of the similarity of the biaryl ratios for the oxidants Ce^{4+}, Fe^{3+}, and DDQ, it is reasonable to conclude that they all operate as electron-transfer agents, as has been well documented for Ce^{4+}. Preferential formation of the bitolyl from either salt clearly implies that the tolyl group undergoes preferred attack, for a random oxidation would yield a 1:1 biaryl mixture. It is understandable that the tolyl group should be more prone than the phenyl to electron loss. If aryl migration from boron to carbon (**60** → **61**) were completely random, then a maximum of 2:1 in the bitolyl:p-methylbiphenyl ratio would be expected. Since the observed ratios (1.4:1.0–2.5:1.0) are within, or only slightly outside of, this maximum, no great migratory preference is revealed in the collapse of **60**.

$$(31)$$

The behavior of the halogen oxidants, however, cannot be accommodated by an electron-transfer mechanism, but clearly require a pathway governed by different electronic factors. Attack by some electrophilic form of halogen, X_n^+, on **57** would produce an electron deficiency α to boron, requisite for migration [cf. Eqs. (6) and (7)], and simultaneously would attain a configuration (**62**) stabilized by the methyl group. The biaryl ratio (>2:1) indicates that iodine not only attacks the m-tolyl group selectively, but also that m-tolyl subsequently migrates preferentially. With the para salt **58**, attack by X_n^+ at either the para or ortho positions of the tolyl group is seriously hindered by the methyl and triarylboryl groups, respectively. As a consequence, electrophilic attack on the more accessible phenyl group is favored, as is the ensuing migration leading to p-methylbiphenyl (**63**).

An electrophilic mechanism for iodinative biaryl couplings clarifies the behavior of **59**. Since ceric ion oxidation involves electron-transfer with little or no aryl migratory preference, there seems to be no bridging (**61**) in the aryl group migrating from boron to carbon. Hence, radical **64** can collapse to yield a quaterphenyl precursor. On the other hand, the attack of I_2 seems to require some bridging of the migrating aryl group (cf. preference of *m*-tolyl migration with **57**). But, because of the perpendicular array of the aryl groups in **59**, a bridging transition state should be highly strained (**65**). Since such bridging would make C—C bond coupling energetically unfavorable, the alternative reaction of simple iododeboration occurs exclusively [Eq. (32)].

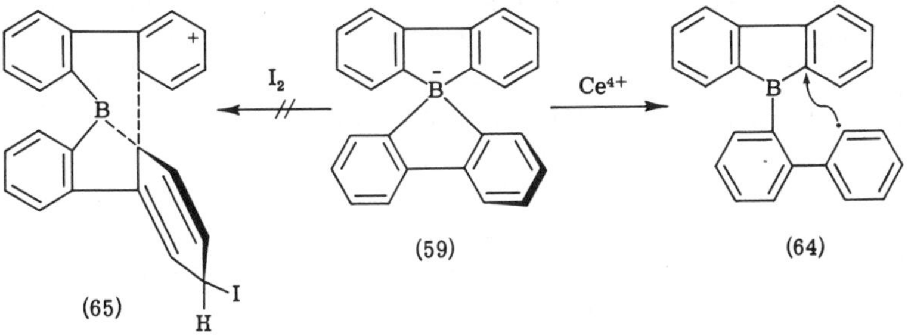

$$(32)$$

Although the ratio of biaryls formed in these halogenative oxidations is somewhat sensitive to the reaction medium, it is noteworthy that the amount of biaryl formation is quite dependent both on the nature of the solvent and the metal ion (L^+, Na^+, or K^+). Although potassium phenyl(tri-*p*-tolyl)borate (**58**) dissolved in chloroform reacted with iodine to yield a 1.4 : 1.0 ratio of biaryl to *p*-iodotoluene, the same reaction in dimethylformamide gave almost exclusively *p*-iodotoluene. Likewise, the sodium analog of **58** gave a biaryl:*p*-iodotoluene ratio of 1 : 32 (*48*).

IV

ALLYLIC ISOMERIZING OR DERIVATIZING REACTIONS OF ORGANOALANES

A. *Isomerization*

Since the preparation and properties of unsaturated organoaluminum compounds have been the subject of a recent critical review (*62*), attention will be directed here to newer evidence bearing on allylic structure and its benzylic analogs. Examination of the NMR spectra of donor solvent-free allylaluminum derivatives show them to exist as mixtures of dynamically interconverting structures in the temperature range of $+30°$ to $+50°C$ (*74, 86*). Purified, dimeric methallyl (dimethyl)aluminum in toluene-d_8 solution displays only one signal for the methylene protons at $+30°C$; these protons appear, however, as broad bands at $-70°C$ (*86*). Likewise, allyl(diethyl)aluminum exhibits an allyl signal of the AX_4 type at $+50°$, but an absorption pattern of the $ABCX_2$ at $-70°C$ (*74*), indicative of a nonexchanging σ-allyl group. Cyclopentadienylaluminum compounds, on the other hand, give only a singlet for the ring protons even down to the limit of measurement, $-90°C$ (*70, 71*). Other substituted allylic-type systems, tribenzylaluminum (*19*), diisobutyl(1,1-dimethyl-3-indanyl)aluminum (*27*) (**66**, see p. 95) and 1-acenaphthenyl(diisobutyl)aluminum (*27*) (**67**), possess NMR spectra at $+25°C$ consistent with the nonequilibrating expected structures. Although their tendency to be associated is not known, there is no significant broadening of the CH_2 protons in $(PhCH_2)_3Al$ down to $-70°C$ (*19*). The rather broad benzylic reasonances of **67** centered at $35°C$ at (δ) 3.43 and 2.75 ppm with relative areas of $2:1$ sharpened into a doublet and a triplet at $112°C$. Such a typical A_2X pattern indicates an averaging of the environment of these protons (*27*) [Eq. (33)]. Whether such inversion at the carbon–aluminum bond occurs via a small amount of dimer is not established.

$$\underset{\text{(67a)}}{H_A\overset{H_B}{\diagdown}\overset{H_X}{\diagup}AlBu_2} \quad \overset{>100°C}{\rightleftharpoons} \quad \underset{\text{(67b)}}{H_B\overset{H_A}{\diagdown}\overset{H_X}{\diagup}AlBu_2} \qquad (33)$$

Since some unsolvated allylic systems exhibit such facile rearrangement, it is evident that Lewis bases may drive the equilibrium toward one form by preferential coordination [Eq. (34)]. If the relative stabilities

$$\underset{\text{(68)}}{\overset{H}{\underset{H}{>}}C=C-C\overset{R'}{\underset{R'}{<}}AlR''} \quad \overset{:D}{\rightleftharpoons} \quad \underset{\text{(69)}}{R''Al\overset{H}{\underset{D}{-}}C-C=C\overset{R'}{\underset{R'}{<}}} \qquad (34)$$

of the allylic forms are similar, then the aluminum may favor that carbon offering less steric hindrance to coordination (i.e., **69**). Thus, in the hydralumination of 1,1-diphenylallene with diisobutylaluminum hydride in the presence of a donor, the apparent proportions of 1- and 3-alumino adducts (as determined by hydrolysis) indicate that a stronger donor favors the formation of more of the 3-alumino adduct (71) (*37, 38*).

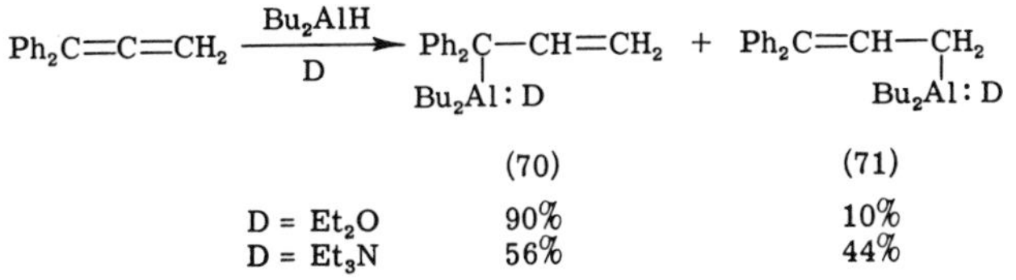

$$Ph_2C{=}C{=}CH_2 \xrightarrow[D]{Bu_2AlH} \underset{\underset{(70)}{Bu_2Al:D}}{Ph_2C{-}CH{=}CH_2} + \underset{\underset{(71)}{Bu_2Al:D}}{Ph_2C{=}CH{-}CH_2} \tag{35}$$

D = Et$_2$O	90%	10%
D = Et$_3$N	56%	44%

In the preceding discussion it was stated that hydrolysis gave "the apparent proportions" of 70 and 71, for the uncertainty always exists that hydrolysis proceeds via an allylic rearrangement. In fact, the mode and reagent used in hydrolyzing allylic aluminum systems can bring about varying amounts of rearrangements (cf. Section IV,B). Hence, when feasible, physical methods should be used to verify the validity of hydrolysis data. Thus, the addition of diisobutylaluminum hydride to 1,1-diphenyl-1,3-butadiene and subsequent hydrolysis with D$_2$O led to a mixture of 73 and 74 (*37, 38*). In this instance, the absence of hydrolytic rearrangement was demonstrable through an NMR examination of the aluminum adducts before hydrolysis; no 72 was observable. Accordingly, the rearrangement leading to 73 occurred during the hydralumination (72 → 75).

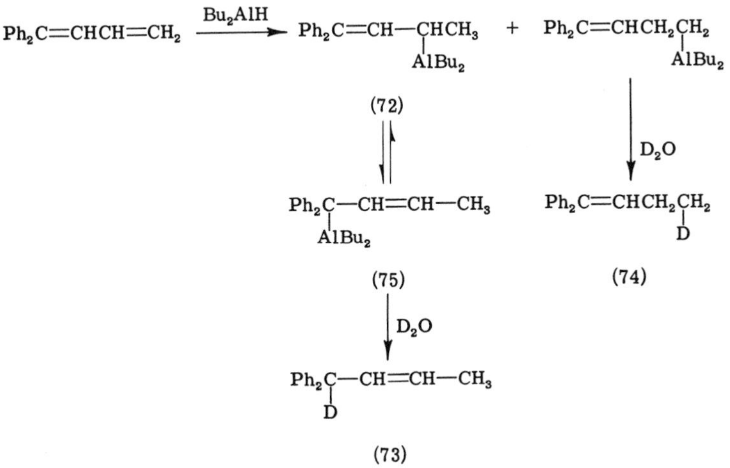

$$Ph_2C{=}CHCH{=}CH_2 \xrightarrow{Bu_2AlH} \underset{\underset{(72)}{AlBu_2}}{Ph_2C{=}CH{-}CHCH_3} + \underset{AlBu_2}{Ph_2C{=}CHCH_2CH_2}$$

(72) ⇅

$$\underset{\underset{(75)}{AlBu_2}}{Ph_2C{-}CH{=}CH{-}CH_3} \qquad \underset{\underset{(74)}{D}}{Ph_2C{=}CHCH_2CH_2} \tag{36}$$

D$_2$O ↓ (from 74)

$$\underset{\underset{(73)}{D}}{Ph_2C{-}CH{=}CH{-}CH_3}$$

On the other hand, to anticipate the discussion in Section IV,B and show how misleading hydrolysis products can be, consider the case of diethyl(3-phenyl-2-propen-1-yl)aluminum etherate (76), whose structure is securely based on infrared (IR) and NMR spectral data. Here, hydrolysis gives almost exclusively the rearranged product (74).

$$Ph—CH=CH—CH_2AlEt_2 \cdot OEt_2O \xrightarrow{H_2O} PH—CH_2CH=CH_2 + Ph—CH=CHMe \tag{37}$$

$$(76) \qquad\qquad\qquad 98\%$$

B. Derivatizing Allylic Rearrangements

The foregoing discussion has stressed the need for reliable structural information for the allylic or benzylic aluminum under consideration, for without such data it is impossible to determine whether a chemical reaction has led to a skin rearrangement. Thus, although a sample of tribenzylaluminum prepared *in situ* was reported to undergo the Tiffeneau–Delange rearrangement (88) shown in Eq. (2) (57), the pure compound has been isolated only recently (19). Tribenzylaluminum undergoes deuterodealumination and insertion of diphenylacetylene without rearrangement; but smaller substrates, carbon dioxide (19) and acetylene (85), give predominantly o-tolyl products (19, 27) (Scheme 7).

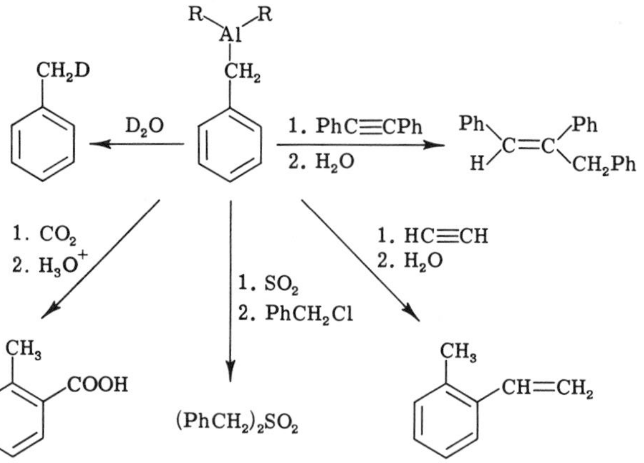

Scheme 7

The behavior of the structurally well-defined 1,1-dimethyl-3-indanylaluminum system (66) toward various reagents has also proved instructive (25–27) (Scheme 8).

Furthermore, the 1-acenaphthenylaluminum system (67), possessing a more reactive aromatic system, was found to respond most readily to various reagents to give extensive amounts of rearrangement products (25–27). Significantly, with this system both kinetic and thermodynamic

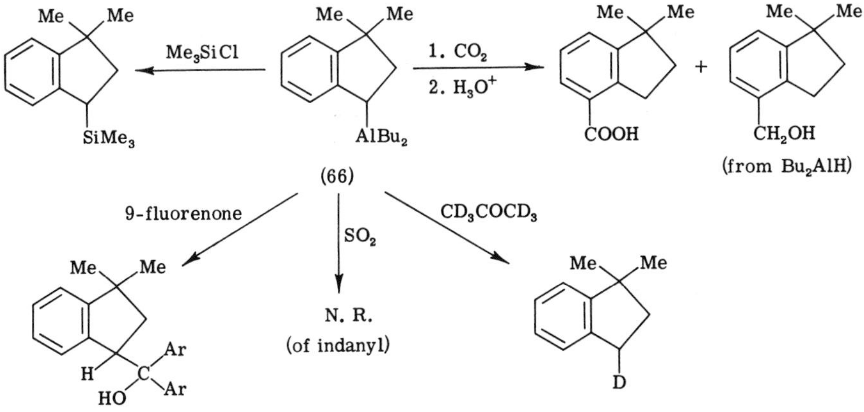

Scheme 8

control of products became noticeable (26). First, the hydrolysis was
shown to be very sensitive to conditions [Tables III and IV; Eq. (38)]. In

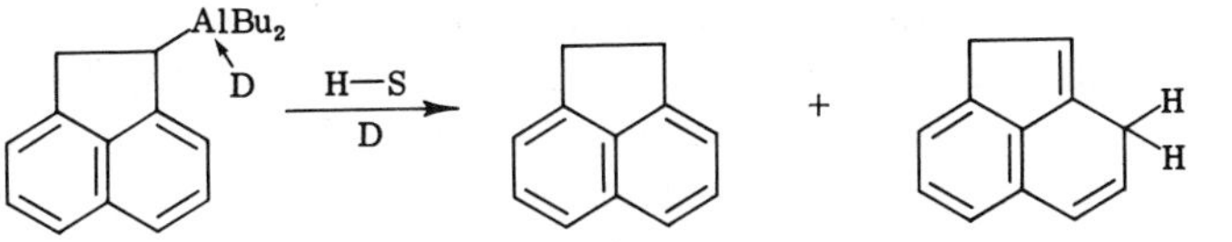

$$(38)$$

TABLE III

THE INFLUENCE OF DONORS ON THE PROPORTIONS OF 1,3-
DIHYDROACENAPHTHYLENE AND ACENAPHTHENE FORMED IN THE
HYDROLYSIS OF 1-ACENAPHTHENYL(DIISOBUTYL)ALUMINUM (67)[a]

Donor	Acenaphthene (77)	1,3-Dihydroacenaphthylene (78)
None	73	27
$(CH_3CH_2)_2O$	90	10
Tetrahydrofuran	92	8
Pyridine	95	5
n-C_4H_9Li	100	0
$(CH_3CH_2)_3N$	73[b]	27

[a] Treatment of 67 in heptane solution with the given donor and
subsequent treatment with H_2O at 25°C.

[b] The presence of $(CH_3CH_2)_3N$ did not alter the ratio of 77:78;
hence, it appears that this sterically hindered amine complexes
very slightly with 67 and thus does not influence the course of
hydrolysis.

TABLE IV

THE INFLUENCE OF PROTON SOURCE ON THE FORMATION OF 1,3-
DIHYDROACENAPHTHYLENE IN THE PROTODEALUMINATION OF 1-
ACENAPHTHENYL(DIISOBUTYL)ALUMINUM (67)[a]

Proton source	Acenaphthene (77)	1,3-Dihydroacenaphthylene (78)
H_2O	76	24
20% HCl	61	39
CH_3OH	66	34
$(CH_3)_2CHOH$	62	38
$(CH_3)_3COH$	57	43

[a] Reaction was conducted at 0°C.

general, treatment with H_2O leads to a higher proportion of acenaphthene as the donor present increases in strength: none < Et_2O < THF
< pyridine < n-butyllithium. This finding suggests that coordination of
the proton source at aluminum in **67** is important for the formation of
1,3-dihydroacenaphthylene (**78**). With hydroxylic proton sources, the
proportions of **78** rises in the order: H_2O < MeOH < i-PrOH < t-BuOH.
This trend may reflect the increasing effective basicity of these Lewis
bases toward aluminum. Finally, and perhaps not surprisingly, the
strongest and less selective proton source used, 20% HCl, gives the
highest amount of **78** (27).

The reactions of **67** with carbon dioxide, acyl chlorides, and, in fact,
aliphatic or aromatic ketones lead to 3-substituted products (Scheme 9)

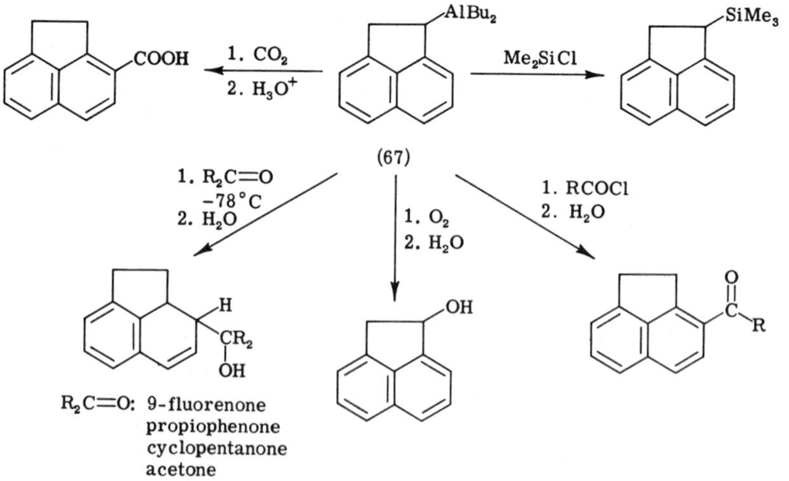

$R_2C=O$: 9-fluorenone
propiophenone
cyclopentanone
acetone

Scheme 9

(25–27). The reactions with oxygen and Me₃SiCl, on the other hand, occurred without allylic rearrangement; it is noteworthy that the presence of ether prevented any reaction of **67** with the chlorosilane (27).

C. *Kinetic versus Thermodynamic Control*

The reaction given in Scheme 9 between the 1-acenaphthenylaluminum system and ketones at −78°C is all the more striking when one learns that the same reagents react at 25°C to yield essentially only the 1-acenaphthenyl products. Clearly one is dealing with a case of kinetically favored adduct **79** rearranging to a more stable system **80** (26). The

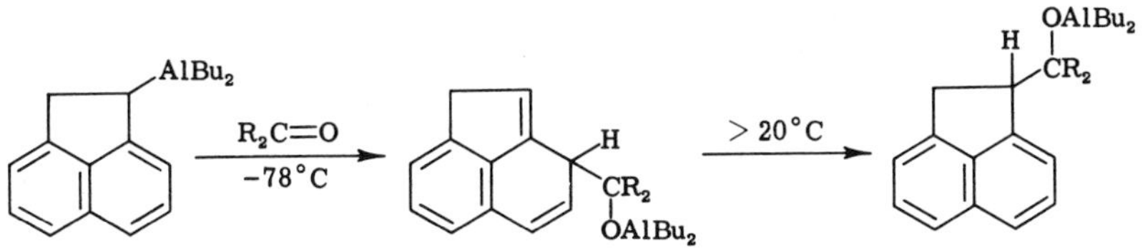

$$(39)$$

failure of 3-substituted products derived from carbon dioxide and acyl chlorides to undergo a similar rearrangement to 1-acenaphthenyl derivatives may be due to their undergoing an irreversible enol salt formation [Eq. (40)]. Such salt formation (**81**) would prevent reversal to the components; the aluminum adducts **79** do not have this capability.

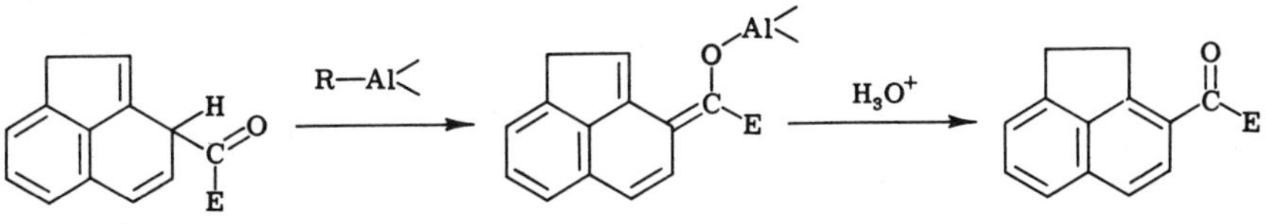

$$(40)$$

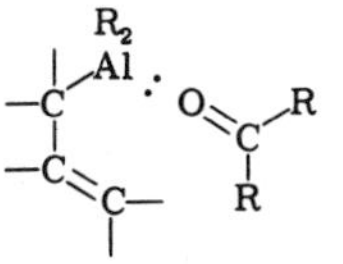

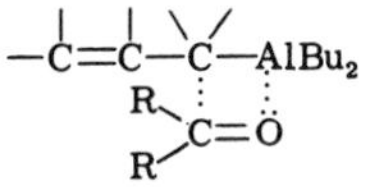

Since **79** is formed more rapidly than **80**, it is attractive to suggest that, where a choice is to be made, a six-membered transition state (**82**) is of lower energy than a four-membered configuration (**83**). A lower energy of **82** may stem from its resemblance to the transition state of an allowed [3, 3] sigmatropic process (cf. **3**); similarly, **83** resembles a disallowed $[_\sigma 2_s + _\pi 2_s]$ interaction.

A similar relationship between kinetic and thermodynamic control appears to exist in the reactions of 3-phenyl-2-propyn-1-ylaluminum derivatives (Scheme 10) (*40*).

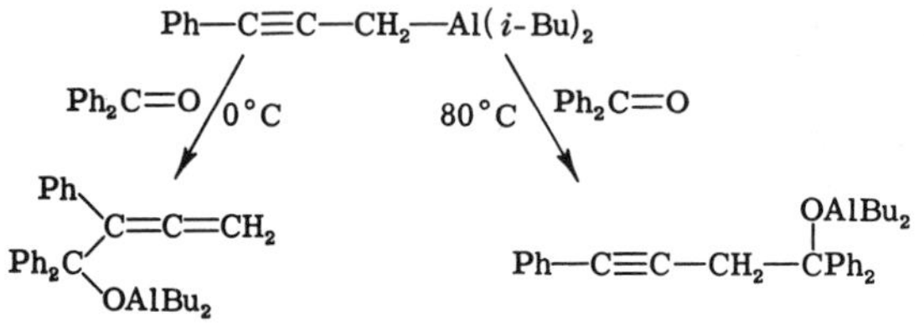

$$Ph—C{\equiv}C—CH_2—Al(\textit{i-}Bu)_2$$

Ph$_2$C$=$O $\diagdown$ 0°C   80°C $\diagdown$ Ph$_2$C$=$O

Ph$\diagdown$
C$=$C$=$CH$_2$
Ph$_2$C$\diagup$
OAlBu$_2$

OAlBu$_2$
|
Ph—C$\equiv$C—CH$_2$—CPh$_2$

Scheme 10

V

CARBALUMINATION VIA INTERMEDIATE COMPLEXES

A. *Behavior of Aluminum-Chiral Carbon Bonds*

The preceding suggestion on the thermally disallowed character of planar four-centered concerted insertions of carbonyl groups into carbon–aluminum bonds (**83**) might be tested with aluminum alkyls having carbon–aluminum bonds of known configuration. Yet the synthesis of such appropriate alkyls of known orientation, either of stereoisomeric or geometric isomeric orientation, must take into account how readily configurational change can occur about the carbon–aluminum bond. Geometric isomers of vinylalanes (**84**) are usually quite stable, and their reactions with halogen, carbon dioxide, carbonyl derivatives, olefins, and acetylenes uniformly occur with retention of configuration (*17, 28, 29, 39, 89, 99, 100*).

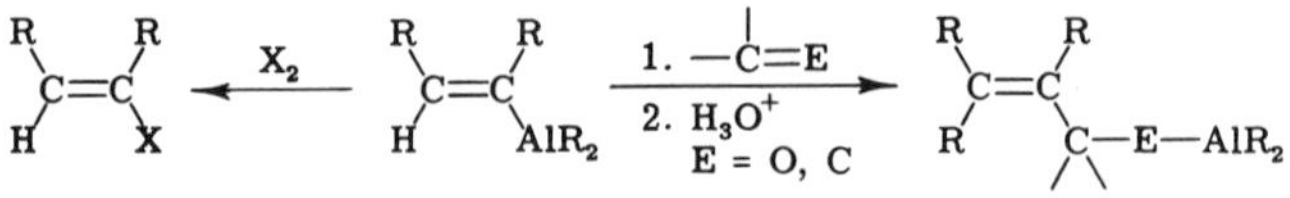

$$\begin{array}{ccccc}
\underset{H}{\overset{R}{\diagdown}}C=C\underset{X}{\overset{R}{\diagup}} & \xleftarrow{\ X_2\ } & \underset{H}{\overset{R}{\diagdown}}C=C\underset{AlR_2}{\overset{R}{\diagup}} & \xrightarrow[\ 2.\ H_3O^+\ ]{\ 1.\ -\overset{|}{C}=E\ } & \underset{R}{\overset{R}{\diagdown}}C=C\underset{\overset{|}{C}-E-AlR_2}{\overset{R}{\diagup}}
\end{array} \tag{41}$$

$$\text{(84)}E = O, C$$

Only in the special instances of β-dialkylamino-, β-alkoxy-, α-trialkylsilyl-, and α-trialkylgermyl-vinylalanes does the configuration (**85**) undergo facile isomerization below 20°C (*35, 42, 43*).

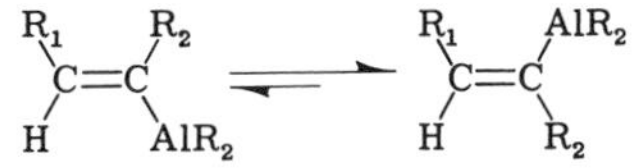

$$(85) \quad R_1 = R_2N, RO; \quad R_2 = R, H, Ar$$
$$or$$
$$R_1 = R, H, Ar; \quad R_2 = R_3Si, R_3Ge$$

(42)

With alkyl carbon–aluminum bonds, NMR studies have shown that configurational inversion occurs more readily than most vinylalanes, namely, between 50 and 100°C *(49, 94)*. Accordingly, although *syn*-hydralumination of suitable olefins might lead to stereochemically defined carbon–aluminum bonds **(86)**, configurational inversion might well occur at a rate (k_2) comparable to hydralumination (k_1) [Eq. (43)].

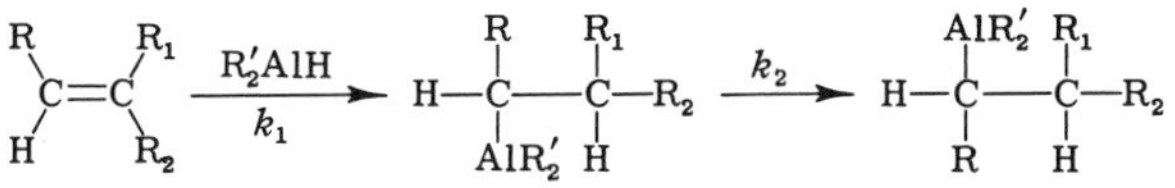

(43)

Thus, hydralumination of 1,1-dimethylindene **(87)** with diisobutylalu-

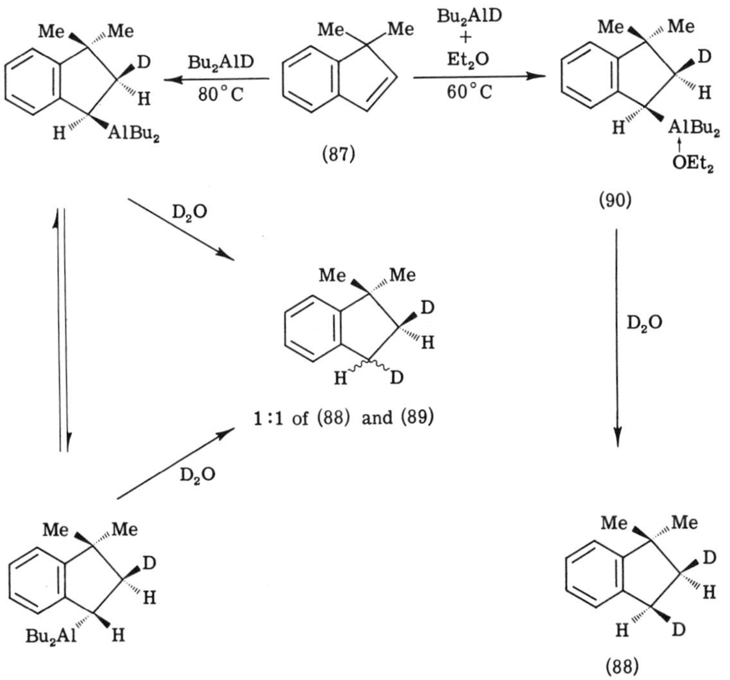

Scheme 11

minum deuteride at 80°C, followed by treatment with D_2O, gave a 1:1 mixture of the cis- and trans-2,3-dideuterio-1,1-dimethylindans (**88** and **89**, distinguishable by IR spectroscopy) (Scheme 11). Hydralumination in the presence of diethyl ether did retard the addition (k_1), but retarded the configurational loss (k_2) more markedly. In this way, pure **88** was obtainable, and hence the configurationally stable aluminum etherate **90** was available for stereochemical studies (25).

Treatment of **90** with 1.5 equivalents of methyllithium and warming at 60°C for 5 hours gave, upon addition of D_2O, an equimolar mixture of **88** and **89**, showing that epimerization of **90** had taken place. Clearly, the aluminate complex **91** must undergo reversible dissociation into a benzylic-like carbanion that suffers inversion:

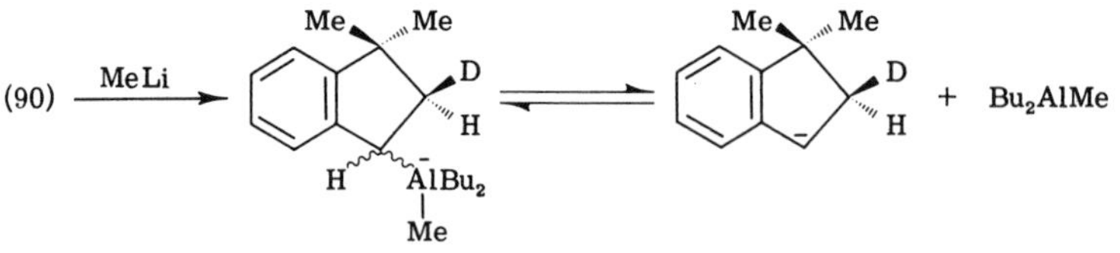

Finally, the behavior of **90** and its deuterated analogs toward unsaturated substrates was examined as to their stereochemical course.[5] Phenylacetylene was found to insert in a syn-manner with retention of configuration at C-3 (**92**); 9-fluorenone caused complete loss of configuration of the indanyl moiety in forming the adduct **93**; and acetone-d_6 underwent insignificant addition to its carbonyl group, but rather was transformed into its enol salt with the stereospecific formation of cis-2,3-dideuterio-1,1-dimethylindan (**88**) (Scheme 12) (25, 26).

Similar deuterated aluminum adducts of acenaphthylene, such as the 2-deuterio isomer of **67**, also suffer loss of configuration when allowed to react with ketones at 20°C [Eq. (39)]. But since the 1-substituted product (**80**) is known to be preceded by the 3-substituted product (**79**), dissociation of **79**, and hence loss of configuration in going to **80**, is to be expected (26, 27).

The various stereochemical outcomes in Scheme 11 merit comment. Possibly the accessibility of a six-membered transition state for the enol state formation with acetone (**93**) is again more favorable than a four-center carbonyl insertion (cf. **82** and **83**). When no alternative is possible, as with 9-fluorenone, clearly the indanyl and $AlBu_2$ fragments must separate (**95**) before collapsing to product. The suggestion that the

---

[5] For convenience in the analysis of products, in some instances **90** was studied, but in other cases the Bu_2AlH adducts of 2-deuterio- or 2,3-dideuterio-3,3-dimethylindene were used. The stereochemical outcomes are illustrated here with **90**.

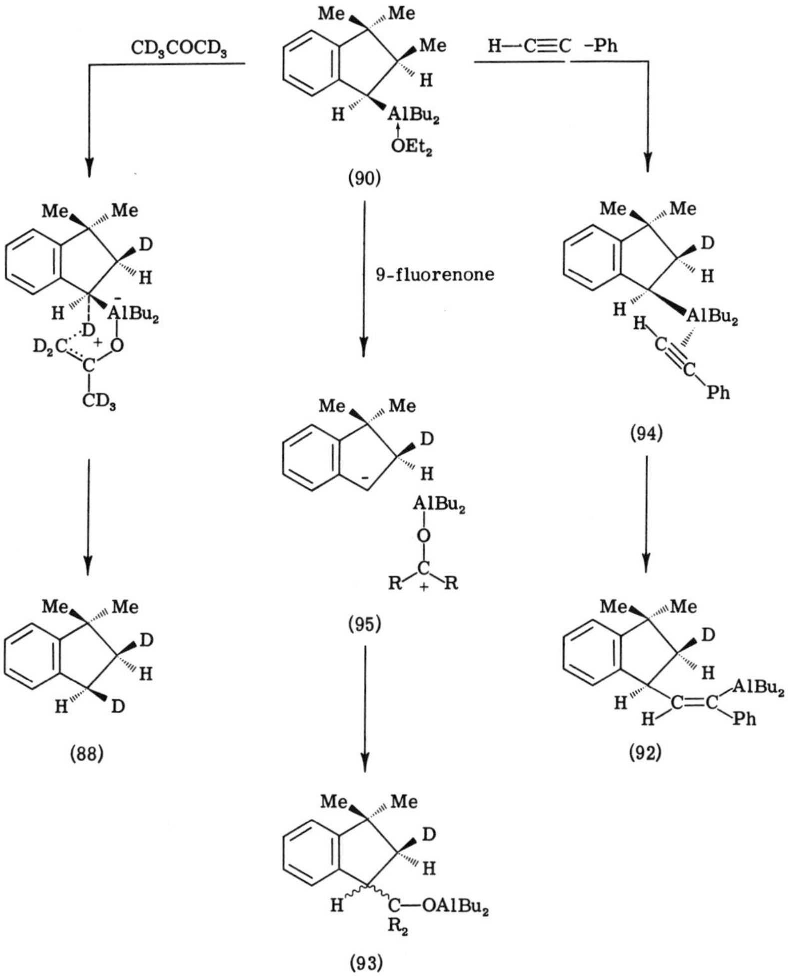

Scheme 12

heterolysis in **95** stems from the coordination with the ketone finds support in the observed loss of configuration of **90** in the presence of methyllithium [Eq. (44)]. Finally, the retention of configuration seen in the insertion of phenylacetylene can be ascribed to the failure of alkynes to form stable complexes with aluminum alkyls, but rather undergo addition via π-complex-like transition states showing relatively little charge separation (**94**) (*17, 21, 42*). Consequently, the possibility of configurational loss, as in **95**, is minimized.

B. *Thermally Induced 1,n-Additions*

Although alkynes and alkenes do not form detectable amounts of complex with aluminum alkyls,[6] various α,β-unsaturated ketones and anils do form such complexes. Such systems, upon heating, do undergo 1,4-carbalumination [Eqs. (45) and (46)]. The rearrangements of com-

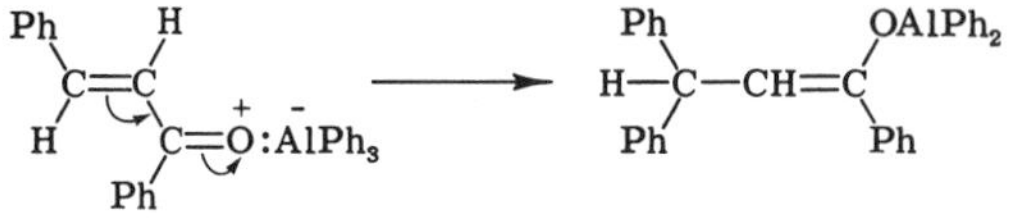

$$(45)$$

$$(46)$$

plexes, such as **96** and **97**, can be viewed as a thermally allowed [$_\sigma 2_s$ + $_\sigma 2_s$ + $_\pi 2_s$] process and hence favored. With aluminum alkyls and α,β-unsaturated cyclic ketones, a six-membered transition state leading to product is not favored, and apparently the reaction then requires free-radical promotion (*66, 67*).

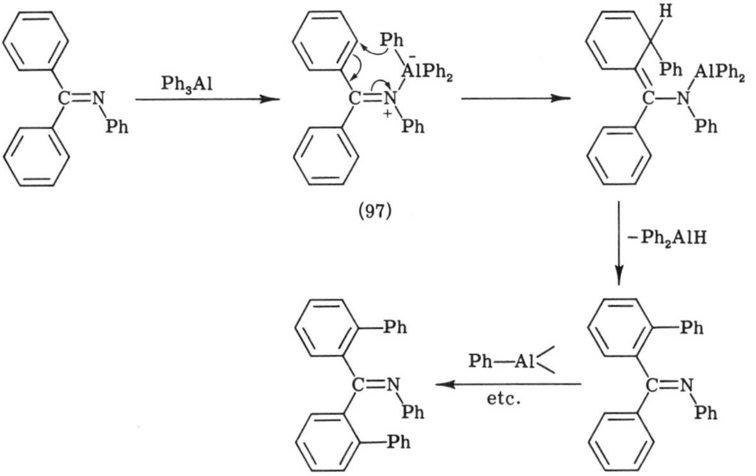

The rearrangement of dimeric diphenyl(phenylethynyl)aluminum (**98**) (cf. refs. *65* and *87*) presents the interesting equilibrium situation where the bridging acetylenic groups are disposed in a π-complex-like configuration about the aluminum centers. The observed intradimer carbaluminations can be viewed as model reactions for the well-known intermolec-

---

[6] There is ample evidence, however, that certain alkynyl- and alkenyl-aluminum compounds do exhibit inter- or intramolecular interactions between their π-bonds and aluminum centers (*62, 87*).

ular carbaluminations of alkynes, whose transition seems to resemble a π-complex (Scheme 13) (87).

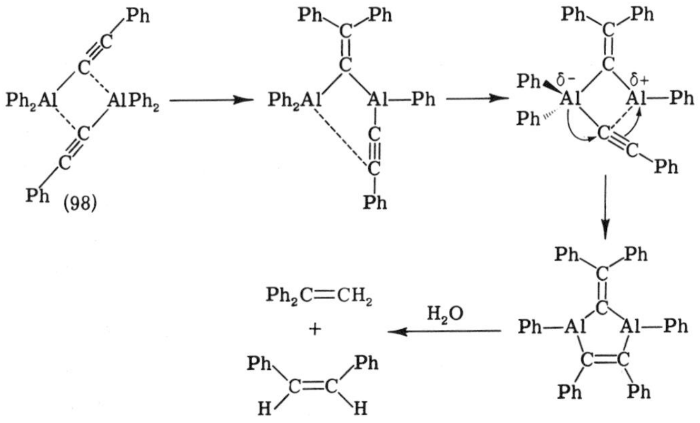

Scheme 13

C. Photorearrangements of Dimeric Aluminum Aryls

The photochemistry of aluminum aryls bears certain features in common with boron aryls: the low reactivity of the monomeric triarylmetallic, the formation of biaryls in tetracoordinate systems having proximate aryl groups and the formation of subvalent metallic products (22–24). Thus, when irradiated at 254 nm in ethyl ether solution, triphenylaluminum ethyl etherate underwent no formation of biphenyl or aluminum metal. On the other hand, the irradiation of triphenylaluminum (99) in benzene or toluene solution, in which the aluminum aryl is largely dimeric, led to the production of biphenyl (45%) and aluminum metal. When such a reaction mixture was filtered to remove the aluminum and the filtrate treated with D_2O, hydrogen gas was evolved that consisted of >98% H—D. The biphenyl now isolated was 20% undeuterated and 80% monodeuterated; further spectral comparisons showed that the deuteron was at the 2-position. These findings indicated the formation of an aluminum-hydride and an o-biphenylyl aluminum bond [Eq. (47)]. The source of the hydrogen in H—D was not the

$$(Ph_3Al)_2 \xrightarrow[\text{254 nm}]{\text{ArH}} Ph\text{---}Ph + Al + \quad \text{(47)}$$

(99)

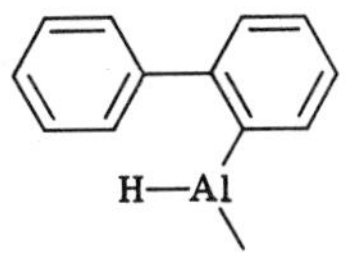

solvent, for when the irradiation was repeated in toluene-d_8, the hydrogen gas obtained upon work-up with D_2O was still >98% H—D. Nor were there any significant amounts of solvent-derived photoproducts with ordinary toluene: the ratio of biphenyl to bibenzyl was 99:1.

To see whether a subvalent organoaluminum precursor to the aluminum metal [Eq. (47)] could be trapped, **99** was irradiated in the presence of diphenylacetylene (**22**). No deposition of aluminum occurred during irradiation and work-up with D_2O. The biphenyl was now found to be a 36:64 mixture of 2-deuterated and undeuterated hydrocarbon. In addition, the *cis*-stilbene isolated was found to consist of 55% of α-deuterio-*cis*-stilbene (**100**) and 45% of α,α'-dideuterio-*cis*-stilbene (**101**). Finally, α-deuteriotriphenylethylene (**102**) was also formed. The formation of **100** and **102** were to be expected, since they would result from the hydralumination [cf. Eq. (47)] and carbalumination, respectively, of diphenylacetylene. The absence of aluminum metal and the detection of **101** point to the syn-addition of some subvalent phenylaluminum reagent to the C≡C bond; such a reagent may be phenylaluminum(I). This subvalent compound need not be free, but may be complexed as in **103**. Moreover, the results do not require that the precursor **104** persist as such; it may undergo a ring-opening dimerization.

$$(Ph_3Al \cdot AlPh_2) \longrightarrow Ph\!-\!Ph \ + \ Ph_3Al \cdot AlPh \xrightarrow{\ PhC\equiv CPh\ } \underset{(104)}{\overset{\displaystyle Ph\diagdown_{\!}C\!=\!C\diagup^{\!Ph}}{\underset{\underset{Ph}{|}}{\diagdown\!\!\underset{Al}{}\!\!\diagup}}}$$

(99)   (103)

$$\Big\downarrow D_2O \qquad\qquad (48)$$

$$\underset{(101)}{Ph\diagdown_{\!}C\!=\!C\diagup^{\!Ph}\atop D\diagup\qquad\diagdown D}$$

That the generation of **103** does not simply involve a 1,2-shift of phenyl from aluminum to C-1 of an adjacent phenyl (as in Scheme 5 for boron) was clearly shown by examining the photomigrations of tri-*m*-tolylaluminum (**105**) and tri-*p*-tolylaluminum (**106**) (*23*). If such a 1,2-shift were operative, **105** and **106** should yield *m,m'*-bitolyl and *p,p'*-bitolyl, respectively. Instead, neither aryl yielded these expected products: the meta isomer gave a 50:50 mixture of 2,3'- and 3,4'-bitolyls, while the para isomer gave only 3,4'-bitolyl (**107**). As with **99**, work-up with D_2O gave both undeuterated and monodeuterated derivatives of these bitolyls. In the case of **106**, the deuteron (and hence the aluminum) was shown to be para to the methyl group in the 3'-tolyl ring.

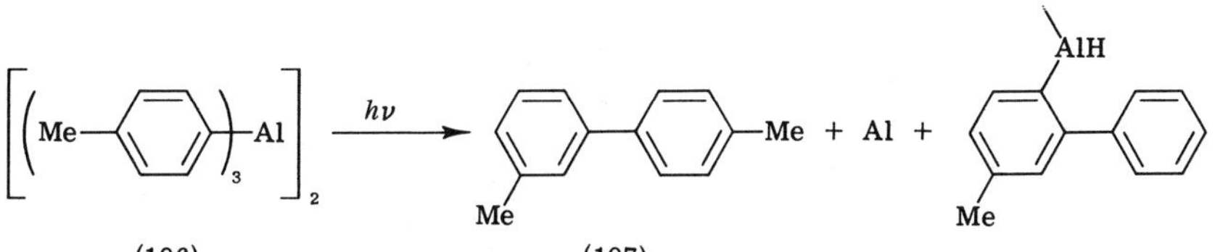

$$(49)$$

These results require the occurrence of a 1,3-shift proceeding through the dimeric aryl. A mechanistic interpretation of these results is offered in Scheme 14.

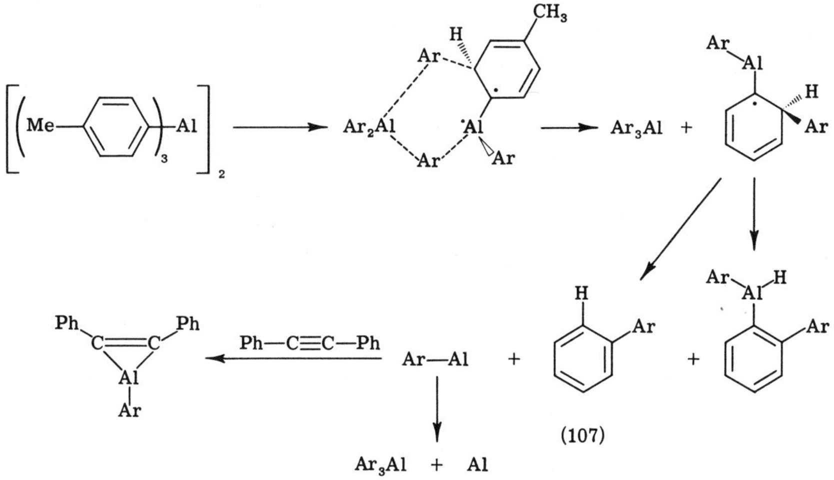

Scheme 14

The photobehavior of the monomeric triphenylgallium and triphenylindium in toluene solution is consistent with this pathway (*24*). The gallium compound underwent photorearrangement very inefficiently under the conditions successful with aluminum aryls. Only 4% of biphenyl was formed, and it was accompanied by 2% of bibenzyl and 2% of methylbiphenyl, products that are definite signs of free-radical involvement of the solvent. The photolysis of triphenylindium gave a 44% yield of biphenyl, but here again comparable amounts of bibenzyl (33%) and the methylbiphenyls (16%) point to intermolecular free-radical attack. Although the photorearrangement of dimeric aluminum aryls may well involve radicals, such intermediates seem to be efficiently scavenged within the rearranging complex.

VI
PROSPECTS FOR FUTURE DEVELOPMENTS

Close examination of various substitution and addition reactions of organoboron and organoaluminum compounds reveals how these processes occur via intermediate complexes, which in turn rearrange with remarkable site- and stereo-selectivity. Having a greater covalent character than lithium and magnesium reagents, boron and aluminum organometallics will generally have a lower reactivity toward organic substrates. Yet the same enhanced covalent character means that these alkyls will often show greater discrimination among competing pathways, such as those in allylic derivatizations, α,β-unsaturated carbonyl additions, 1,n-migrations, and stereospecific insertions into carbon–metal bonds. All these synthetic advantages should eventually find their just recognition in novel preparative procedures for the construction and functionalization of complex carbon skeletons.

It is further evident that the rearrangements of unsaturated organoboranes and organoalanes will prove to be a fertile area in which to explore the reaction mechanisms involving the chirality of carbon–metal bonds, and to probe the implications and applicability of the principles of pericyclic reactions. As was suggested in the Introduction (Section I,A), there are already strong indications that mechanistic insight gained from a scrutiny of these *intramolecular*, unsaturated organometallic rearrangements will aid our comprehension of how *intermolecular* reactions take place. The principles governing allylic isomerizations (Section IV,A) and 1,n-migrations (Section V,A) in complexes, for example, are clearly pertinent to the course of carbaluminations with unsaturated substrates.

Finally, the rearrangements encountered in these studies have opened up some fresh vistas on unusual reactions, on the generation of subvalent group III organometallics, and on the synthesis of metallocycles of great theoretical interest.

ACKNOWLEDGMENTS

The author is grateful for the zeal and dedication of his many graduate and postdoctoral students, the fruits of whose labors he is here privileged to present. Their individual contributions are designated in the References; what is not adequately conveyed there is the zest of the day-to-day intellectual fellowship. The research support stemmed largely from grants from the National Science Foundation and the National Institute of General Medical Sciences, whose recognition of our research has been deeply appreciated.

References

1. Abley, P., and Halpern, J., *J. Chem. Soc.* D, 1238 (1971).
2. Binger, P., *Angew. Chem.* **79,** 57 (1967).
3. Binger, P., and Köster, R., *Tetrahedron Lett.* 1901 (1965).
4. Braye, E. H., Hübel, W., and Caplier, I., *J. Am. Chem. Soc.* **83,** 4406 (1961).
5. Brown, H. C., "Organic Syntheses Via Boranes," p. 77. Wiley (Interscience), New York, 1975.
6. Brown, H. C., and Krishnamurthy, S., *J. Am. Chem. Soc.* **95,** 1669 (1973).
7. Bruno, G., *J. Org. Chem.* **30,** 623 (1965).
8. Bubnov, Y. N., Frolov, S. I., Kiselev, V. G., Bogdanov, V. S., and Mikhailov, B. M., *Organometal. Chem. Syn.* **1,** 37 (1970).
9. Buckton, G. B., and Odling, W., *Proc. Roy. Soc. London* **14,** 19 (1865).
10. Clark, G. M., Hancock, K. G., and Zweifel, G., *J. Am. Chem. Soc.* **93,** 1308 (1971).
11. Courtois, G., and Miginiac, L., *J. Organometal. Chem.* **69,** 1 (1974).
12. Curtin, D. Y., and Koehl, W. J., Jr., *J. Am. Chem. Soc.* **84,** 1967 (1962).
13. Doty, J. C., Grisdale, P. J., Evans, T. R., and Williams, J. L. R., *J. Organometal. Chem.* **32,** C35 (1971).
14. Eisch, J. J., *Trans. N.Y. Acad. Sci.* [2] **27,** 450 (1965).
15. Eisch, J. J., *Ann. N.Y. Acad. Sci.* **239,** 292 (1974).
16. Eisch, J. J., *Ind. Eng. Chem., Proc. Res. Develop.* **14,** 11 (1975).
17. Eisch, J. J., and Amtmann, R., *J. Org. Chem.* **37,** 3410 (1972).
18. Eisch, J. J., Amtmann, R., and Foxton, M. W., *J. Organometal. Chem.* **16,** P55 (1969).
19. Eisch, J. J., and Biedermann, J. M., *J. Organometal. Chem.* **30,** 167 (1971).
20. Eisch, J. J., and Biedermann, J. M., Unpublished studies (1970).
21. Eisch, J. J., Burlinson, N. E., and Bolelawski, M., *J. Organometal. Chem.* **111,** 137 (1976).
22. Eisch, J. J., and Considine, J. L., *J. Am. Chem. Soc.* **90.** 6257 (1968).
23. Eisch, J. J., and Considine, J. L., *J. Organometal. Chem.* **26,** C1 (1971).
24. Eisch, J. J., and Considine, J. L., Unpublished dissertation studies (1972).
25. Eisch, J. J., and Fichter, K. C., *J. Am. Chem. Soc.* **96,** 6815 (1974).
26. Eisch, J. J., and Fichter, K. C., *J. Am. Chem. Soc.* **97,** 4772 (1975).
27. Eisch, J. J., and Fichter, K. C., Unpublished dissertation studies (1975).
28. Eisch, J. J., and Foxton, M. W., *J. Organometal. Chem.* **11,** P50 (1968).
29. Eisch, J. J., and Foxton, M. W., *J. Org. Chem.* **36,** 3520 (1971).
30. Eisch, J. J., and Galle, J. E., *J. Am. Chem. Soc.* **97,** 4436 (1975).
31. Eisch, J. J., and Galle, J. E., *J. Organometal. Chem.* **127,** C9 (1977).
32. Eisch, J. J., and Galle, J. E., Unpublished studies (1976).
33. Eisch, J. J., and Gonsior, L. J., *J. Organometal. Chem.* **8,** 53 (1967).
34. Eisch, J. J., and Gonsior, L. J., Unpublished studies (1964).
35. Eisch, J. J., Gopal, H., and Rhee, S. G., *J. Org. Chem.* **40,** 2064 (1975).
36. Eisch, J. J., Hota, N. K., and Kozima, S., *J. Am. Chem. Soc.* **91,** 4575 (1969).
37. Eisch, J. J., and Husk, G. R., *J. Organometal. Chem.* **4,** 415 (1965).
38. Eisch, J. J., and Husk, G. R., *J. Organometal. Chem.* **64,** 41 (1974).
39. Eisch, J. J., and Kaska, W. C., *J. Am. Chem. Soc.* **88,** 2213 (1966).
40. Eisch, J. J., and Komar, D., Unpublished studies (1976).

41. Eisch, J. J., and Rakowsky, A., Unpublished studies (1969).
42. Eisch, J. J., and Rhee, S. G., *J. Am. Chem. Soc.* **97**, 4673 (1975).
43. Eisch, J. J., and Rhee, S. G., *Justus Liebigs Ann. Chem.* 565 (1975).
44. Eisch, J. J., and Stucky, G. D., In preparation (1977).
45. Eisch, J. J., and Tamao, K., Unpublished studies (1974).
46. Eisch, J. J., Tamao, K., and Wilcsek, R. J., *J. Am. Chem. Soc.* **97**, 895 (1975).
47. Eisch, J. J., and Wilcsek, R. J., *J. Organometal. Chem.* **71**, C21 (1974).
48. Eisch, J. J., and Wilcsek, R. J., Unpublished studies (1974).
49. Fraenkel, G., Dix, D. T., and Carlson, M., *Tetrahedron Lett.* 589 (1968).
50. Frankland, E., *Justus Liebigs Ann. Chem.* **71**, 171 (1849).
51. Frankland, E., *Justus Liebigs Ann. Chem.* **85**, 347, 360 (1853).
52. Frankland, E., *Justus Liebigs Ann. Chem.* **95**, 33 (1855).
53. Frankland, E., and Duppa, D. F., *Justus Liebigs Ann. Chem.* **115**, 319 (1860).
54. Geske, D. H., *J. Phys. Chem.* **66**, 1743 (1962).
55. Gill, G. B., and Willis, M. R., "Pericyclic Reactions." Chapman & Hall, London, 1974.
56. Gilman, H., and Kirby, R. H., *J. Am. Chem. Soc.* **63**, 2046 (1941).
57. Gilman, H., and Nelson, J. F., *J. Am. Chem. Soc.* **61**, 741 (1939).
58. Grisdale, P. J., Williams, J. L. R., Glogowski, M. E., and Babb, B. E., *J. Org. Chem.* **36**, 544 (1971).
59. Hancock, K. G., and Kramer, J. D., *J. Am. Chem. Soc.* **95**, 6463 (1973).
60. Hancock, K. G., and Kramer, J. D., *J. Organometal. Chem.* **64**, C29 (1974).
61. Hennion, G. F., McCusker, P. A., Ashby, E. C., and Rutkowski, A. J., *J. Am. Chem. Soc.* **79**, 5190 (1957).
62. Henold, K. L., and Oliver, J. P., in "Organometallic Reactions" (E. I. Becker and M. Tsutsui, eds.), Vol. V, p. 387. Wiley (Interscience), New York, 1975.
63. Hooz, J., and Linke, S., *J. Am. Chem. Soc.* **90**, 5936, 6891 (1968).
64. Ioffe, D. V., and Mostova, M. I., *Russ. Chem. Rev. Eng. Transl.* **42**, 56 (1973).
65. Jeffery, E. A., Mole, T., and Saunders, J. K., *Aust. J. Chem.* **21**, 137 (1968).
66. Kabalka, G. W., Brown, H. C., Suzuki, A., Honma, S., Arase, A., and Itoh, M., *J. Am. Chem. Soc.* **92**, 710 (1970).
67. Kabalka, G. W., and Daley, R. F., *J. Am. Chem. Soc.* **95**, 4428 (1973).
68. Köbrich, G., and Merkle, H. R., *Chem. Ber.* **100**, 3371 (1967).
69. Köster, R., *Prog. Boron Chem.* **1**, 289 (1964).
70. Kroll, W. R., McDivitt, J. K., and Naegele, W., *Inorg. Nucl. Chem. Lett.* **5**, 973 (1969).
71. Kroll, W. R., and Naegele, W., *Chem. Commun.* 246 (1969).
72. Lehmkuhl, H., *Angew. Chem.* **76**, 817 (1964).
73. Lehmkuhl, H., *Justus Liebigs Ann. Chem.* **719**, 40 (1968).
74. Lehmkuhl, H., and Reinehr, D., *J. Organometal. Chem.* **23**, C25 (1970).
75. Leusink, A. J., Drenth, W., Noltes, J. G., and van der Kerk, G. J. M., *Tetrahedron Lett.* 1263 (1967).
76. Lustgarten, R. K., Brookhart, M., and Winstein, S., *J. Am. Chem. Soc.* **94**, 2347 (1972).
77. Matteson, D. S., *Prog. Boron Chem.* **3**, 117 (1970).
78. Mikhailov, B. M., *Organometal. Chem. Rev. A* **8**, 1 (1972).
79. Mikhailov, B. M., *Eur. Conf. Organometal Chem., 1st, Abstr.*, p. 29 (1976).
80. Mikhailov, B. M., Negrebetskii, V. V., Bogdanov, V. S., Kessenikh, A. V., Bubnov, Y. N., Baryshnikov, T. K., and Smirnov, V. N., *J. Gen. Chem. USSR, Engl. Transl.* **44**, 1844 (1975).

81. Nesmeyanov, A. N., Sazonova, V. A., Liberman, G. S., and Emelyanova, L. I., *Izv. Akad. Nauk SSSR, Otd. Khim. Nauk.* 48 (1955).
82. Onak, T., "Organoborane Chemistry," Chapters 3 and 4. Academic Press, New York, 1975.
83. Razuvaev, G. A., and Brilkina, T. G., *Zh. Obshch. Khim.* **24,** 1415 (1954).
84. Schlesinger, H. I., and Walker, A., *J. Am. Chem. Soc.* **57,** 621 (1935).
85. Stefani, A., and Consiglio, G., *Helv. Chim. Acta* **55,** 117 (1972).
86. Stefani, A., and Pino, P., *Helv. Chim. Acta* **55,** 1110 (1972).
87. Stucky, G. D., McPherson, A. M., Rhine, W. E., Eisch, J. J., and Considine, J. L., *J. Am. Chem. Soc.* **96,** 1941 (1974).
88. Tiffeneau, M., and Delange, R., *C. R. Acad. Sci., Paris* **137,** 573 (1903).
89. Wilke, G., and Müller, H., *Justus Liebigs Ann. Chem.* **629,** 222 (1960).
90. Williams, J. L. R., Doty, J. C., Grisdale, P. J., Regan, T. H., Happ, G. P., and Maier, D. P., *J. Am. Chem. Soc.* **90,** 53 (1968).
91. Williams, J. L. R., Doty, J. C., Grisdale, P. J., Searle, R., Regan, T. H., Happ, G. P., and Maier, D. P., *J. Am. Chem. Soc.* **89,** 5153 (1967).
92. Williams, J. L. R., Grisdale, P. J., and Doty, J. C., *J. Am. Chem. Soc.* **89,** 4538 (1967).
93. Winstein, S., and Sonnenberg, J., *J. Am. Chem. Soc.* **83,** 3235, 3244 (1961).
94. Witanowski, M., and Roberts, J. D., *J. Am. Chem. Soc.* **88,** 737 (1966).
95. Woodward, R. B., and Hoffmann, R., "The Conservation of Orbital Symmetry," Verlag Chemie-Academic Press, Weinheim, 1971.
96. Zakharkin, L. I., *Izv. Akad. Nauk SSSR., Ser. Khim.* 539 [*Chem. Abstr.* **57,** 10997 (1962)].
97. Zakharkin, L. I., and Okhlobystin, O. Y., *Izv. Akad. Nauk SSSR, Ser. Khim.* 1278 [*Chem. Abstr.* **53,** 4115 (1958)].
98. Ziegler, K., *Justus Liebigs Ann. Chem.* **629,** 1 ff (1960).
99. Zweifel, G., and Steele, R. B., *J. Am. Chem. Soc.* **89,** 5085 (1967).
100. Zweifel, G., and Whitney, C. C., *J. Am. Chem. Soc.* **89,** 2753 (1967).

Rearrangements of Organoaluminum Compounds and Their Group III Analogs

JOHN P. OLIVER

Department of Chemistry
Wayne State University
Detroit, Michigan

I

INTRODUCTION

Rearrangement and exchange reactions of main group organometallic compounds offer a rich variety of processes important both to the inorganic and organic chemist. These processes have provided examples of electrophilic substitution, electron transfer, and of concerted multicentered processes, all of which are of both theoretical and practical importance.

In this review discussion is restricted to those systems that involve a group III metal bound to various organic ligands and deals principally with simple exchange processes, but also includes some discussion of the more complex rearrangements, especially those in which multiple bonds are involved.

Many of the older studies have been reviewed in classic works, such as the monograph by Coates (8), the treatise by Mole and Jeffery (35) on aluminum chemistry, and that on reaction mechanisms by Matteson (33).

Recently, a review of the structural aspects of main group electron deficiently bound systems has appeared (*39*), and the chemistry and exchange reactions of both group I (*3–5*) and group II (*19, 38*) have been covered.

In addition, the studies on exchange reactions prior to about 1970 for group III alkyls have been reviewed by Oliver (*38*), Mole (*34*), and Ham and Mole (*19*). The chemistry of unsaturated organoaluminum compounds also has been reviewed by Henold and Oliver (*22*). No attempt is made to cover in detail the work covered in the earlier reviews, but the reader will be referred to these references for the detailed discussion and the primary references.

II

STRUCTURES AND BONDING IN GROUP III DERIVATIVES

A very brief summary of the salient points with regard to the structures, bonding, and their influence on the mechanisms of exchange and rearrangement process for aluminum and its heavier congeners follows, prior to the discussion of specific systems. The structures for several of the important aluminum derivatives are shown in **1–6**. All of

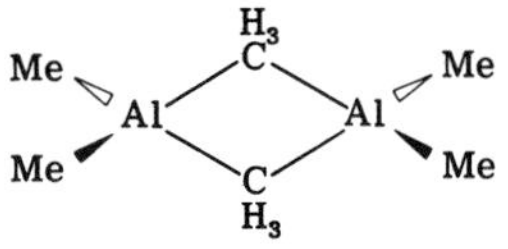

(refs.: *23, 28, 54*)

(1)

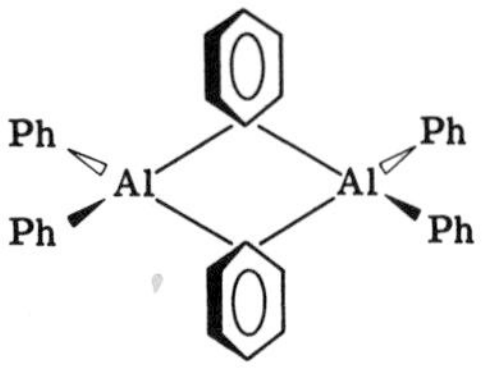

(ref.: *31, 32*)

(2)

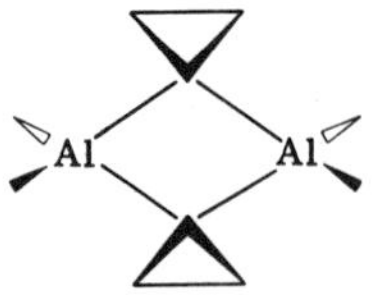

(ref.: *36*)

(3)

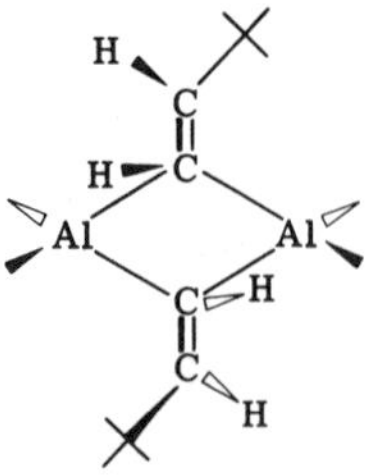

(ref.: *2*)

(4)

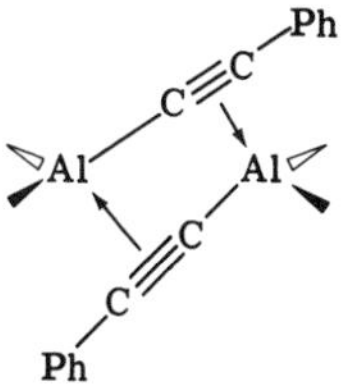

(ref.: 52)   (refs.: 29,30)

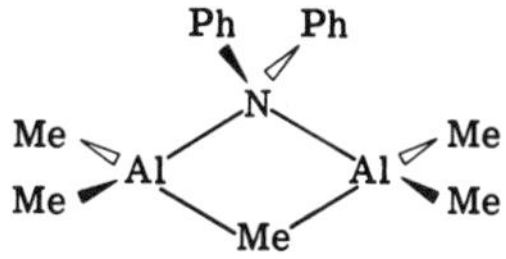

(5)   (6)

these structures show the dimeric electron-deficient bridged structure characteristic of organoaluminum compounds.

Structure of the heavier congeners of group III have not been as extensively investigated; those that have been determined in the solid state do not have bridging alkyl or aryl groups, but are found to exist either as monomers or loosely associated aggregates (8, 39). The structures for these monomeric derivatives can be represented by 7 with

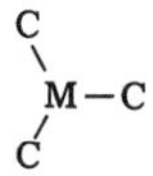

(7)

the metal and three α-carbon atoms in the same plane. These molecules are further characterized by the presence of a vacant p-orbital of the metal atom perpendicular to the plane, which can readily be used as an electron-pair acceptor.

Evidence for formation of dimeric gallium derivatives with bridging vinyl groups has been reported (40, 53), and both phenylethynylgallium and indium also have been shown to be dimeric (24).

Formation of more stable bridge bonds by unsaturated groups, including vinyl, ethynyl, or aryl groups, or the saturated cylopropyl derivatives appears to be common, not only for the gallium and indium species, but also for aluminum derivatives. This will be discussed further in Sections III and VI. Further, it should be noted that hydride bridging groups and all the bridging groups formed with an oxygen, a nitrogen, or a halogen atom tend to be very stable relative to the electron-deficient carbon-bridged systems.

The bonding present within the bridged molecules can be treated in a simple manner, making use of two orbitals from the two metal atoms and one from the bridging group for each bridge bond. This is shown schematically in **8** assuming that the orbitals provided by all atoms are sp^3 hybrid orbitals.

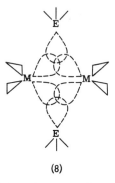

(8)

Construction of simple (although approximate) molecular orbital (MO) schemes from this system leads to an energy level diagram having three levels, bonding, nonbonding, and antibonding, as shown in **9a, 9b,** and **9c.**

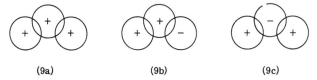

(9a)   (9b)   (9c)

In the case where only two electrons are available, one from the bridging group and one from the metal atoms, only the bonding orbital is occupied, giving rise to the single three-centered (or electron-deficient) bond. If additional orbitals are available of appropriate symmetry, such as in the vinyl, phenyl, or ethynyl systems, as shown in **10**, these may

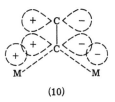

(10)

interact with the unoccupied nonbonding orbitals to lower the energy of the system, hence increasing the stability of the dimeric species as observed in vinyl (*7*), phenyl (*31, 32, 51*), and cyclopropylaluminum (*46*) or leading to formation of dimers as observed for vinyl (*40, 53*) and phenylethynylgallium (*24*) or phenylethynylindium (*24*) derivatives or even lead to a change in the mode of bonding, as suggested for phenylethynylaluminum (*52*). These features are all depicted in structures **1–5**. In structure **6**, which contains an NR_2 bridging group, we find "normal" 2 electron-2 center bonding present for one of the bridges. This can be extended if both bridging groups present are formed by electron-rich centers, thus leading to the higher stability of these species.

III

MONOMER–DIMER EQUILIBRIA

As indicated in Section II, the compounds of interest may form either monomers or dimer. Further, it was noted that alteration in either the substituents or in the metal may cause substantial changes in the structure of the molecule or in the position of the equilibrium indicated in Eq. (1). The monomer form has a vacant orbital available that

$$\tag{1}$$

represents the active site for most reactions involving these derivatives, thus it is crucial for the aluminum dimers to undergo dissociation or at least partial dissociation to provide a low-energy path for reaction. For the monomeric species such as Me_3Ga or Me_3In, this is unnecessary and these derivatives undergo facile exchange even at very low temperatures (*38*).

If we examine **1–6** we find that the stability of the dimers increases in approximately the order given, the dialkyls being least stable and the mixed bridged species with the Ph_2N bridging group being most stable.

Quantitative data are not available for the majority of these systems, but several studies have appeared that deal with the relative stabilities of the simple alkyl dimers (*48–50*) and of a few olefinic derivatives (*10*). Some of the pertinent data are collected in Table I. These data clearly establish that increasing the chain length decreases the stability of the dimer shifting the equilibrium to the right. Further chain branching causes a similar effect.

Examination of these data leads to an additional conclusion: introduction of a terminal double bond results in a dramatic shift in the equilibrium from dimer to monomer. The implication of this with regard to the reactions is discussed more fully in Section VIII, however it should be noted that this indicates stabilization of the monomer toward

TABLE I
COLLIGATIVE PROPERTIES AND HEATS OF DISSOCIATION FOR SELECTED
ORGANOALUMINUM COMPOUNDS

Compound	β^a	$\bar{n}^b$	ΔH_d (kcal/mole)
$Me_2Al_6{}^{c,d}$	0.00029	1.97^e	19.40
$Et_2Al_6{}^d$	0.00501	—	16.94
$(n\text{-}Pr)_6Al_2{}^f$	0.0253	—	15.40
$(n\text{-}Bu)_6Al_2{}^f$	0.0376	1.96^e	15.02
$(i\text{-}Bu)_6Al_2{}^f$	0.164^g		8.1
$[H_2C\!\!=\!\!CH(CH_2)_2]_3Al^e$	—	0.98	—
$[CH_3(CH_2)_4]_6Al_2{}^e$	—	1.80	—
$[H_2C\!\!=\!\!CH(CH_2)_3]_3Al^e$	—	0.99	—
$[CH_3(CH_2)_7]_6Al_2{}^f$	0.0501	—	14.68

^{a} The fraction dissociated at 20°C in 0.05 mole fraction solution in cyclohexane.
^{b} The average degree of aggregation in freezing cyclohexane.
^{c} Smith (48).
^{d} Smith (49).
^{e} Dolzine and Oliver (10).
^{f} Smith (50).
^{g} 40°C.

reaction with a second molecule and does not necessarily imply that these species should be more reactive.

If we turn to the other extreme, the case involving the Ph_2N bridge, we see that complete dissociation of this molecule does not occur until elevated temperatures are reached (42). This suggests that alternative mechanisms for reaction will be required for this type of system. These might include partial dissociation, formation of a 5 or 6 coordinate metal atom, or a radical, or even an ionic path. These items are explored in later sections.

IV

EXCHANGE PROCESSES INVOLVING DISSOCIATED SPECIES

A. Alkylaluminum Exchange Reactions

The first and most extensively studied alkyl exchange process is that for trimethylaluminum dimer (6, 25, 27, 37, 41, 51, 55). The mechanism favored for this exchange process is the dissociation as indicated in Eq.

(1), followed by a rapid recombination step. This is one of the mechanisms originally proposed, but was initially rejected on the basis that the observed activation energy for this process was greater than the dissociation energy for the Me_6Al_2 dimer measured in the gas phase (*41*).

Further studies on this system, and studies on the exchange between Me_6Al_2 and Me_3Ga in which random recombination and exchange occurs as shown in Eq. (2), provide convincing evidence that dissociation occurs as the rate-determining step (*6, 25, 27, 51*).

$$Me_2Al_6 + Me_3Ga \rightleftharpoons 2Me_3Al + Me_3Ga \qquad (2)$$

Studies on the exchange of Et_6Al_2 and $(n\text{-}Pr)_6Al_2$ by ^{13}C NMR spectroscopy also have lead to the conclusion that the dominant path for exchange is a slow dissociation step followed by rapid recombination (*56*). Further, these studies show that the activation energy falls off as the chain length increases in the same way that the stability of the dimer decreases as discussed in Section III. The extreme cases for this occur when the metal alkyl is in the monomeric form, as discussed elsewhere for the $Me_3Ga\text{-}Me_3In$ or Me_3Tl systems in which the recombination step is rate controlling and the activation process involves not dissociation, but simple diffusion of the species together with a small reorganization energy (*38*).

B. *Main Group Metal–Transition Metal Bonded Compounds*

An interesting new development in this area has resulted from the observation which has been made on compounds that have main-group metal–transition metal bonds as well as carbon–metal bonds (*9, 44*). Only preliminary exchange studies have appeared on these systems, but it has been shown that compounds of the type $[\pi\text{-}C_5H_5(CO)_3W]_n GaMe_{3-n}$ (n = 0–3) can be observed in equilibrium systems. Two of the transition metal derivatives have been obtained in the solid state, and their structures were determined. One of these is the tris(transition metal)gallium compound $[\pi\text{-}C_5H_5(CO)_3W]_3Ga$ with three direct W—Ga bonds (*9*), and the second is the mono(transition metal)gallium derivative, $\pi\text{-}C_5H_5(CO)_3WGaMe_2$, which has been shown to have the structure indicated in **11** with both W—Ga and Ga—C bonds present. Both research groups have reported preliminary NMR data indicating that fast exchange occurs under normal conditions for alkyl groups between Me_3Ga and $TmGaMe_2$. Further, these studies show that methyl group exchange may occur on a slower time scale between Tm_2GaMe and $TmGaMe_2$, and of course exchange between Tm_3Ga and alkylgallium

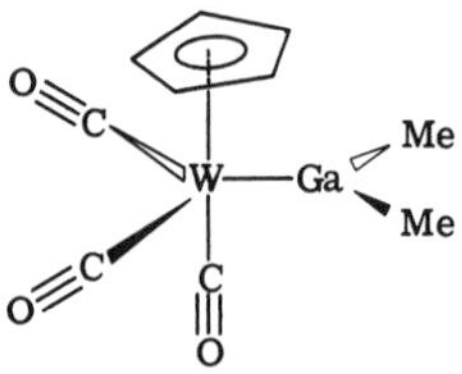

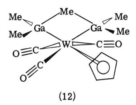

(11)

compounds must occur in order to obtain the mixed species. This implies that bridge transition states that involve species such as that in **12** can be formed in which both an alkyl and a transition metal enter into bridging sites. The relatively slow rate of exchange of the transition metal moiety to that of the alkyl groups, where exchange occurs with two alkyls in the bridged transition state, suggests that the transition metal enters the bridging position with difficulty.

(12)

A number of questions with regard to these systems remain unanswered. Certainly one must consider why the transition metal does not enter more effectively into the bridge position. Is it a result of the steric interference of the substitution on the transition metal? Is it a result of the size of the transition metal? Or does it result from the electronic nature of the metal? Similarly one must ask these questions concerning the exchange of the methyl groups between the various species, since this exchange appears to proceed far more slowly than in the simple alkyls. Further, one should note that the qualitative data suggests the activation energy is substantially increased and that the rate-determining step in these systems involves the reorganization of the groups surrounding the main group metal. Quantitative data must be obtained on these systems to answer the many questions surround these systems.

V

EXCHANGE REACTIONS OF GROUP III ALKYL ADDITION COMPOUNDS

As indicated in Sections II and III, it is desirable to have a free monomer for reaction. Thus, we have seen that dissociation of dimers occurs as the rate-determining step in many of the simple alkyl exchange processes. However, if a base is present forming the group III adduct or as the solvent, then, the group III species is present as a monomer, but is prevented from reacting by formation of a complex, as indicated in Eq. (3).

$$R_6Al_2 + 2S \rightleftharpoons 2R_3Al \cdot S \tag{3}$$

For reaction or exchange to occur, either of two paths are possible, as previously discussed for the Me_3Ga amine systems (38). The first of these is described by Eq. (4) for the $Me_3Ga-PMe_3$ system (21), in which the activation step is the dissociation of the adduct followed by exchange

$$Me_3GaPMe_3 \rightleftharpoons Me_3Ga + PMe_3 \tag{4}$$

and the second is represented by Eq. (5), in which the adduct Me_3AlPMe_3 undergoes exchange with Me_3Al (1).

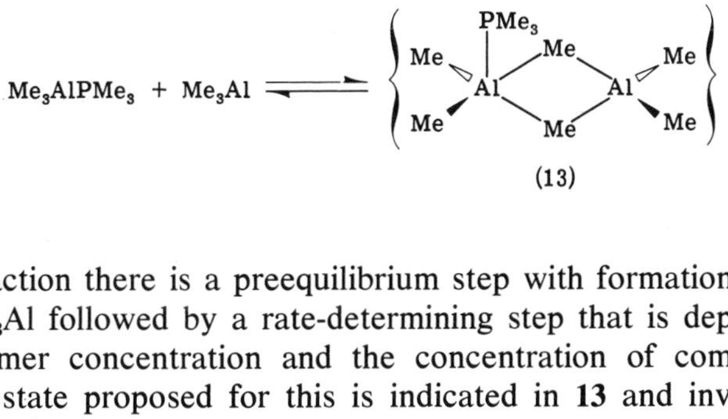

$$\tag{5}$$

(13)

In this reaction there is a preequilibrium step with formation of monomeric Me_3Al followed by a rate-determining step that is dependent on the monomer concentration and the concentration of complex. The transition state proposed for this is indicated in 13 and involves a 5-coordinate aluminum atom.

The difference in activation energy between the two processes is small (E_a = 16 kcal/mole for the $Me_3Ga \cdot PMe_3$; E_a = 14.8 kcal/mole for the $Me_3Al \cdot PMe_3$ with ΔH_d = 21 kcal/mole) with the lower energy for the bimolecular aluminum exchange.

The differences in the two processes are determined by the higher dissociation energy for the $Me_3Al \cdot PMe_3$, which arises from the higher acidity of the Me_3Al and from the greater tendency of this species to enter into electron-deficient bonds as seen in the transition state. These two reactions illustrate some of the differences that occur in these systems as the central metal atom is changed while holding all other factors constant.

VI

INTRAMOLECULAR EXCHANGE PROCESSES

A. *Exchange in the Unsymmetrically Bridged μ-Me-μ-Ph$_2$NAl$_2$Me$_4$ and Related Systems*

Several systems, such as μ-Me-μ-Ph$_2$NAl$_2$Me$_4$ shown in **6** have been investigated. These derivatives have one strong bond that remains intact while the second bridging group undergoes rapid exchange with the terminal positions (*43*). This reaction reported by Magnuson and Stucky (*28, 29*) has been examined in detail recently by Rie and Oliver (*43*). The reaction was shown to be first order in the concentration of dimer. Further, no dissociation occurs via reaction (6), which would lead to the

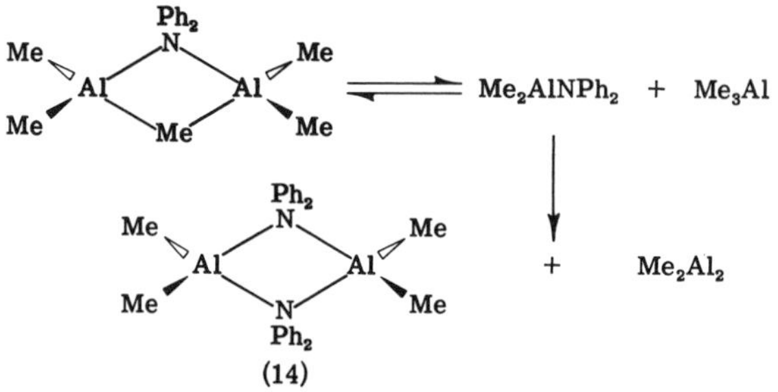

$$\text{(6)}$$

production of the stable bridged dimer **14** (*42*). These findings lead to the proposed mechanism shown in Eq. (7), in which the methyl bridge bond

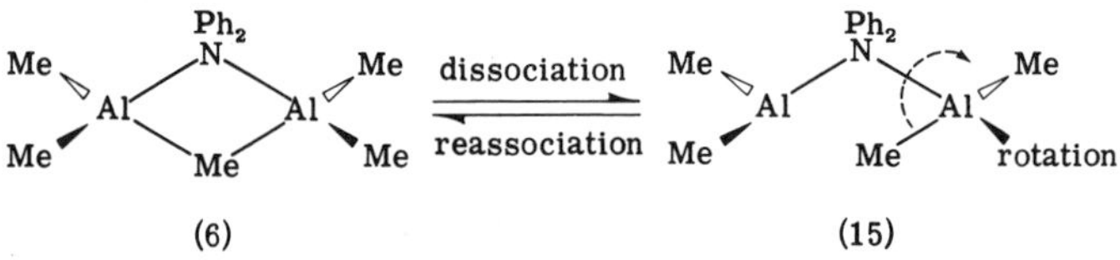

$$\text{(7)}$$

opens initially followed by rotation about the Al—N bond and finally reassociation to the bridge derivative. This process is consistent with the observed concentration dependence and activation parameters for the reaction (see Table II).

Similar processes have been suggested for the hemialkoxides, such as μ-(t-BuO)-μ-MeAl$_2$Me$_4$ (*26*) and for the imide formed on addition of Me$_6$Al$_2$ to phenylnitryl (*26*) shown in **16**. A study of these systems reveals that the bridge terminal exchange is greatly dependent on the bridging group. Further, in **16** the exchange rate for the A methyl groups with the bridge methyl is substantially greater than for the B methyl groups. The suggested reason for this is the stabilization of the transition

TABLE II

ACTIVATION PARAMETERS FOR EXCHANGE REACTIONS OF BRIDGED ALUMINUM
DERIVATIVES

Reactions Intramolecular bridge terminal exchange reactions	E_a (kcal/mole)	$\Delta H\ddagger$ (kcal/mole)	$\Delta S\ddagger$ (e.u.)
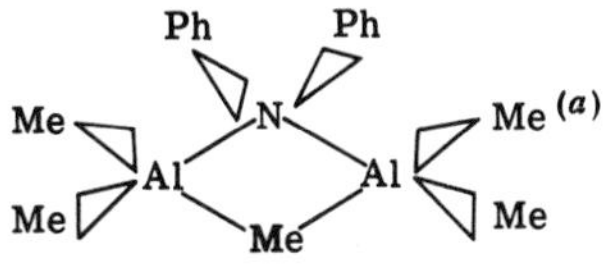	13.67 ± 0.45	13.0 ± 0.6	−5.8 ± 3.6 @ 290°C
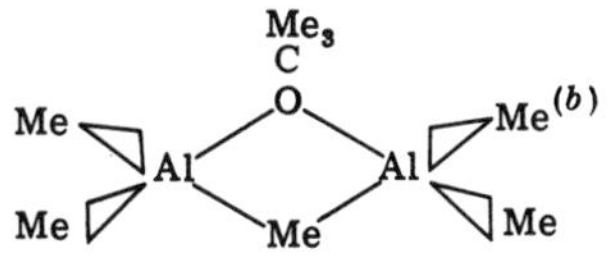	~ 20 ± 3		
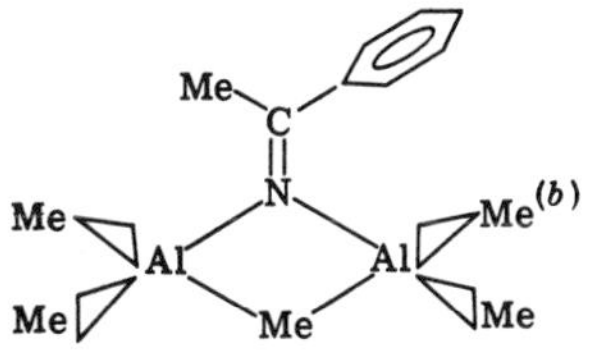	12.7 ± 2.0 for A methyls ~ 20 for B methyls		
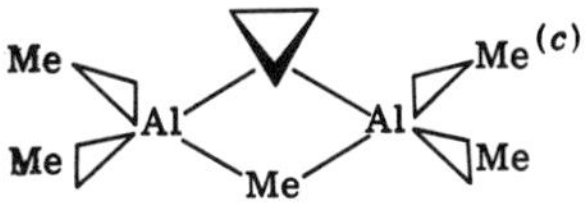	15.8 ± 0.7	15.1	−7.6
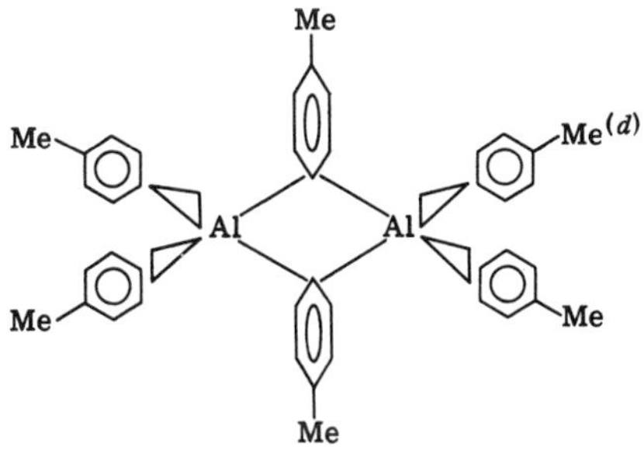	9.7 ± 1.0	9.2	−9 @ 225°C

[a] Rie and Oliver (43).
[b] Jeffery and Mole (26).
[c] Sanders et al. (47).
[d] Stanford and Henold (51).

TABLE II—*(Continued)*

Reactions *Intramolecular bridge terminal exchange reactions*	E_a (kcal/mole)	$\Delta H\ddagger$ (kcal/mole)	$\Delta S\ddagger$ (e.u.)
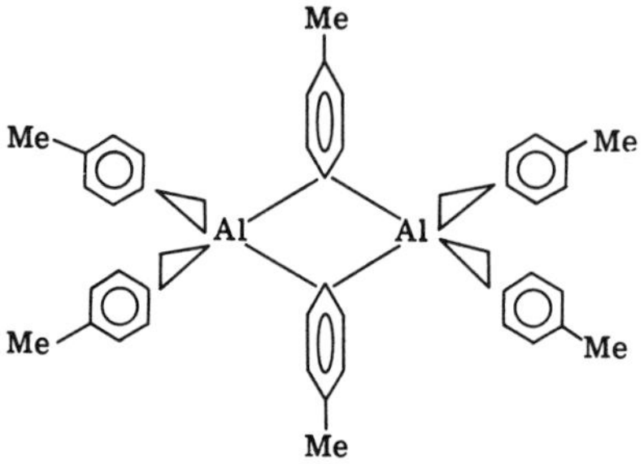 (structure: two Al centers bridged, with four Me-phenyl groups)	8.0 ± 1	7.6	-13 @ 225 °C

Intermolecular Exchange Reactions

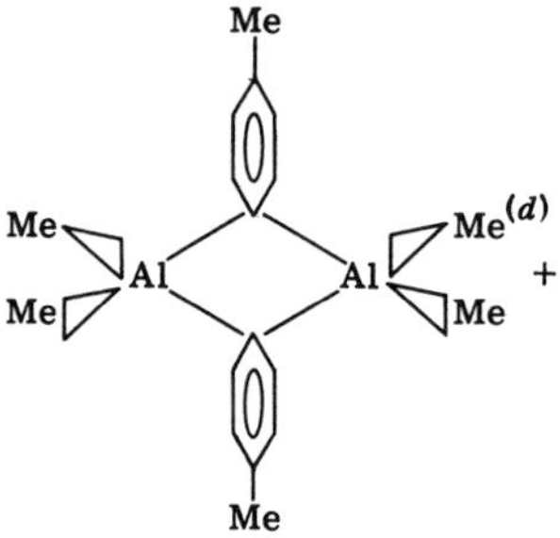

(structure) $+$ Al$_2$Me$_6$	$11.1\pm$	10.6	$-11.$ @ 225 °C
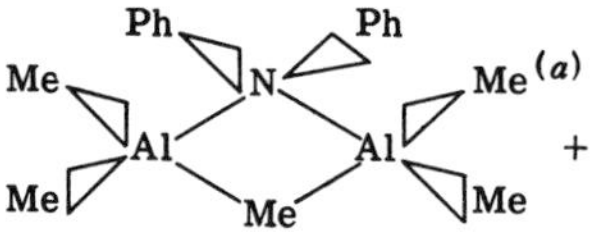 (structure) $+$ GaMe$_3$	20.93 ± 0.8	20.35 ± 0.8	-20.6 ± 4.2 @ 290 °C

state [Eq. (8)] by the interaction with the phenyl groups as shown in **17**.

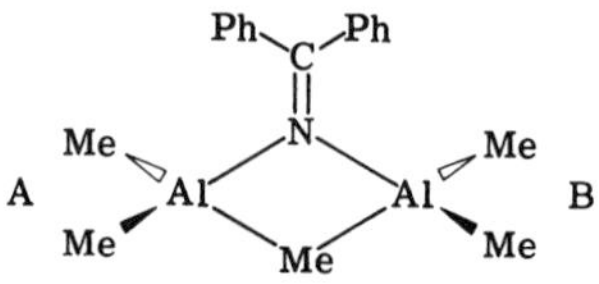

$$\begin{array}{c} Ph\diagdown \underset{C}{}\diagup Ph \\ \| \\ N \\ A\ \ Me\diagdown Al\diagup N\diagdown Al\diagup Me\ \ B \\ Me \diagup\ \ Me\ \ \diagdown Me \end{array}$$

(16)

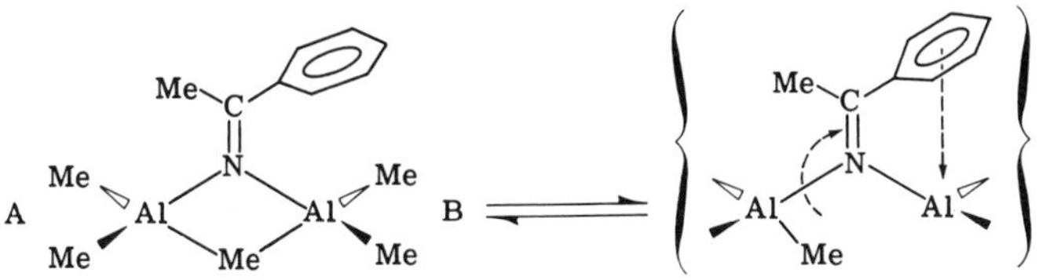

$$(8)$$

(17)

B. Exchange in Systems with Electron-Deficient Bonds That Differ in Stability

The limiting cases for this type of exchange process occur as the relative stabilities of the two bridging groups approach each other. Two systems investigated in which this condition exists are represented in **18** (*47*) and **19** (*51*), in which the strong bridging group is either cyclopropyl

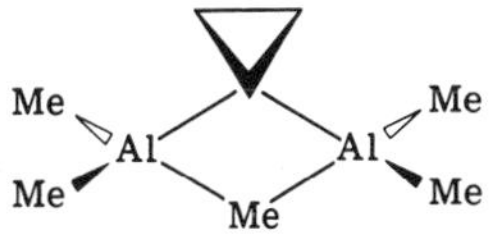

(18) 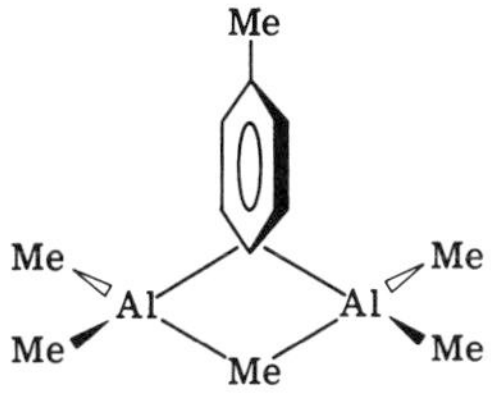   (19)

or *p*-tolyl. These systems both undergo rapid bridge-terminal methyl group exchange, and the observed kinetic parameters (Table II) are consistent with a single bridge-opening process similar to that described for the diphenylamino bridged system.

C. Exchange Reactions in Systems with Identical Bridging Groups

Examination of the parent tri(*p*-tolyl)aluminum system provides another interesting result (*51*). The activation parameters for the bridge-terminal exchange in this system shows that the process requires less energy for exchange than the exchange observed for trimethylaluminum. This, along with the kinetics, is consistent with the proposed mechanism shown in Eq. (9) with a single bridge opening (*51*). This system does differ from the diphenylamino bridged system since the bridge bonds are of the same energy, thus allowing either bridge to open. This type of exchange may prevail in other derivatives, such as those indicated in **3, 4,** and **5,** but there is insufficient evidence to either prove or disprove this possibility.

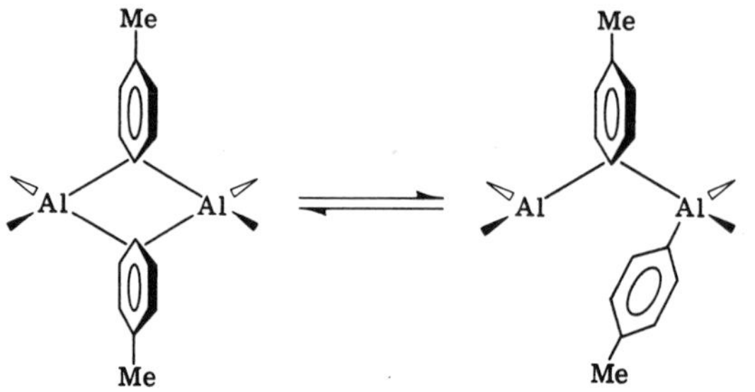

$$(9)$$

The final intramolecular process which has been examined is illustrated in **20,** Eq. (10), and in **21a** and **21b** (*36, 47*). This process has been represented as a rotational process taking the bridging cyclopropyl

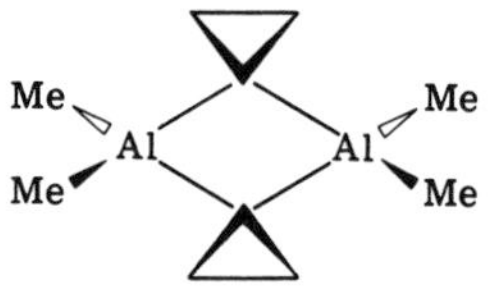

(20)

$$(10)$$

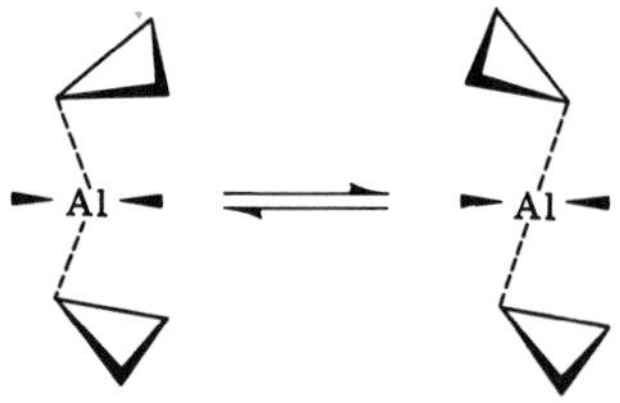

syn	*anti*
(21a)	(21b)

groups from the syn position observed in the solid state, in which the terminal methyl groups are nonequivalent, to the anti configuration, in which all methyl groups are equivalent. The equilibrium between the two forms has not been determined, but the barrier to rotation has been measured from variable-temperature NMR studies (Table II) and has also been calculated by complete neglect of differential overlap (CNDO/2) methods (*45*). These results, although approximate in nature, support the proposed rotational process and give results fortuitously close to the value obtained experimentally.

VII

INTERMOLECULAR EXCHANGE INVOLVING PARTIAL DISSOCIATION

The bridge-terminal intramolecular exchange of methyl groups in μ-Me-μ-Ph$_2$NAl$_2$Me$_4$ has been described in Section VI,A. This material also may undergo intermolecular exchange with a second organometallic species, such as Me$_3$Ga (43). The reaction between these two species is second order, first order in each component, and proceeds without disruption of the Al—N—Al bridge bond. The activation parameters are given in Table II.

At least two possible mechanisms can be proposed that are compatible with these results. The first, originally proposed by Mole (34, 35), for systems containing two strong bridges, is given in Eq. (11) and has a 5-

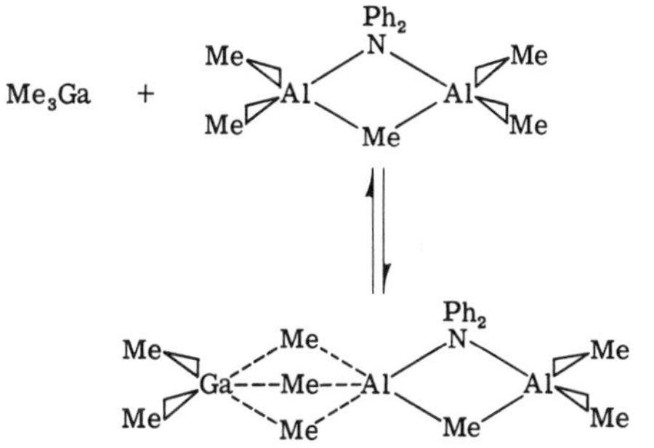

(11)

(22)

coordinate aluminum atom in the transition state similar to that suggested for the exchange of Me$_3$Al with Me$_3$Al·PMe$_3$ (Section V). The second mechanism involves the initial opening of the alkyl bridge followed by reaction with the Me$_3$Ga in a rate-determining step, as suggested in Eq. (12). The transition state may take the cyclic form

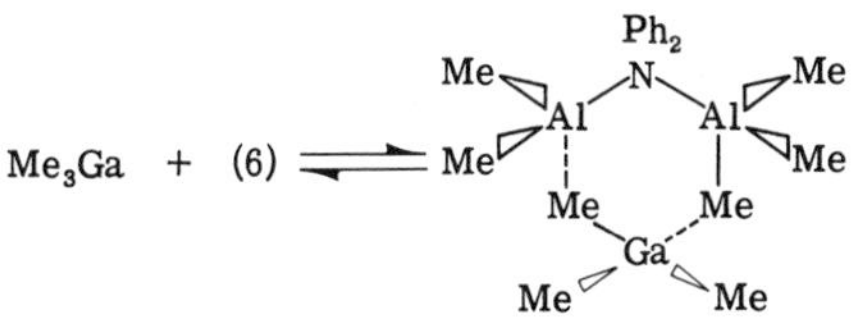

(12)

(23)

indicated in **23** or may form some other transition state involving only one of the two aluminum centers.

The most recent work on these systems favors the mechanism described in Eq. (12). This mechanism also has been suggested for the intermolecular methyl exchange in the Me_6Al_2-p-tolyl·Me_5Al_2 system (*51*).

Unfortunately, it is not possible in the present studies with similar bridging and terminal groups to assign a mechanism unequivocally, but only to show that it is consistent with the observed kinetics for the system. Certainly additional work must be carried out in order to establish paths completely and to explore fully the influence of the bridging group on the mode of reaction.

VIII

PROCESSES INVOLVING METAL–MULTIPLE BOND INTERACTIONS

The final type of process which is to be dealt with here is the interaction of a vacant metal orbital with an unsaturated organic groups, such as a double or triple bond. The interactions possible are of two types. The first involves the interaction of a filled multiple bond on an organic moiety with the vacant nonbonding orbitals generated on formation of the three-centered MO scheme described in Section II. This type of interaction has been confirmed by X-ray structural data for the *trans-t*-butylvinyl-, cyclopropyl-, and phenyl-bridged aluminum derivatives.

The second class of metal–multiple bond interactions has been postulated both for interaction of olefins (*10, 11, 13, 14, 20, 21, 45*) and for acetylenes (*12, 15–18*) as the initial step in metal–carbon or metal–hydrogen addition across the multiple bond. The complexes proposed are illustrated in **24** and **25**.

$$
\begin{array}{cc}
\begin{array}{c}
R \diagdown \\
R \!\rightarrow\! M \!\Leftarrow\! \\
R \diagup
\end{array}
\begin{array}{c}
\diagdown \diagup \\
C \\
\| \\
C \\
\diagup \diagdown
\end{array}
&
\begin{array}{c}
R \diagdown \\
R \!\triangleleft\! M \!\Leftarrow\! \\
R \diagup
\end{array}
\begin{array}{c}
| \\
C \\
\|\| \\
C \\
|
\end{array}
\\
(24) & (25)
\end{array}
$$

The overall reactions for these additions are indicated in Eqs. (13) and

(14), and lead to either a new saturated or olefinic-metal derivative, depending upon the substrate.

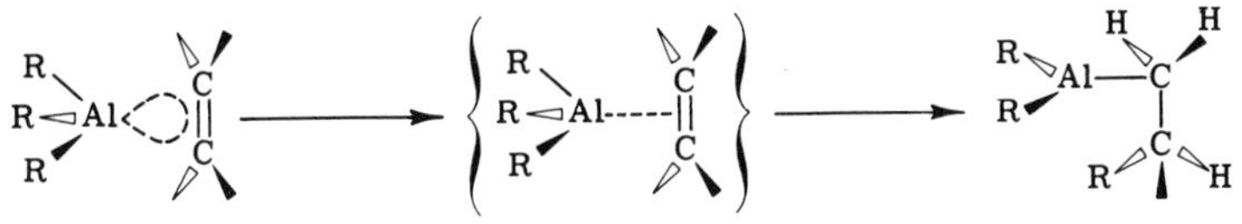

(13)

(20)

rearrangement

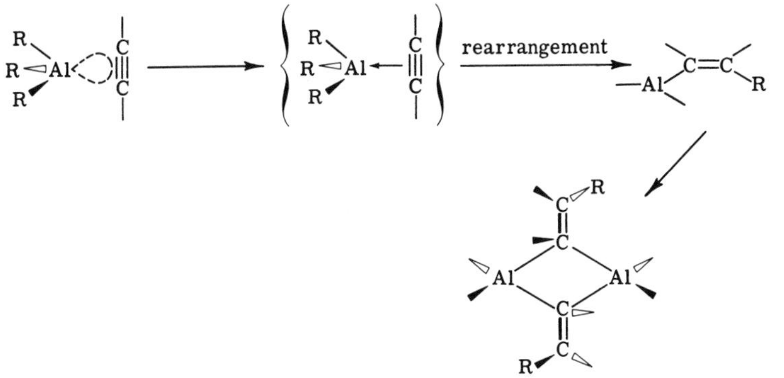

(14)

Eisch has developed a substantial amount of evidence based on the kinetics of reaction and on the stereochemistry of the products for the aluminum–triple bond interaction (*12, 15–18*). Further, structure of the phenylethynyl bridged dimer, recently determined by X-ray crystallography (*52*), provides strong evidence for this metal ← triple bond interaction with bridging groups clearly oriented appropriately for complex formation, as seen in **26**.

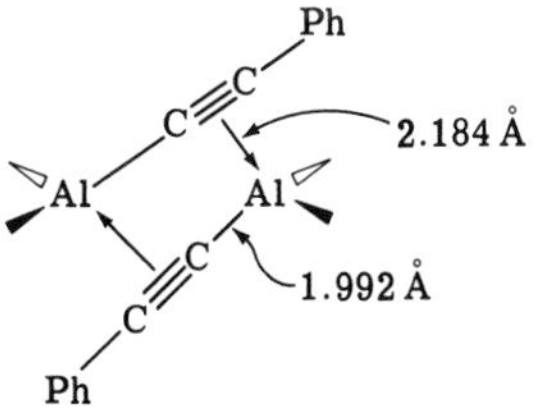

(26)

If we now turn our attention to the interaction of the metal center with olefinic derivatives, we find two classes of reactions. The first of these is the basis for the olefin polymerization or simple addition reactions, as illustrated in Eq. (13). A wide variety of data have been provided with regard to this reaction, which supports the formation of the complex (*13, 14, 21, 35*).

The second class of reactions to be examined is that involving intramolecular processes. A variety of metal derivatives that contain an organic moiety with a terminal double bond undergo a facile cyclization process, as seen in Eq. (15) (*10, 11, 20, 45*). This is particularly true for

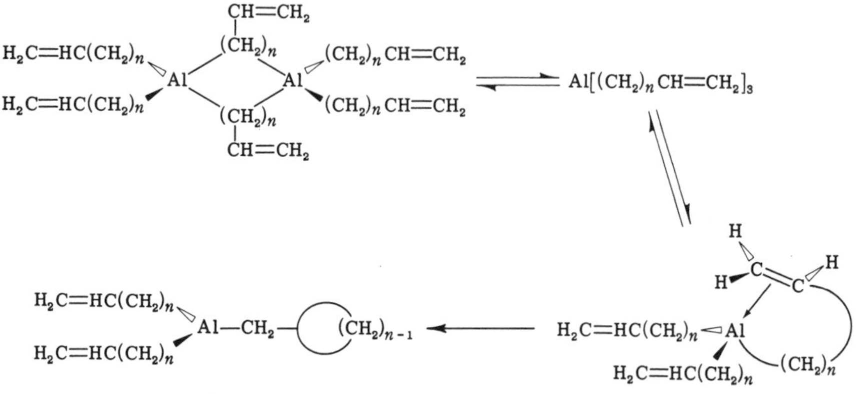

$$(15)$$

those derivatives that contain a 6-carbon chain and decreases for longer groups. Since this 6-carbon chain can most readily be oriented to form a metal ← double bond interaction, this observation supports the proposed mechanism (*10, 11, 45*).

The most convincing arguments in favor of complex formation, however, come from examination of the physical properties of these olefinic derivatives (*10*). To obtain this evidence, a series of unsaturated metal derivatives were prepared with increasing numbers of methylene units between the metal and double bond: $Al[(CH_2)_nCH=CH_2]_3$; $n = 2$, 3, 4, 5. The system may then be described by the various steps indicated in Eq. (15). When $n = 2$ or 3, colligative property measurements show that the monomer (**27**) is clearly favored over the dimer (Table I, Section III). With $n = 4$, the final cyclicalized product (**28**) is formed rapidly. The shift in the equilibrium from the favored dimeric form to the monomer indicates that the metal–double bond interaction is sufficiently strong to disrupt 2 Al—C—Al three-centered bond. We know that the dissociation energy for these bridge bonds is of the order of 12–16 kcal/mole; thus we can set a lower limit on the individual metal–olefin interaction of 6–8 kcal/mole, since two of these interactions may result on dissociation of each dimer.

Investigation of these systems by variable-temperature NMR spectroscopy, where all groups are of the same type, clearly shows even at −80°C that the butenyl or pentenyl groups are all equivalent. This implies that the three groups present on the aluminum derivative are in rapid equilibrium between the free and complexed states. This rapid exchange suggests that the activation energy for the process is low. Since it seems probable that the mechanism for equilibrating the R groups requires initial dissociation of the metal ← olefin interaction, an upper limit of 10 or 11 kcal/mole for this may be set.

Unfortunately, quantitative data for these systems are not yet available, either with regard to the structure or with regard to the exchange phenomena, and a full treatment of these systems must therefore await additional experimental information.

REFERENCES

1. Alaluf, E., Alford, K. J., Bishop, E. O., and Smith, J. D., *J. Chem. Soc., Dalton Trans.* 669 (1974).
2. Albright, M., Butler, W. M., Anderson, T. J., Glick, M. D., and Oliver, J. P., *J. Am. Chem. Soc.* **98**, 3995 (1976).
3. Brown, T. L., *Adv. Organometal. Chem.* **3**, 365 (1965).
4. Brown, T. L., *Rev. Pure Appl. Chem.* **23**, 447 (1970).
5. Brown, T. L., *Acc. Chem. Res.* **1**, 23 (1968).
6. Brown, T. L., and Murrell, L. L., *J. Am. Chem. Soc.* **94**, 378 (1972).
7. Clark, G. M., and Zweifel, G., *J. Am. Chem. Soc.* **93**, 527 (1971).
8. Coates, G. E., "Organometallic Compounds," 3rd ed., Vol. I. Methuen, London, 1967.
9. Conway, A. J., Hitchcock, P. B., and Smith, J. D., *J. Chem. Soc., Dalton Trans.* 1945 (1975).
10. Dolzine, T. W., and Oliver, J. P., *J. Am. Chem. Soc.* **96**, 1737 (1974).
11. Dolzine, T. W., and Oliver, J. P., *J. Organometal. Chem.* **78**, 165 (1974).
12. Eisch, J. J., *Ann. N. Y. Acad. Sci.* **239**, 292 (1974).
13. Eisch, J. J., Burlinson, N. E., and Boleslawski, M., *J. Organometal. Chem.* **111**, 137 (1976).
14. Eisch, J. J., and Burlinson, N. E., *J. Am. Chem. Soc.* **98**, 753 (1976).
15. Eisch, J. J., Gopal, H., and Rhee, S. G., *J. Org. Chem.* **40**, 2064 (1975).
16. Eisch, J. J., and Rhee, S. G., *J. Am. Chem. Soc.* **96**, 7276 (1974).
17. Eisch, J. J., and Rhee, S. G., *J. Am. Chem. Soc.* **97**, 4673 (1975).
18. Eisch, J. J., and Rhee, S. G., *J. Organomet. Chem.* **86**, 143 (1975).
19. Ham, N. S., and Mole, T., *Prog. NMR Spectrosc.* **4**, 91 (1969).
20. Hata, G., *Chem. Commun.* 7 (1968).
21. Henold, K. L., Deroos, J. B., and Oliver, J. P., *Inorg. Chem.* **8**, 2035 (1969).
22. Henold, K. L., and Oliver, J. P., *Organomet. React.* **5**, 387 (1975).
23. Hoffman, J. C., and Streib, W. E., *J. Chem. Soc., D* 911 (1971).
24. Jeffery, E. A., and Mole, T., *J. Organomet. Chem.* **11**, 393 (1968).
25. Jeffery, E. A., and Mole, T., *Aust. J. Chem.* **22**, 1129 (1969).
26. Jeffery, E. A., and Mole, T., *Aust. J. Chem.* **23**, 715 (1970).
27. Jeffery, E. A., and Mole, T., *Aust. J. Chem.* **26**, 739 (1973).
28. Lewis, P. H., and Rundle, R. E., *J. Chem. Phys.* **21**, 986 (1953).
29. Magnuson, V. R., and Stucky, G. D., *J. Am. Chem. Soc.* **90**, 3269 (1968).
30. Magnuson, V. R., and Stucky, G. D., *J. Am. Chem. Soc.* **91**, 2544 (1969).

31. Malone, J. F., and McDonald, W. S., *J. Chem. Soc., Dalton Trans.* 2646 (1972).
32. Malone, J. F., and McDonald, W. S., *J. Chem. Soc., Dalton Trans.* 2649 (1972).
33. Matteson, D. S., "Organometallic Reaction Mechanisms." Academic Press, New York, 1974.
34. Mole, T., *Organomet. React.* **1,** 1 (1970).
35. Mole, T., and Jeffery, E. A., "Organoaluminum Compounds." Elsevier, Amsterdam, 1972.
36. Moore, J. W., Sanders, D. A., Scherr, P. A., Glick, M. D., and Oliver, J. P., *J. Am. Chem. Soc.* **93,** 1035 (1971).
37. Muller, N., and Pritchard, D. E., *J. Am. Chem. Soc.* **82,** 248 (1960).
38. Oliver, J. P., *Advan. Organomet. Chem.* **8,** 167 (1970).
39. Oliver, J. P., *Advan. Organomet. Chem.* **15,** 235 (1977).
40. Oliver, J. P., and Stevens, L. G., *J. Inorg. Nucl. Chem.* **24,** 953 (1962).
41. Ramey, K. C., O'Brien, J. F., Hasegawa, I., and Borchert, A. E., *J. Phys. Chem.* **69,** 3418 (1965).
42. Rie, J. E., and Oliver, J. P., *J. Organomet. Chem.* **80,** 219 (1974).
43. Rie, J. E., and Oliver, J. P., *J. Organomet. Chem.,* **00,** 000 (1977).
44. St. Denis, J. N., Butler, W., Glick, M. D., and Oliver, J. P., *J. Organomet. Chem.* **129,** 1 (1977).
45. St. Denis, J., Dolzine, T., and Oliver, J. P., *J. Am. Chem. Soc.* **94,** 8260 (1972).
46. Sanders, D. A., and Oliver, J. P., *J. Am. Chem. Soc.* **90,** 5910 (1968).
47. Sanders, D. A., Scherr, P. A., and Oliver, J. P., *Inorg. Chem.* **15,** 861 (1976).
48. Smith, M. B., *J. Phys. Chem.* **76,** 2933 (1972).
49. Smith, M. B., *J. Organomet. Chem.* **70,** 13 (1974).
50. Smith, M. B., *J. Organomet. Chem.* **22,** 273 (1970).
51. Stanford, T. B., Jr., and Henold, K. L., *Inorg. Chem.* **14,** 2426 (1975).
52. Stucky, G. D., McPherson, A. M., Rhine, W. E., Eisch, J. J., and Considine, J. L., *J. Am. Chem. Soc.* **96,** 1941 (1974).
53. Visser, H. D., and Oliver, J. P., *J. Am. Chem. Soc.* **90,** 3579 (1968).
54. Vranka, R. G., and Amma, E. L., *J. Am. Chem. Soc.* **89,** 3121 (1967).
55. Williams, K. C., and Brown, T. L., *J. Am. Chem. Soc.* **88,** 5460 (1966).
56. Yamamoto, O., Hayamiza, K., and Yanagisawa, M., *J. Organomet. Chem.* **73,** 17 (1974).

Organomagnesium Rearrangements

E. ALEXANDER HILL

Department of Chemistry
University of Wisconsin-Milwaukee
Milwaukee, Wisconsin

I

INTRODUCTION

To those for whom the term *rearrangement* implies the Wagner–Meerwein 1,2-shift of carbonium ion chemistry, organomagnesium chemistry (along with most of organometallic chemistry) may seem to provide rather meager fare. The simple 1,2-shift of alkyl or hydrogen is a "forbidden" process on orbital symmetry grounds, requiring the occupancy of an antibonding orbital in the transition state (*68, 69*), and is not observed.

However, other rearrangement types are possible, and a rather impressive variety of rearrangements has been reported. The principal general class of rearrangement involves ring formation by intramolecular addition of the carbon–magnesium bond to a carbon–carbon multiple bond, the reverse reaction of β-cleavage with rupture of a strained ring, or the sequential combination of these reactions shown in Eq. (1).

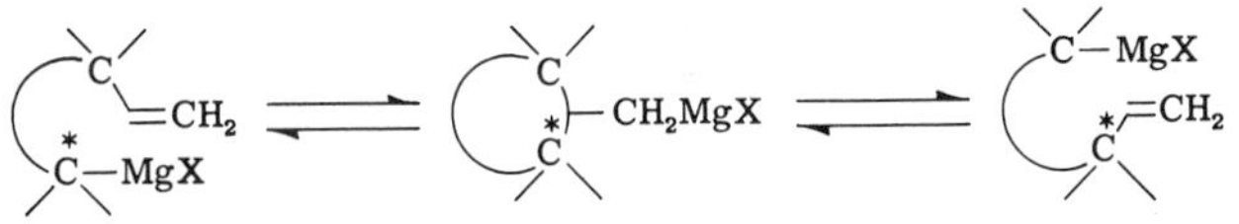

$$\underset{\overset{|}{C}-MgX}{\overset{|}{C}}=CH_2 \;\rightleftharpoons\; \overset{|}{\underset{\overset{|}{C}}{C}}\!\!-CH_2MgX \;\rightleftharpoons\; \overset{\overset{|}{C}-MgX}{\underset{\overset{|}{C}}{}}=CH_2 \tag{1}$$

Where the ring generated is a cyclopropane ring, the latter process produces the same net result as a 1,2-shift of a neighboring vinyl group. If the carbon–carbon unsaturation should be a phenyl group, rearrangement would correspond to phenyl migration, found principally with the more polar organoalkali compounds. Finally, in the formal limit of a two-membered ring, this becomes the allylic shift.

In addition to rearrangements in which one organomagnesium compound is converted to another, rearrangements may occur in the process of formation of the organometallic, or in the course of reaction of the organometallic with another reactant. In many cases, there is evidence that such rearrangements occur in a free-radical intermediate in the reaction.

A critical discussion of Grignard rearrangements has been published recently (29). For this reason, the present review will concentrate on some selected aspects of organomagnesium rearrangement chemistry, rather than attempting to be comprehensive. The most recent literature will be summarized in some detail, but older results will be used more selectively.

II

INTRAMOLECULAR ADDITION AND RING-CLEAVAGE REARRANGEMENTS

A. *Historical Survey*

In 1950, Smith and McKenzie reported an attempt to prepare cyclo-propylacetic acid (65). They carbonated the Grignard reagent derived from a bromide made from cyclopropylmethanol and PBr_3. However, the product turned out instead to be allylacetic acid [Eq. (2)]. Since

$$\triangleright\!\!-CH_2OH \xrightarrow{PBr_3} \xrightarrow[\text{ether}]{Mg} \xrightarrow{CO_2} \diagup\!\!\diagdown\!\!\diagup\!\!\diagdown_{COOH} \qquad (2)$$

carbonium ion rearrangements in cyclopropylmethyl systems were known, they concluded that the ring-cleavage rearrangement had most likely occurred in the step where the alcohol was converted to the bromide. The next year, Roberts and Mazur reported that, on the basis of spectroscopic and solvolytic data, the bromide from cyclopropylme-thanol and PBr_3 is mostly saturated (57). Cyclopropylmethyl bromide comprises the principal component. They found that Grignard formation from cyclopropylmethyl halides (but not cyclobutyl halides) gave the anilide derivative corresponding to ring-cleaved Grignard [Eq. (3)]. Roberts had also noted in 1950 that products derived from the Grignard reagent from dehydronorbornyl chloride have the nortricyclyl structure

(58). At that stage, it was uncertain whether rearrangement had occurred *during formation* of the Grignard, *after formation* but before its reaction, or *during reaction* with the carbon dioxide or phenyl isocyanate.

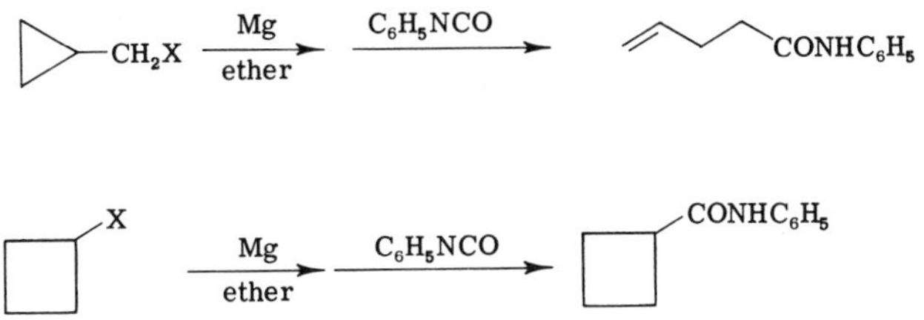

(3)

In 1965, in work published after additional examples of Grignard and other organometallic rearrangements had been uncovered, Patel, Hamilton, and Roberts settled this question (52). Grignard reagents prepared from cyclopropylmethyl halides gave at most traces of methylcyclopropane on hydrolysis. However, reaction of the halide with magnesium in the presence of a proton donor, such as benzoic acid, led to as much as 50% of methylcyclopropane. It may be concluded that about half of the halide led to rearranged Grignard **2** directly in the process of Grignard formation. Unrearranged Grignard **1** that was formed must then have been rapidly converted to **2**, unless trapped immediately by protolysis [Eq. (4)]. It was also found that cyclopropylmethyl magnesium bromide

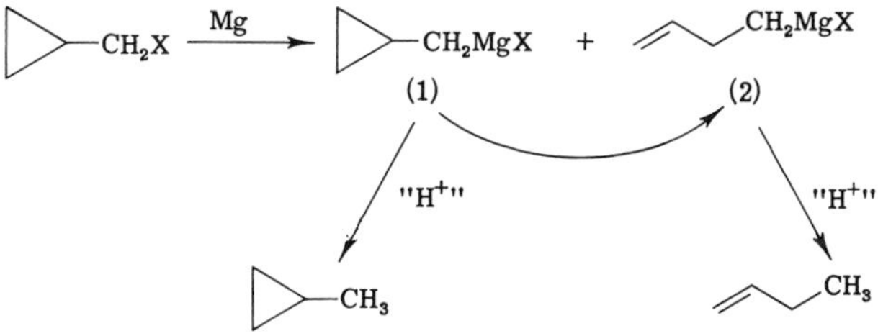

(4)

could be prepared in refluxing methyl ether at −24°C. It rearranged slowly enough that methylcyclopropane was again found in about 50% yield on hydrolysis after Grignard formation was complete. Rearrangement at −24°C occurred with a half-life of about 2 hours.

Meanwhile, Roberts' group had also reported in 1960 that the cyclopropylmethyl–3-buten-1-yl Grignard rearrangement could be approached from the side of ring-opened Grignard (63). In an attempt to synthesize 4-amino-1-pentene by reaction of the Grignard reagent of the corresponding chloride with methoxyamine, 4-amino-3-methyl-1-butene was the unexpected major product. A Grignard rearrangement via the sequence of cyclization and cleavage shown in Eq. (5) conveniently explained the result. Further exploration clearly demonstrated rear-

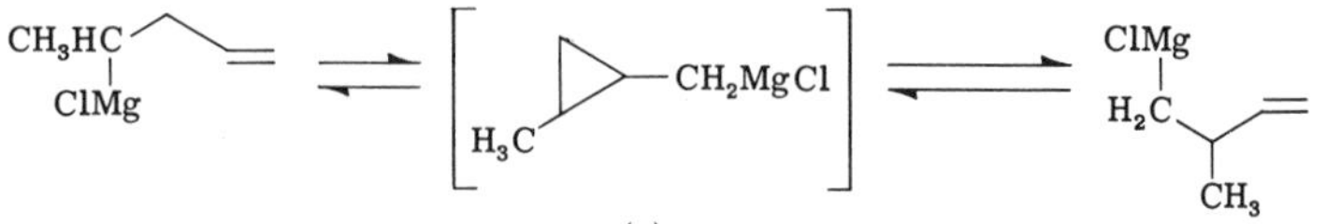

$$\text{(5)}$$

rangement of an organomagnesium intermediate with deuterated 3-buten-1-yl Grignard [Eq. (6)]. Grignard observed immediately after formation lacked the high-field NMR signal for protons α to the magnesium. Rearrangement to an equilibrium mixture, followed by appearance of this resonance, occurred with half-life of 30 hours at 27°C and 40 minutes at 55°C.

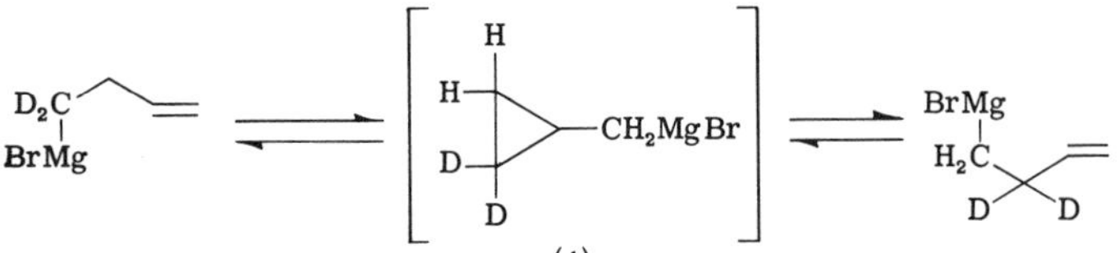

$$\text{(6)}$$

Equations (5) and (6) presume that the cyclic Grignards **3** and **4** are intermediates in the formal 1,2-vinyl migrations. The subsequent finding (*52*) that the cyclopropylmethyl Grignard is preparable, though rapidly rearranging, supports this presumption. Further proof of its intermediacy is derived from experiments by Maercker (*50, 50a*) and by Hill (*33*), shown in Eqs. (7) and (8) respectively. In Eq. (7), cis-trans equilibration

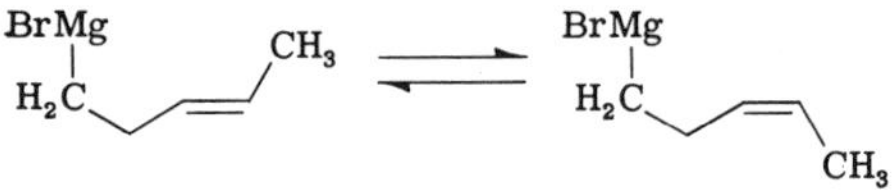

$$\text{(7)}$$

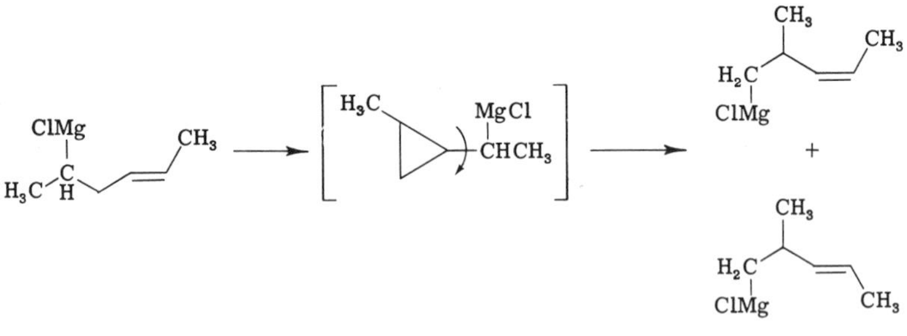

$$\text{(8)}$$

at the double bond occurred at a rate similar to isotopic scrambling of the two methylene groups, measured in separate experiments. This result indicates an intermediate with free rotation about the bond to the ring. The same conclusion is reached from the results in Eq. (8), in which the reaction is essentially irreversible because a secondary Grignard is converted to a primary one by the rearrangement. No cis-trans equilibration of the starting Grignard was observed.

Since the original reports by Roberts and co-workers of the cyclopropylmethyl–3-butenyl Grignard interconversion, numerous examples of related organomagnesium rearrangements have been reported. These have involved cleavages of three-, four-, and five-membered rings and cyclizations forming three- through six-membered rings. The cyclizations have included additions to double and triple bonds and to the cumulated double bond of an allene. Cyclizations and cleavages have been observed in bicyclic and tricyclic carbon skeletons. In the following section, work which has been published more recently than 1973 will be summarized in some detail. From rearrangements included in that section, plus examples noted elsewhere in this review, a good cross section of organomagnesium rearrangement chemistry may be viewed. Reference (29) contains a more complete survey of prior literature.

B. Recently Reported Studies of Organomagnesium Rearrangements

Richey and Veale (55) have reported a kinetic study of the cyclization of 5-hexen-1-yl organomagnesium solutions in THF [Eq. (9)]. Rates for

$$
\text{(structure: hexenyl-CH}_2\text{MgBr)} \longrightarrow \text{(cyclopentyl)}-CH_2MgBr \tag{9}
$$

the dialkylmagnesium compound, for the alkylmagnesium bromide, and for mixtures of the latter with additional magnesium bromide were determined. The rates found for various mixtures could best be fit on the assumption of a Schlenk equilibrium [Eq. (10)] with an equilibrium

$$R_2Mg + MgBr_2 \rightleftharpoons 2RMgBr \tag{10}$$

constant of about 30 to 40 at 100°C, and independent parallel rearrangement of alkyl groups in monomeric RMgBr and RMgR. Rearrangement of RMgR was about 45 times as rapid as that of RMgBr. The observation of good first-order kinetics at various concentrations and compositions was taken to indicate that exchange equilibria among the components RMgBr, R_2Mg, R'MgBr, R'RMg, and R_2'Mg are rapid and not affected greatly by any differences between the original and cyclized alkyl groups (R and R', respectively).

Richey and Veale have also reported a study of the effect of phenyl substitution on the rate of Grignard cyclizations to yield three- and five-membered rings (56). The phenyl substitution in Grignard reagents **5** and **6** led to rates of deuterium label equilibration [analogous to Eq. (6)]

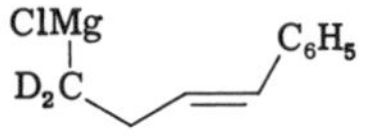

(5) 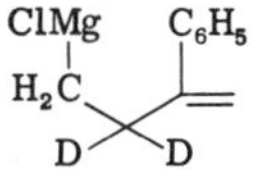   (6)

which were retarded by factors of 1600 and 3300, respectively. This result contrasts with Roberts' earlier report that the two phenyl groups in **7** accelerate the cyclization to the extent that rearrangement was

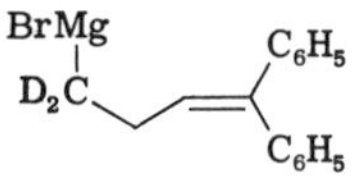

(7)

complete by the time of the first observation (40). Reinvestigation of that reaction showed that equilibration was substantial but not complete immediately after formation of the reagent. Subsequent complete equilibration was very slow, the second phenyl producing a further rate decrease by a factor of roughly 30 relative to **5**. Radical cyclization during Grignard formation is probably responsible for the initial equilibration. The phenyl substitution in **8** produced a modest 7-fold rate

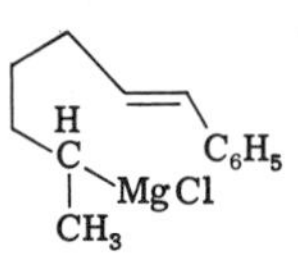

(8)

increase in cyclization to a five-membered ring. The authors suggested that the different effects of phenyl substitution may result from a rate-enhancing electronic effect and a rate-retarding steric effect. Significant observations on the stereochemistry of this cyclization will be discussed in Section II,F.

The most impressive array of new Grignard cyclization-cleavage rearrangement examples has come from studies on the *intermolecular* addition of Grignard reagents to alkenes. In 1975, Lehmkuhl and co-workers published extensive descriptions of their work in this area (46, 47). If an allylic Grignard reagent adds to an alkene, the addition product has organomagnesium and olefinic functions appropriately situated for an intramolecular addition to yield a substituted cyclobutylmethyl Grignard. In the reactions of crotyl and prenyl Grignard reagents with 1-octene [Eq. (11), R=H, CH$_3$, respectively], the isolation of substituted

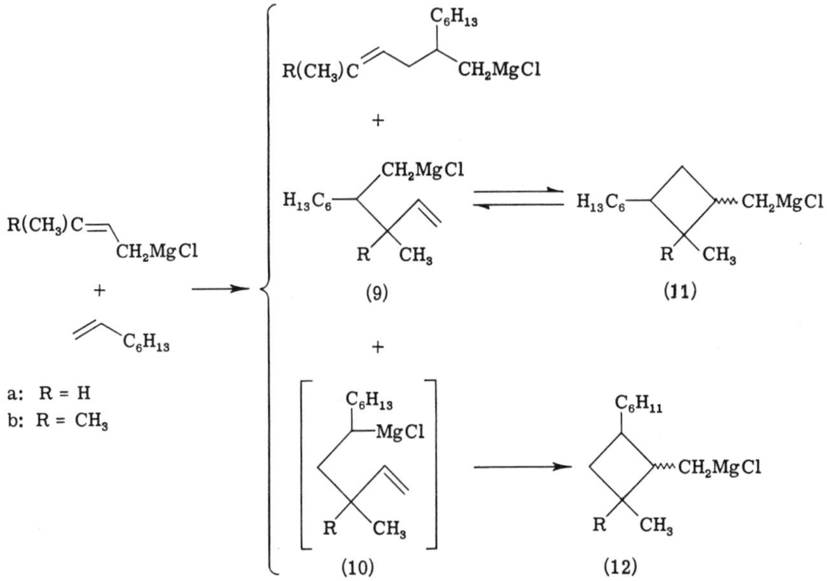

$$
\begin{aligned}
&R(CH_3)C{=} \\
&\qquad CH_2MgCl \\
&\quad + \\
&\diagup\diagup C_6H_{13}
\end{aligned}
\longrightarrow
\left\{
\begin{array}{c}
\overset{C_6H_{13}}{R(CH_3)C{=}}\diagup\diagdown CH_2MgCl \\[2pt]
+ \\[4pt]
(9) \rightleftharpoons (11) \\[4pt]
+ \\[4pt]
[(10)] \longrightarrow (12)
\end{array}
\right.
$$

a: R = H
b: R = CH$_3$

(11)

cyclobutane derivatives on hydrolysis demonstrated that cyclization to form stable cyclobutylmethyl Grignard reagents **11b**, **12a**, and **12b** had occurred. In the reaction of allylmagnesium chloride with 1-octene [Eq. (12)], cyclization to a six-membered ring took place after addition of a

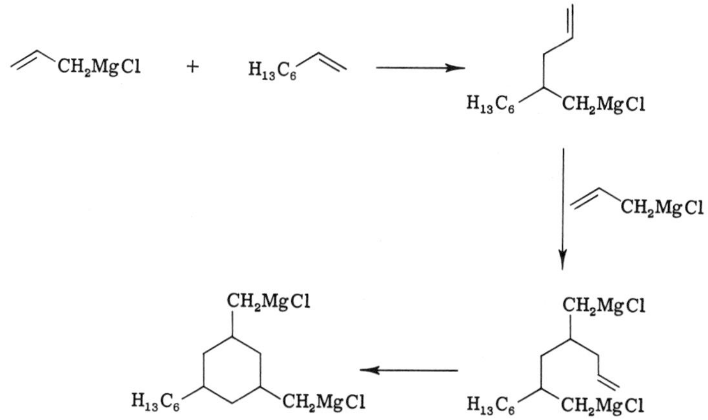

$$
\diagup\diagdown CH_2MgCl \;+\; H_{13}C_6\diagdown\diagup \longrightarrow \underset{H_{13}C_6 \quad CH_2MgCl}{\diagup\diagdown}
$$

(12)

second mole of Grignard to the original adduct. Similar cyclizations to cyclobutylmethyl structures were observed in the additions of crotyl and methallyl Grignards to norbornene [Eq. (13)]. In both instances, hydrol-

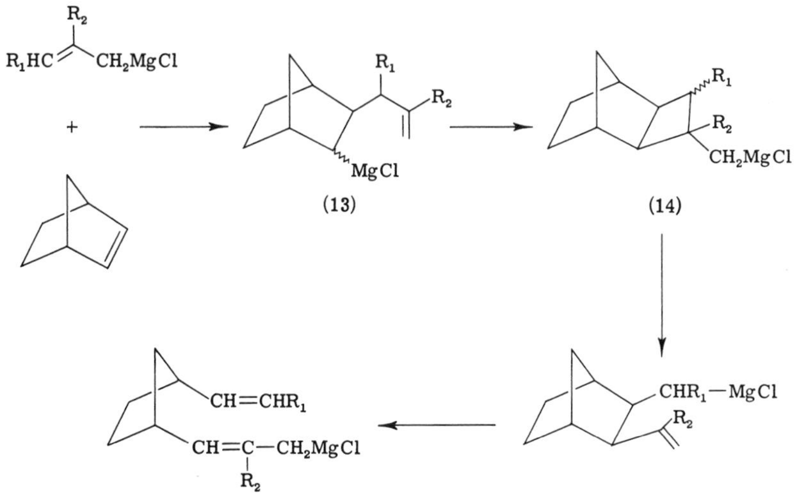

$$(13)$$

a: $R_1 = CH_3$; $R_2 = H$
b: $R_1 = H$, $R_2 = CH_3$

ysis products from the initial adduct (13) could be isolated after shorter or milder reaction periods. Further heating of this adduct before hydrolysis led to products derived from rearranged Grignard reagents 14a from the crotyl Grignard (as a mixture of two stereoisomers) and 15b from the methallyl Grignard. In both reactions, more vigorous heating led to rupture of one of the five-membered rings of the norbornane skeleton. In the presence of excess norbornene, the allylic Grignard product 16 adds to an additional molecule of alkene, and the sequence of cyclization and cleavages is repeated.

Rearrangement of the adducts of crotylmagnesium chloride to styrene [Eq. (14)] was also reported (46). The cyclization appears to follow an unusual course in producing the larger of two possible rings. In previously reported organometallic (29) or radical (43) cyclizations, the smaller-ring cycloalkylmethyl skeleton is the kinetically determined product (i.e., cyclobutylmethyl rather than cyclopentyl, or cyclopentyl-methyl rather than cyclohexyl).

A different variety of rearrangement was also noted (45) in the reaction of allyl and methallyl Grignards with styrene [Eq. (15)]. On long

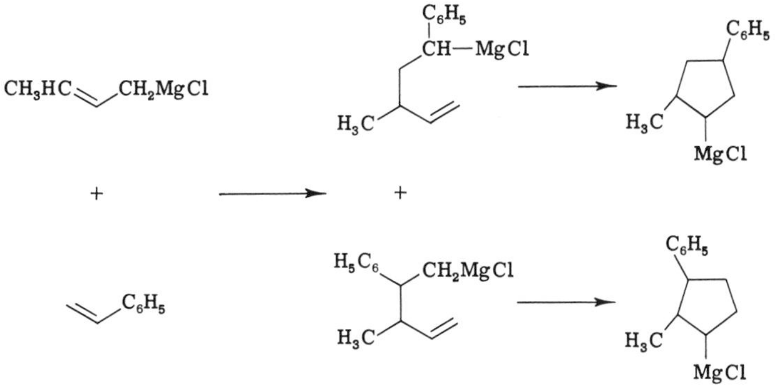

$$(14)$$

heating, the ratio of benzylic Grignard **17** to primary Grignard **18** increased slightly. It was proposed that this occurs via reversal of the addition (i.e., a cleavage-addition sequence). A similar mechanism is proposed for isomerization of the products of some allylic Grignard reagents with carbonyl compounds (*7, 51*).

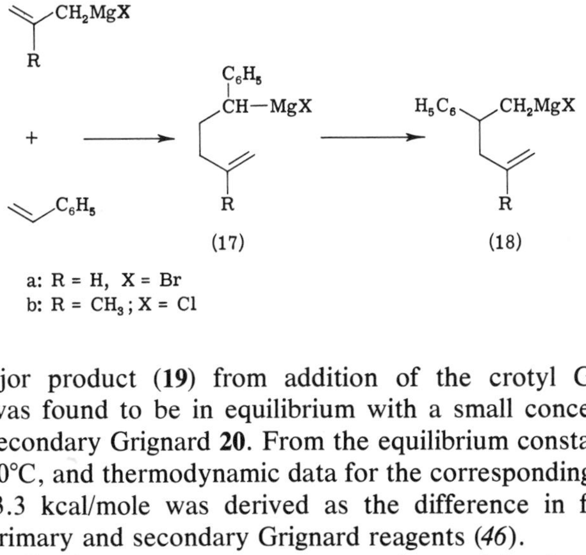

$$(15)$$

a: R = H, X = Br
b: R = CH$_3$; X = Cl

The major product (**19**) from addition of the crotyl Grignard to ethylene was found to be in equilibrium with a small concentration of isomeric secondary Grignard **20**. From the equilibrium constant [**19**]/[**20**] = 290 at 70°C, and thermodynamic data for the corresponding alkenes, a value of 3.3 kcal/mole was derived as the difference in free energy between primary and secondary Grignard reagents (*46*).

Other cyclization rearrangements were observed in the addition of allyl and methallyl Grignard reagents to butadiene (*47*). Intramolecular addition occurs after addition of a second butadiene [Eq. (17)]. With allylmagnesium bromide, three of four possible stereoisomeric hydrolysis products of the cyclized Grignard **21a** were identified. With methallyl magnesium chloride, a cyclization product incorporating three butadiene molecules was also isolated. Further details have also been furnished

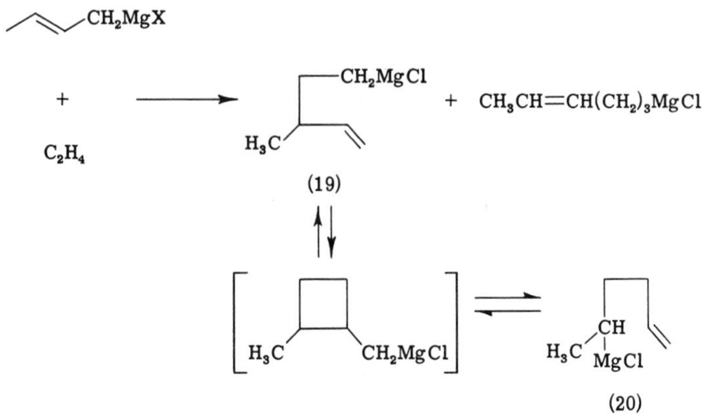

$$\text{CH}_2\text{MgX}$$

$$+ \longrightarrow \overset{\text{CH}_2\text{MgCl}}{\underset{\text{H}_3\text{C}}{\Big|}} + \text{CH}_3\text{CH}=\text{CH}(\text{CH}_2)_3\text{MgCl}$$

$$\text{C}_2\text{H}_4$$

(19)

(16)

$$\left[\underset{\text{H}_3\text{C}}{\square} \underset{\text{CH}_2\text{MgCl}}{} \right] \rightleftharpoons \underset{\text{H}_3\text{C}}{\overset{\text{CH}}{\underset{\text{MgCl}}{}}}$$

(20)

(*46*) on the previously reported rearrangement products from reaction of ethylene with the bisorganomagnesium dimers of butadiene and isoprene.

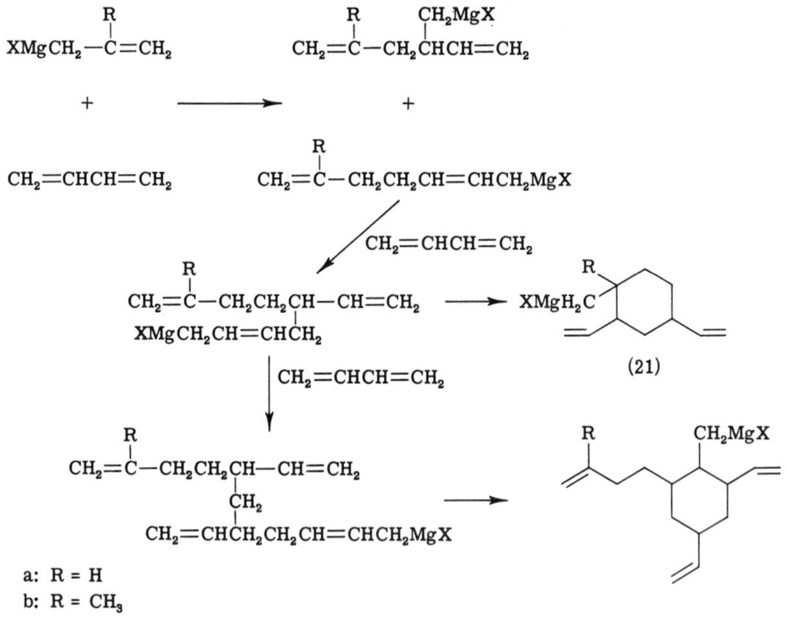

$$\underset{\text{XMgCH}_2-\overset{\text{R}}{\underset{|}{\text{C}}}=\text{CH}_2}{} \qquad \text{CH}_2=\overset{\text{R}}{\underset{|}{\text{C}}}-\text{CH}_2\overset{\text{CH}_2\text{MgX}}{\underset{|}{\text{CH}}}\text{CH}=\text{CH}_2$$

$$+ \longrightarrow +$$

$$\text{CH}_2=\text{CHCH}=\text{CH}_2 \qquad \text{CH}_2=\overset{\text{R}}{\underset{|}{\text{C}}}-\text{CH}_2\text{CH}_2\text{CH}=\text{CHCH}_2\text{MgX}$$

$$\diagdown \text{CH}_2=\text{CHCH}=\text{CH}_2$$

$$\text{CH}_2=\overset{\text{R}}{\underset{|}{\text{C}}}-\text{CH}_2\text{CH}_2\overset{}{\underset{\text{XMgCH}_2\text{CH}=\text{CHCH}_2}{\text{CH}}}-\text{CH}=\text{CH}_2 \longrightarrow \text{XMgH}_2\text{C}\overset{\text{R}}{\diagup}$$

(21)

$$\Big| \text{CH}_2=\text{CHCH}=\text{CH}_2$$

$$\text{CH}_2=\overset{\text{R}}{\underset{|}{\text{C}}}-\text{CH}_2\text{CH}_2\overset{}{\underset{\overset{|}{\text{CH}_2}}{\text{CH}}}-\text{CH}=\text{CH}_2$$

$$\text{CH}_2=\text{CHCH}_2\text{CH}_2\text{CH}=\text{CHCH}_2\text{MgX}$$

$$\longrightarrow$$

a: R = H
b: R = CH$_3$

(17)

Derocque and Sundermann (*16*) have reported that reaction of magnesium with bromide (**22**) leads to a mixture of unrearranged (**23**) and ring-cleaved (**24**) Grignard reagents. Subsequent to formation, an apparently irreversible ring-cleavage rearrangement of **23** to **24** occurred. The authors favored a radical process for rearrangement during Grignard formation and a cyclic mechanism (see Section II,E) for rearrangement of the formed Grignard. In previous work (*17*) it had been found that the bromide corresponding to **24** reacts with magnesium to form a mixture containing **24** and the cyclobutenyl Grignard **25**. Once formed, these

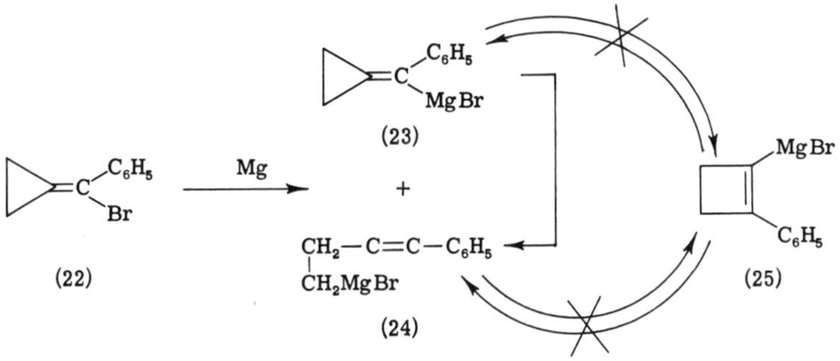

Grignard reagents were not interconverted. No rearranged products were formed in the reaction of the bromide corresponding (**25**) with magnesium or with tri-*n*-butyltin hydride, or in reaction of the corresponding mercuric bromide derivative with sodium–potassium alloy.

Gerard and Miginiac (*27*) have published a study of Grignard reagents of the structure **26**, where R = ethyl, phenyl, and vinyl.

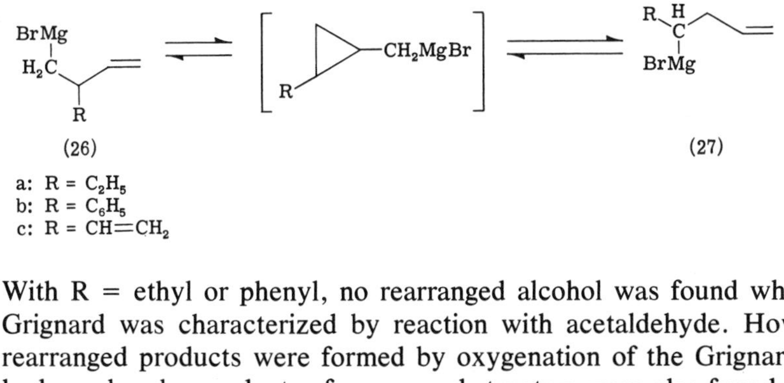

With R = ethyl or phenyl, no rearranged alcohol was found when the Grignard was characterized by reaction with acetaldehyde. However, rearranged products were formed by oxygenation of the Grignard, and hydrocarbon by-products of rearranged structure were also found. When R = vinyl, reaction of the Grignard with acetaldehyde gave both rearranged and unrearranged alcohols. The product mixture was not changed if the Grignard was refluxed 3 hours before reaction with the aldehyde. A Grignard formed from the bromide corresponding to **27c** gave only unrearranged Grignard product. The rearranged side products probably resulted from radical rearrangement during formation and oxygenation; under the conditions studied, no rearrangement of the formed Grignard occurred. In no case was any product with a cyclopropane ring detected.

Hill and co-workers (*31*) have published the details of a study of the 2-methylcyclobutylmethyl Grignard **28**. Both cis and trans isomers rear-

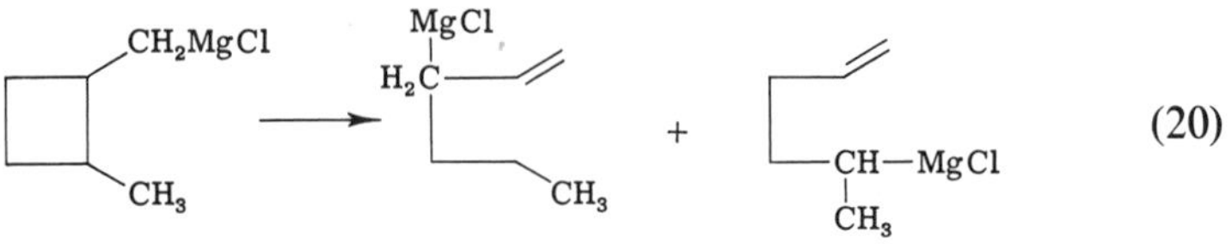

$$(20)$$

range at a rate slightly less than half that of the parent cyclobutylmethyl Grignard. Rearrangement yields almost entirely (>98%) the primary Grignard product **19**. Rearrangement of the corresponding free radical during tributyltin hydride reduction of the chloride occurred principally in the other direction. The results were cited as strong evidence against a radical mechanism for Grignard rearrangement (see Section II,E).

Maercker and Streit have also published details of their study of the rearrangement of **29b–29d**, which provides further evidence against a

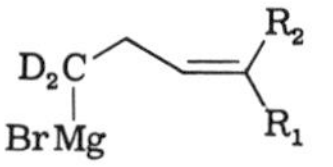

radical mechanism. Reaction was followed by equilibration of the methylene groups. Introduction of the one methyl group in **29b** markedly slowed the rearrangement, and a further decrease in rate was produced by the second methyl in **29c**. Compound **29d**, with a *tert*-butyl substituent, was still slower. The rearrangement rate of **29b** was strongly concentration-dependent, increasing linearly with molarity of the Grignard in the concentration range from 0.9 to 2.1 M. The reaction was about 5.5 times faster in diethyl ether than in THF. The decrease in rate with substitution on the double bond is inconsistent with a radical mechanism, since such substitution should stabilize the cyclized radical, and might be expected to increase the rate of cyclization. The authors concluded that the substitution, concentration, and solvent effects are most consistent with a bimolecular carbanionic mechanism (see Section II,E).

Results have also been published on the cleavage reactions of some Grignard reagents which incorporate the bicyclo[3.2.0]heptyl skeleton (*37*). The 2-bicyclo[3.2.0]heptyl Grignard reagent (**30**) cleaves to a mixture of primary (**31**) and secondary (**32**) Grignard reagents in which the primary cyclopentenylethyl Grignard (31) predominates by a ratio of about 10:1. The cycloheptenyl Grignard 32 rearranged slowly on further heating to 31. Since the corresponding bicyclic radical cleaves as shown

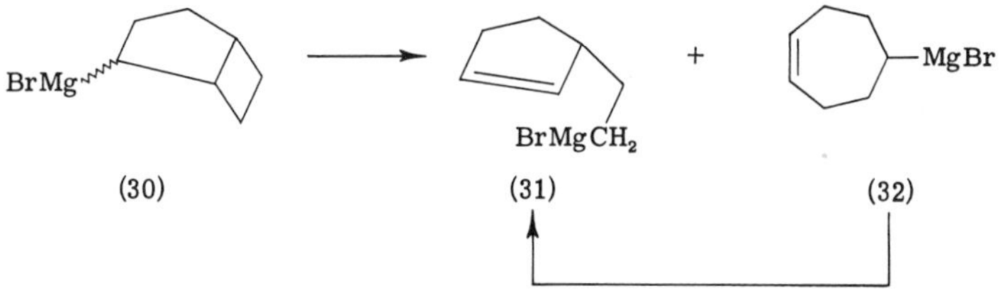

(21)

in Eq. (22), formation of the cycloheptenyl Grignard product does not result from a competing radical process. The rates of the cleavage rearrangements, and their implications for the rearrangement mechanism will be discussed below (Section II,F).

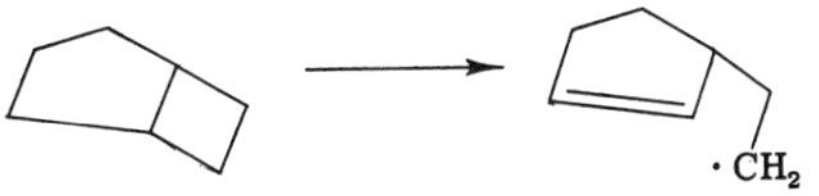

(22)

The tricyclic Grignard 33 undergoes cleavage in a similar fashion, except that no products with a cycloheptene ring were found [Eq. (23)].

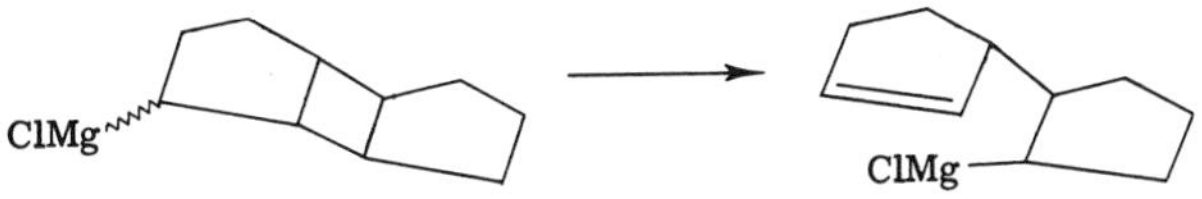

(23)

With Grignard reagents 34, the corresponding cleavage reaction would produce a β-alkoxy Grignard reagent; this should rapidly eliminate magnesium alkoxide, leading to dienes in an overall fragmentation process [Eq. (24)]. Such a fragmentation was indeed observed, with the

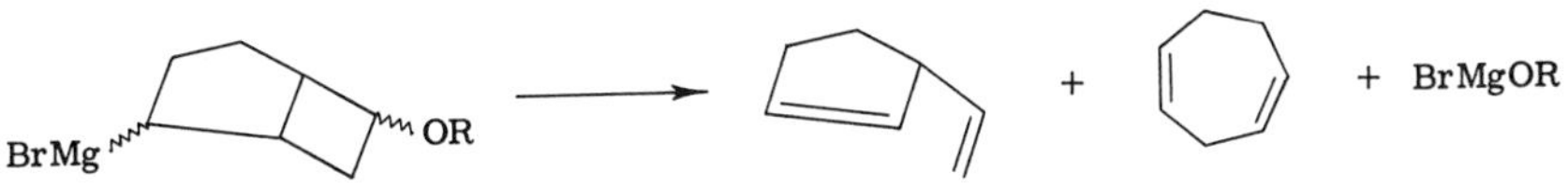

R = CH_3, C_2H_5

(24)

3-vinylcyclopentene and 1,4-cycloheptadiene formed in a ratio between 20:1 and 5:1. Similar products were formed from isomers with the alkoxy group exo or endo. Both isomers reacted at accelerated rates compared with the Grignard cleavage rearrangement of 30. The rate for the *endo* isomer was particularly rapid, about 10^3 times as fast as Eq.

(21), and it formed a larger proportion of vinylcyclopentene. It was suggested that this result might support a concerted fragmentation.

A cyclized product was isolated from the reaction of the 3-buten-1-yl Grignard reagent with thiobenzophenone [Eq. (25)] (14).

$(C_6H_5)_2C=S$

+

$BrMgCH_2CH_2CH=CH_2$

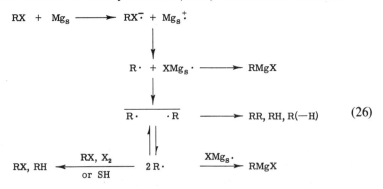

(25)

Radicals corresponding in structure to addition of an alkyl radical to the sulfur were detected by EPR in similar reactions, and other products of probable radical origin were formed. Therefore, it is not certain whether the cyclization resulted from a radical or an organometallic rearrangement.

Radical intermediates in the formation of Grignard reagents have been studied, using the rearrangement of the 5-hexen-1-yl radical as a probe (9). Reaction of 6-bromo-1-hexene with magnesium leads to a mixture of unrearranged and cyclized (cyclopentylmethyl) Grignard reagents. The proportion of rearranged Grignard was greatest in less polar solvents or mixtures. Coupling by-products, containing rearranged and unrearranged alkyl groups, were also identified. Although rearrangement of Grignard reagent does occur according to Eq. (9), this cyclization is too slow to account for a significant fraction of that observed shortly after formation of the reagent. Chemically induced dynamic nuclear polarizations (CIDNP) of the rearranged and unrearranged Grignard reagents formed were also observed. The results were interpreted to support a mechanism for Grignard reagent formation similar to that shown in Eq. (26). Formation of Grignard reagent via the radical pair, detected by its spin polarization, increased with decreasing solvent basicity and viscosity, and on dilution of tetrahydrofuran (THF) solvent with benzene.

$$RX + Mg_s \longrightarrow RX^{\bar{.}} + Mg_s^{\overset{+}{.}}$$

$$R\cdot + XMg_s\cdot \longrightarrow RMgX$$

$$R\cdot \quad \cdot R \longrightarrow RR, RH, R(-H) \qquad (26)$$

$$RX, RH \xleftarrow[\text{or SH}]{RX, X_2} 2\,R\cdot \xrightarrow{XMg_s\cdot} RMgX$$

(Mg_s, Mg_s^+, and $XMg_s\cdot$ are surface-bound species)

A Grignard reagent cyclization has also been observed in the dimerization of 1,3-butadiene brought about by propylmagnesium bromide and catalytic amounts of a nickel complex (23). As shown in Eq. (27),

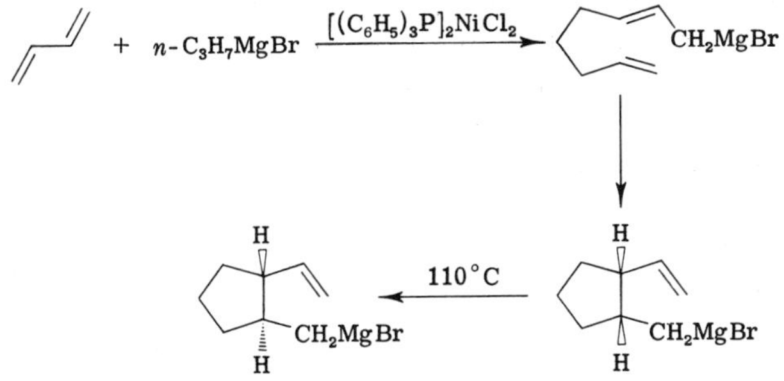

(27)

following the catalytic dimerization, the initial Grignard reagent product cyclizes stereospecifically to yield the cis isomer. On more vigorous heating, this is transformed to the trans isomer, very probably via reversal of the cyclization. The latter two reactions also occur in the absence of the nickel complex. It was not reported whether these steps were accelerated by the catalyst.

C. Equilibria in Grignard Cyclization–Cleavage Rearrangements

Of the Grignard reagent rearrangements noted thus far, some, such as those in Eqs. (9) and (12) proceed essentially to completion in the direction of cyclization. Others, for instance Eqs. (4) or (20), favor ring cleavage. The position of the equilibrium depends upon several factors. Two dominant ones are ring strain and the enthalpy change in the addition process:

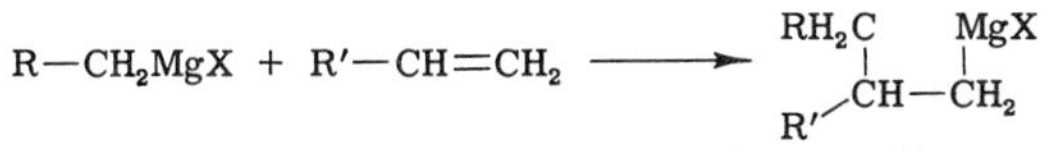

$$R-CH_2MgX + R'-CH=CH_2 \longrightarrow \begin{array}{c} RH_2C \quad MgX \\ | \quad\quad | \\ R'^{\diagup}CH-CH_2 \end{array}$$

For a primary organometallic adding to a vinyl group, the change in enthalpy may be estimated as -21.7 kcal/mole for formal conversion of a carbon–carbon double bond to two single bonds (8, 35). Countering this is the strain energy of the ring formed, which may be taken as Benson's additivity rule corrections of 27.6, 26.2, and 6.3 kcal/mole for three-, four-, and five-membered rings (8). The decreased entropy of the cyclic compound also favors the open-chain isomer [to an extent estimated as 10.4 cal/mole-deg for methylcyclobutane $\rightleftharpoons$ 1-pentene (35)]. The predicted result is that ring closure to a cyclopentane ring should be

rather strongly favored, while cleavage should be favored for the three-
and four-membered rings.

The above prediction should be modified by any structural feature that
affects the stability of either the Grignard or the ring system. For
example, the increased strain of the bicyclic norbornane skeleton makes
the five-membered ring more susceptible to cleavage. However, neither
the norbornyl nor norbornylmethyl Grignard reagents cleave (25, 26, 38).
But, if the Grignard reagent formed on cleavage is additionally stabilized
as an allylic organometallic, cleavage may occur. Two examples were
previously illustrated in Eq. (13) (46, 47), and another is shown in Eq.
(28) (32).

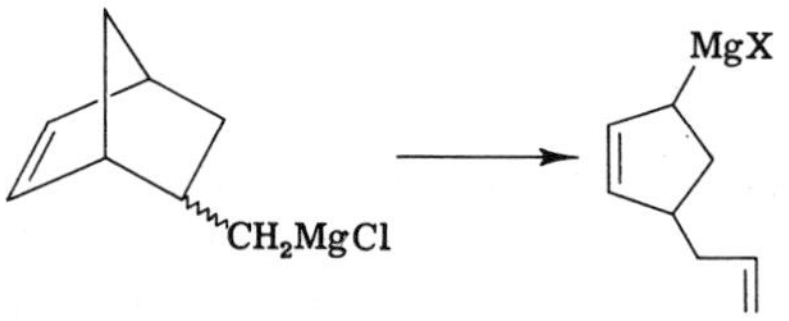

$$(28)$$

At the opposite extreme, the cyclopropylmethyl Grignard reagent **35** is
present at equilibrium to an extent of greater than 99.9% (49). Two

$$(29)$$

(35)

factors favor its formation. First, the open-chain Grignard isomer is
destabilized, since it is tertiary. Second, the cyclic isomer is probably
stabilized by the Thorpe-Ingold "*gem*-dimethyl" effect. With Eq. (30),

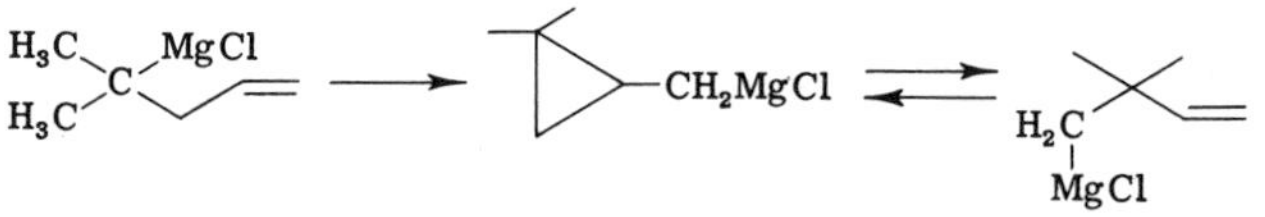

$$(30)$$

one of the open-chain isomers may be primary, so the equilibrium now
lies to the open-chain side. The isolation of about 0.07% of cyclic
hydrolysis product may reflect *gem*-dimethyl stabilization. Other instances where the three-membered ring is stable at equilibrium include
Eqs. (31) and (32), where it is incorporated into a nortricyclyl skeleton

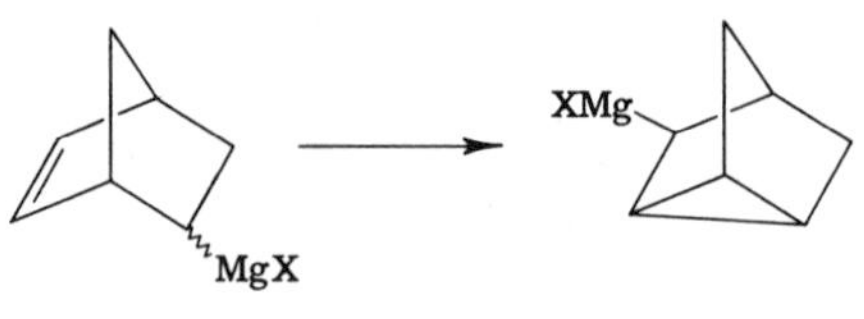

$$(31)$$

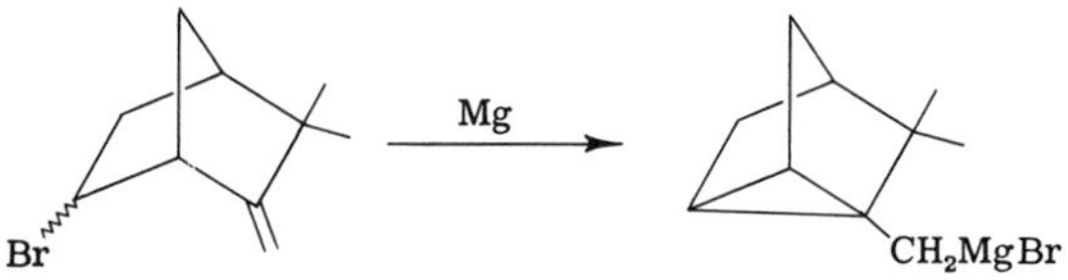

$$(32)$$

(*12, 15, 58*). With **36**, the cyclic Grignard is stabilized by being a vinyl

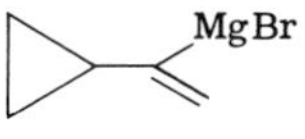

organometallic (*53, 62*). However, the cyclic intermediates in Eqs. (18) and (33) are apparently not stable enough to be isolated at equilibrium (*16, 54*). Interestingly, phenyl conjugation in **37** does not stabilize the

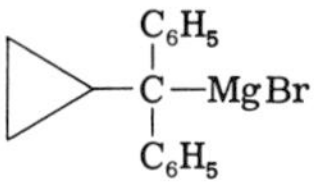

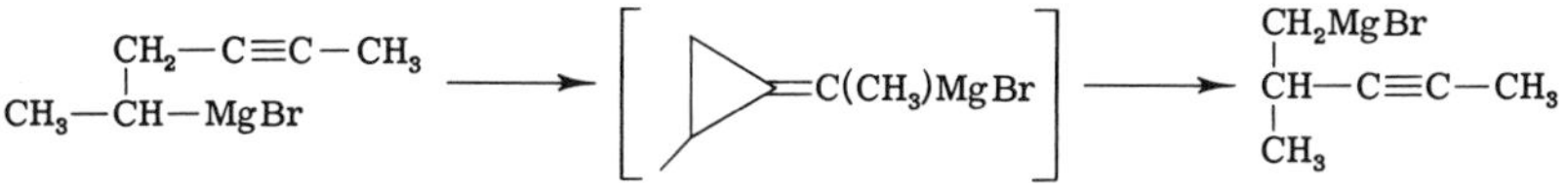

$$(33)$$

cyclic reagent sufficiently for it to lead to isolable amounts of hydrolysis products (*40*), although the corresponding cyclic alkali metal compounds are stable.

The four-membered ring may be similarly stabilized. In reactions shown previously in Eqs. (11) and (13), the combination of ring alkylation and secondary or tertiary open-chain isomers stabilizes cyclic Grignards **11b, 12a, 12b,** and **14a** to the extent that their hydrolysis products dominate (*46, 47*). In a slightly simpler system [Eq. (34)], the equilibrium constant in 1 *M* ether solution has a value of about 3, favoring the cyclized Grignard (*34*). The equilibration was studied at

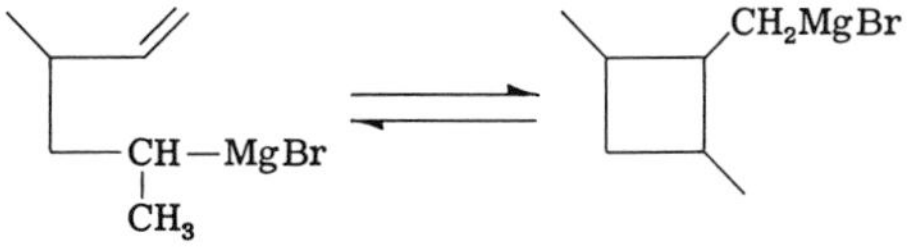

$$(34)$$

100°C, and exhibited little dependence on temperature. The major cyclic isomer has both methyl groups trans to the organomagnesium function, but slower equilibration with other isomers occurs.

Another equilibrium question, which may be approached through the Grignard rearrangement, is the difference in stability between primary and secondary organomagnesium functions. As noted above, an estimate that the primary Grignard is 3.3 kcal/mole more stable (ΔG) was derived from the equilibrium of Eq. (16). A slightly larger preference of 3.7 kcal (ΔH) or 5.0 kcal (ΔG) has been estimated (48) from the rearrangement of Eq. (35).

$$\text{(structure)} \rightleftharpoons \left[\text{(structure)} \right] \rightleftharpoons \text{(structure)} \quad (35)$$

D. *General Features of the Rearrangement*

A survey of quantitative and qualitative data on the Grignard rearrangement (29) leads to several generalizations:

1. Rearrangement occurs more rapidly in less polar solvents. Consistently, rearrangements are somewhat more rapid in diethyl ether than in THF, and are generally increased in rate by addition of a nonpolar diluent. Opposing this trend may be a rate increase produced by stronger solvent coordination to the magnesium.

2. Rearrangements follow first-order kinetics, and the rate is largely independent of total organomagnesium concentration in dilute solutions (<0.5–1.0 M). At higher concentration, rate increases are found, often linear with concentration.

3. The rate is relatively insensitive to magnesium purity. In nearly every case where this has been investigated, increases in rate from sublimed magnesium to "Grignard grade" magnesium have been less than 30%, and frequently negligible.

4. The dialkylmagnesium compound is more reactive than the corresponding alkylmagnesium halides. A recent detailed study has been noted above (55). Of the alkylmagnesium halides, the chloride has generally been more reactive than the bromide or iodide.

5. Relative reactivities of cycloalkylmagnesium halides to ring cleavage are $C_3 > C_4 \gg C_5$, C_6. For ring closure to various ring sizes, reactivities follow the order $C_3 > C_5 > C_4 > C_6$.

6. Alkyl group substitution on either end of the double bond appears to decrease the rate of ring closure (R_1, R_2, or R_3 in 38). An unusual

$$\text{(structure with } R_1, R_2, R_3, R_4, R_5, \text{C—MgX)}$$

(38)

order is found in cyclization for methyl substitution at R_4 and R_5 on the

Grignard carbon; relative reactivities are $1° < 2° > 3°$. In ring cleavage, alkyl substitution at R_1 appears to have little effect on rate, while alkyl at R_4 decreases the rate. Halogen at R_1 or R_2 or phenyl at R_3 decreases the rate of cyclization. Phenyl at R_1, R_2 produces a sizable decrease in cyclization rate for three-membered ring formation, but a slight increase for the five-membered ring.

7. When a phenyl group is substituted at the R_1 position, electron-withdrawing substituents on the benzene ring increase the cyclization rate. However, the substituent effect is rather small; $\rho = +0.9$ and 1.4 for **39** and **40** (*13, 36*).

$$\begin{array}{c} \langle\!\!\!\!\!\!\!\!\!\!\begin{array}{c} -C{\equiv}C-Ar \\ \\ -CH_2MgBr \end{array} \\ (39) \end{array} \qquad \begin{array}{c} Ar-\langle\bigcirc\rangle-MgCl \\ \\ (40) \end{array}$$

E. *Mechanism of Rearrangement*

As noted above, a wide variety of organomagnesium cyclization and cleavage rearrangements have been observed. Because of the similarities in behavior summarized in Section II,D, which appear to be rather consistently followed for rearrangements that have been studied in some detail, we will presume the importance of a single general mechanism throughout the following discussion. In particular, one should note that cyclization and cleavage are the same process in opposite directions, and so for one system, must involve precisely the same mechanism and transition state. However, the rearrangements studied do cover a wide range in conditions, including some that require heating at up to 150°C and occur only in competition with attack upon the ether solvent. It would not be unreasonable if other mechanisms should be utilized in the more extreme cases or for rearrangements of organometallics of other metals. Another mechanistic uncertainty may arise from rearrangements formulated as a cyclization–cleavage sequence, which are formally a 1,2 or 1,3 vinyl group migration. However, as noted in Section II,A, there is good reason to believe that these do indeed involve a cycloalkylmethyl Grignard as intermediate.

Four mechanisms for Grignard rearrangement which have received serious consideration are illustrated in Fig. 1. Mechanisms a and b involve a simple cleavage of the carbon–magnesium bond—homolytically or heterolytically—followed by rearrangement of the radical or carbanion intermediate and recombination to give the product. In

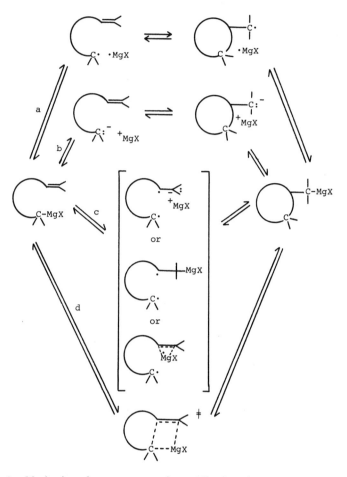

FIG. 1.   Mechanisms for organomagnesium cyclization–cleavage rearrangements.

mechanism c, considered in the direction of addition, the first step is an electron transfer from the carbon–magnesium bond to the olefinic function. This might occur alternatively as a metal transfer. The last mechanism is a concerted four-center mechanism, in which the structure shown is represented as a transition state, rather than an intermediate, and could possibly involve a π-complex as a stable intermediate species between the transition state shown and the open-chain Grignard. The evidence for and against the various mechanisms has been discussed in considerable detail, with the conclusion that the last of these, the concerted mechanism, survives "by default" (29). The mechanistic discussion will not be reproduced in detail, but some salient points will be noted.

The carbanion mechanism encounters serious difficulty from solvent effects. As noted in Section II,D, the rearrangements appear generally to be accelerated slightly by decreased polarity. A detailed explanation of these solvent effects would be quite complex; it would have to include the effect of basicity of the solvent, and the solvent's effect on association and exchange equilibria. However, the small observed effect does not appear to be compatible with the very much larger solvent effects in the opposite direction found for carbonium ion formation and certain other ion-pair forming reactions. Substituent effects also paint a picture that is not consistent with a carbanion process. Although cyclizations of **39** and **40** are favored by electron-withdrawing substituents in the aryl rings, these substituent effects are very small for reactions generating a carbanion center α to the aryl group (*13, 36*). The failure of phenyl or chlorine substituted on the double bond to significantly activate the double bond to addition is also incompatible with known carbanion behavior.

The radical mechanism avoids the problems of the solvent effect, but still runs into substituent effect problems. A most crucial substituent effect difficulty for the radical mechanism is in the ring cleavage reaction of the 2-methylcyclobutylmethyl Grignard in Eq. (20) (*34*). The corresponding radical was shown to cleave mainly in the direction that forms the secondary product radical [Eq. (36)]. The result observed for the

$$(36)$$

Grignard cleavage is precisely the opposite—almost exclusive formation of the primary Grignard. There appears to be no way in which the radical mechanism can be made compatible with this result (*29, 34*).

The electron-transfer or magnesium-transfer mechanism is analogous to mechanisms that have been popular in recent discussions of Grignard addition to carbonyl groups. However, the Grignard rearrangement appears to be relatively insensitive to magnesium purity, while electron transfer between Grignard reagents and carbonyl compounds is extremely sensitive to the presence of trace amounts of transition metals in the magnesium (*1*). It may also be difficult to see why the cyclization should not be accelerated by aryl or chlorine substitution on the double bond. A critical result again appears to be the cleavage of the 2-methylcyclobutylmethyl Grignard. Application of the electron-transfer (or magnesium transfer) mechanism to this reaction is shown in Eq. (37).

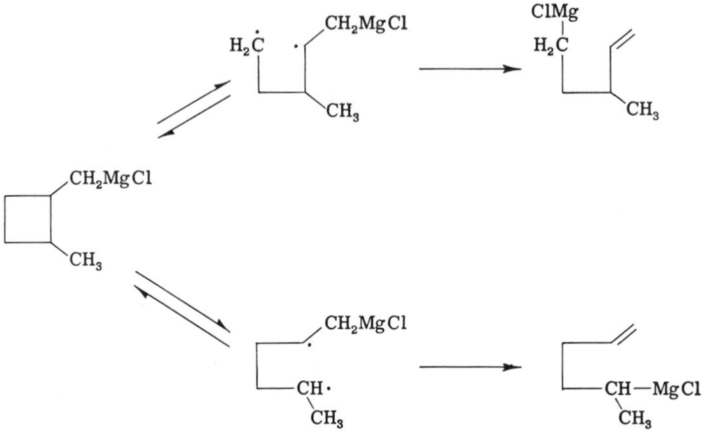

(37)

If the first step, ring cleavage, is rate determining, then cleavage to the secondary Grignard should be preferred. If the second step, magnesium or electron transfer, is rate-determining, then the primary Grignard product might be formed preferentially from the low concentration of the less stable and more reactive primary diradical intermediate. However, in that event the starting Grignard reagent should experience cis-trans isomerization, which is not observed. On this basis, the electron- (or magnesium-) transfer mechanism must also be discarded.

At this point, we are left with the requirement that the transition state should not be significantly more polar than the starting Grignard, that most alkyl, phenyl, or chlorine substitution on the double bond should decrease the rate of intramolecular addition, that cyclization of primary, secondary, and tertiary Grignards should follow the order 1° < 2° > 3°, and that an electron-withdrawing electronic effect of a substituent at the remote end of the double bond should accelerate the reaction. The concerted mechanism appears quite compatible with this pattern. As drawn, the mechanism has no formal charge separation. However, in the transition state, one carbon is in the process of conversion from a tetrahedral carbon α to magnesium to an ordinary saturated carbon, while another carbon is progressing from an ordinary olefinic carbon to $-C_\alpha-Mg$. With the expected polarity of a carbon magnesium bond, we might expect some significant electronic effects, though to a smaller extent than in a carbanion reaction. Finally, the concerted transition state clearly increases congestion about the reactive centers and might be expected to be sensitive to steric hindrance from substituents, regardless of their electronic effect. The irregular trend for primary, secondary, and tertiary Grignards in the addition may be a consequence of opposing effects of steric repulsion and Grignard stability.

An objection that may be raised to a concerted four-center mechanism is the possibility of "violation" of orbital symmetry rules (68, pp. 65–78). Formally, the cyclization and cleavage may be viewed as forbidden $[_\sigma 2_s + _\pi 2_s]$ and $[_\sigma 2_s + _\sigma 2_s]$ cycloadditions (see transition state structure **41**). The polarity of the bonds involved may alleviate the situation somewhat (21, 41). However, if the magnesium atom utilizes an additional orbital as in **42** (formally occupied partially by binding a solvent

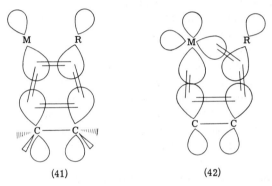

(41)   (42)

molecule), then orbital symmetry constraints are inapplicable. Because of orthogonality of the two orbitals on magnesium, there is no longer the necessary cyclic array of overlapping orbitals.

A variant on the concerted mechanism is one that includes a π-complex intermediate [Eq. (38)]. Such intermediates appear commonly

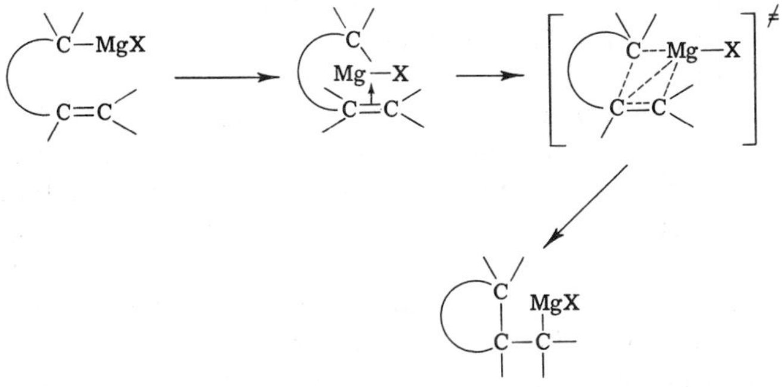

(38)

in discussions of mechanistic organometallic chemistry; they are known stable compounds in transition metal organometallic chemistry; and there is firm experimental evidence for such intramolecular interactions of a double bond with nontransition metal organometallic functions. (For recent references on metal–double bond interactions, see *18, 19, 60, 64*.) NMR spectroscopic evidence has been obtained by Oliver and co-workers for a weak metal–double bond interaction in di(4-penten-1-yl) zinc in hydrocarbon solution (*60*). The interaction is not detected in the presence of bases, or in the corresponding 3-butenyl or 5-hexenyl derivatives. It was not observed for dialkenylmagnesium compounds, which could only be studied in basic solvents. Since interaction of zinc and other organometallic functions with a double bond is shown to occur in several instances, it seems reasonable to postulate such interaction in the course of addition of the organomagnesium function to the double bond. However, if such a π-complex intermediate occurs along the reaction coordinate in the Grignard cyclization, it would probably have only a small equilibrium concentration because the double bond does not compete favorably with the ether solvent for a coordination site on magnesium. Even if the π-complex does not exist as a stable potential minimum along the reaction coordinate, it is likely that the rearranging Grignard species may pass through a configuration resembling the complex en route to the transition state, and a "π-complex-like" transition state might be one possible description for the concerted addition. It might be noted that a π-complex mechanism should avoid orbital symmetry restrictions (*29, 42*).

An alternative mechanistic description to that given here has been proposed by Maercker and co-workers (*48, 50a*). They have found that a number of Grignard rearrangements show a rather strong dependence of rate upon concentration. Rearrangement rates at low organometallic concentrations exhibit kinetics that are first order in unrearranged alkyl group, but essentially independent of concentration. However, at higher concentrations (above about 0.5–1 M) the first-order rate constant increases linearly with concentration, as one might expect for second-order behavior. Based heavily on this finding, Maercker and co-workers argue for a "bimolecular carbanionic" mechanism (Fig. 2). The decreased rate in THF is attributed to its stronger basicity, which makes desolvation of magnesium in the $RMgS_3^+$ species more difficult. The Lewis acid function of $RMgS_3^+$ could be served by other magnesium-containing species, giving the first-order dependence of the rate on total Grignard or magnesium concentration. It is suggested that different mechanisms may apply in the two concentration ranges (*50a*).

Two points might be made that may suggest that the concentration dependence does not necessarily implicate a bimolecular mechanism. First, it is known (*1a, 67*) that alkylmagnesium halides in ether solution are associated. Alkylmagnesium bromides are found to be monomeric only at very low concentrations (<0.1 M), and the average degree of

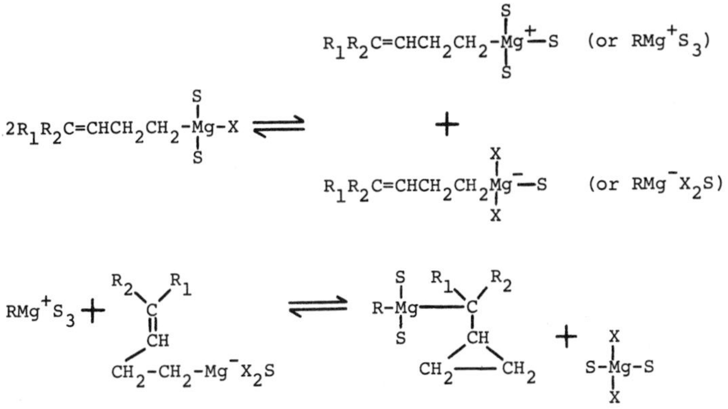

$$R_1R_2C=CHCH_2CH_2-\overset{\overset{S}{|}}{\underset{\underset{S}{|}}{Mg}}{}^{+}-S \quad (\text{or } RMg^+S_3)$$

$$2R_1R_2C=CHCH_2CH_2-\overset{\overset{S}{|}}{\underset{\underset{S}{|}}{Mg}}-X \;\rightleftharpoons\; + $$

$$R_1R_2C=CHCH_2CH_2\overset{\overset{X}{|}}{\underset{\underset{X}{|}}{Mg}}{}^{-}-S \quad (\text{or } RMg^-X_2S)$$

FIG. 2. Bimolecular carbanionic mechanism of Grignard cyclization. S = solvent.

association in the range of 1 to 2 M appears to be two or greater (67). Therefore, a transition state containing two magnesium species (as in Fig. 2) should lead to approximately first-order behavior in the high concentration range. A second problem is that a kinetic scheme involving concurrent first- and second-order processes should not become independent of total Grignard concentration in dilute solutions.

Aside from the concentration dependence, most of the mechanistic discussion could be applied equally well to either the concerted four-center mechanism or the bimolecular carbanionic mechanism, since the latter appears to be described as essentially a concerted addition. It may be that a more complex kinetic scheme, or one with a tetrameric or more highly associated transition state, could fit the concentration dependence.

In the absence of other compelling evidence, and in light of the rather similar structural and medium effects on rate in the two ranges, the present author prefers to interpret the concentration dependence tentatively as a "medium effect" (possibly related to solvent activity).

F. Transition-State Geometry

A further aspect of the rearrangement that we would like to explore is the geometrical requirement of the transition state. If a concerted mechanism is accepted, a first notion of the transition state might be one in which the reacting centers are located in a coplanar, approximately rectangular array, essentially as drawn in Fig. 1,d.

Richey and Veale (56) have pointed out, however, that stereochemical results in the cyclization of substituted hexenyl Grignards (44, 56) are incompatible with such a transition state [Eq. (39)]. During the process

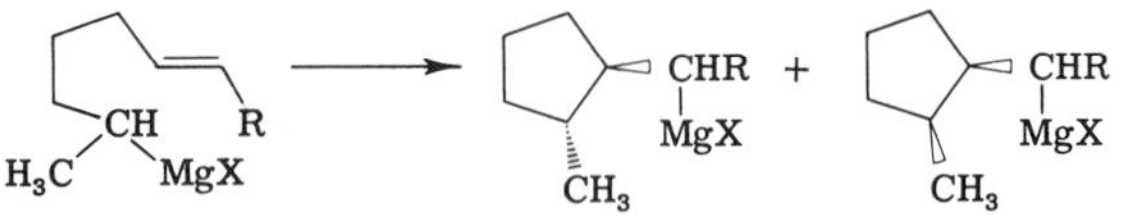

$$
\text{(43)} \longrightarrow \text{(44)} + \text{(45)} \tag{39}
$$

(43)   (44)   (45)

of Grignard formation, some cyclized reagent was produced. This rearranged material was relatively rich in the cis isomer (78% in the case where R = phenyl). The cis preference is similar to that reported for related radical cyclizations (3) and is consistent with cyclization of radical intermediates during Grignard formation. However, rearrangement of Grignard reagent after formation produced a predominance of trans isomer. Again in the case of R = phenyl, this preference was to the extent of 25:1 or greater. Examination of models, or less satisfactorily, Fig. 3, shows that one of the two groups attached to C_α (labeled R_2 in Fig. 3) must be oriented toward the double bond. If it is methyl, or anything else more bulky than hydrogen, it will generate severe steric repulsions. If reaction occurs with retention of configuration at C_α, then R_2 will occupy a ring position trans to the CHR_1MgX group in the adduct, and R_3 will be cis. Then, the less hindered "rectangular coplanar" transition state should lead to the cis isomer. Preference for the cis isomer would be eliminated only in a very late transition state, in which C_α is rotated to position for bonding and the olefinic carbon approaches tetrahedral geometry (56). An apparent preference for trans isomer in excess of the equilibrium proportion may also be found in cyclizations to a four-membered ring (34, 46). The notion of a coplanar transition state therefore appears to be incompatible with experimental results.

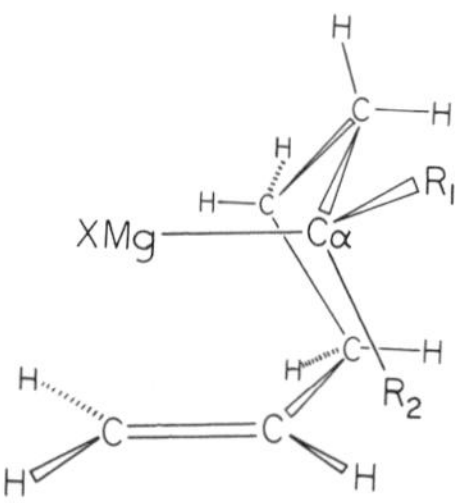

Fig. 3.   Coplanar transition state model for "hexenyl Grignard" cyclization.

Another approach to learning about transition state geometry is to incorporate the cycloalkylmethyl group into a more rigid polycyclic skeleton, in which there is restriction of conformational freedom. The rearrangements of Eqs. (40)–(42) have been investigated with this

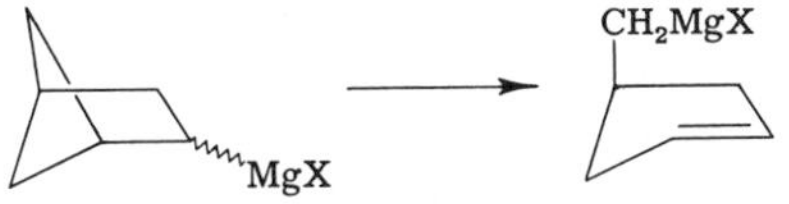

$$(40)$$

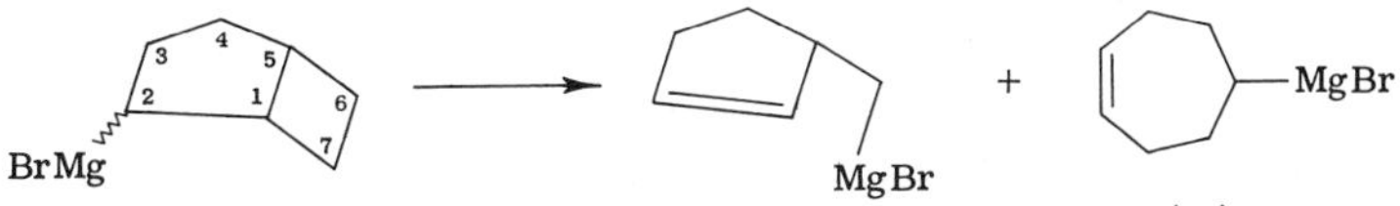

$$(41)$$

$$(30) \qquad (31) \qquad (32)$$

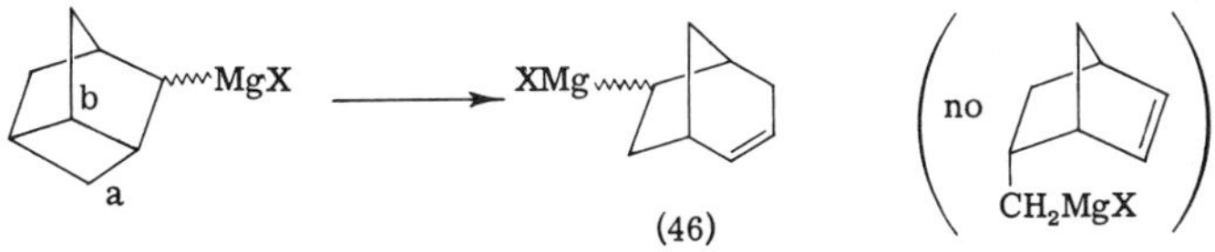

$$(42)$$

$$(46)$$

purpose in mind (39, 37, 32). In all instances, the bi- or tricyclic Grignard rearranges somewhat more slowly than the monocyclic analog of Eq. (43) (chosen as a model since it involves conversion of a secondary

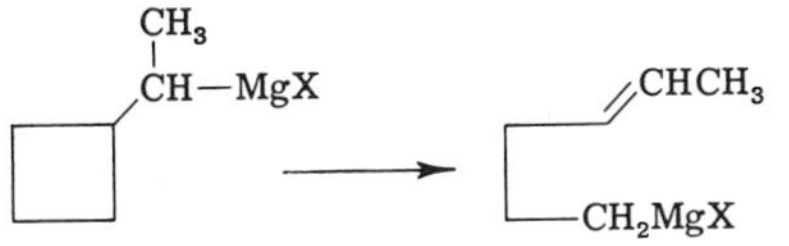

$$(43)$$

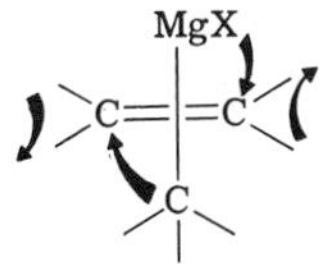

$$(47)$$

Grignard to a primary one). Approximate relative rates are <0.06 for Eq. (40), 0.06 and 0.005 for formation of 31 and 32, in Eq. (41), and 0.015 in Eq. (42). The decrease in rate appears to come in all cases despite added strain in the initial Grignard, and probably a greater relife of strain in the course of the ring cleavage. The decrease has been attributed to geometrical restriction that prevents the rearranging molecule from adopting the optimum transition state configuration.

Two of the above examples warrant some special comment. In the rearrangement of Eq. (41), the secondary cycloheptenyl Grignard product is produced to a significant extent ($\sim$10%), although it is considerably less stable than the primary. Although the ring cleavage rate to the major, primary Grignard product 31 appears to be retarded, the cleavage process to the cycloheptenyl product may be "normal" for a secondary → secondary cleavage. It might then be concluded that the transition state for the latter cleavage is closer to optimum transition state geometry. However, this cleavage may not even approximately approach a coplanar transition state geometry (see Fig. 4a). Examination from the direction of *addition* (Fig. 4b) shows that the initial approach of the organomagnesium function to the double bond must be in a *perpendicular* fashion. On this basis, it has been suggested that addition may preferentially involve perpendicular approach of the C—Mg and C≡C bonds, possibly with π-complex interaction between the metal and the π-orbital (37). This would be followed by twisting of the double bond and a rotary motion of the C—Mg bond, leading to addition as in 47. Cleavage would involve the reverse process.

Grignard 46 in Eq. (42) is related to 30 by the introduction of a one-carbon bridge between carbons 3 and 6. This modification increases the strain and "freezes" the four- and five-membered rings into one puckered conformation. Cleavage of the bond designated a in the four-membered ring would lead to a primary Grignard. However, no trace of this product is observed. Instead, cleavage of bond b occurs cleanly to yield the secondary Grignard. The carbon–magnesium bond is again nearly perpendicular to the cleaving ring bond.

In another case of interest, in the cyclization of the 3-cyclohexen-1-yl Grignard reagent (48) shown in Eq. (35), a nearly perpendicular geometry would appear to be enforced between the C—Mg and C≡C bonds. It appears by comparison with the rearrangement in Eq. (8), which also involves rate-determining secondary → secondary cyclization to form a three-membered ring (33), that this geometric constraint does not significantly reduce the rate.

In conjunction with rearrangement studies in bi- and tricyclic Grig-

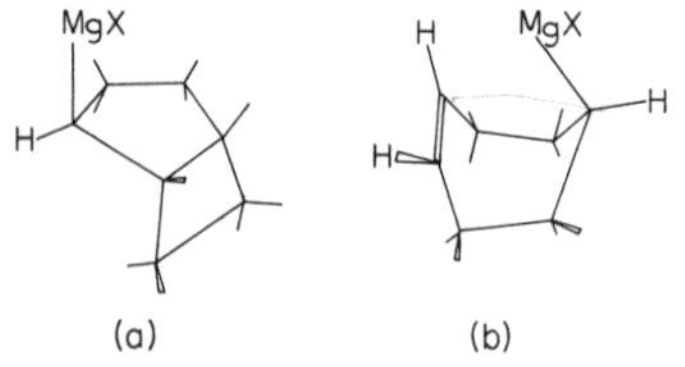

(a)   (b)

FIG. 4. Geometrical relationship of reacting bonds in (a) the 2-bicyclo[3.2.0]heptenyl, and (b) the cyclohepten-5-yl Grignard reagents.

nards, rearrangements of the corresponding radicals have been investigated *(30)*. Tributyltin hydride reductions of the halides corresponding to Grignard reagents in Eqs. (40)–(43) were studied, and related radical cleavage rates were estimated from the proportions of unrearranged and rearranged hydrocarbons. It has been postulated that the transition state for addition of a radical to a double bond, or the reverse process of β-cleavage, should preferentially have a coplanar rearrangement of the three reacting orbitals *(2)*. Radical cleavage reactions have been reported in which stereoelectronic control is exercised by this requirement that the cleaving bond be approximately in the same plane as the half-filled orbital of the radical (for leading references, see *4*). Radical cleavages to the primary radicals corresponding to the major products in Eqs. (40) and (41) were retarded relative to that of Eq. (43) by a factor of roughly five, probably as a consequence of the balance between increased strain and less satisfactory geometry. Neither of the cleavages to a secondary radical, corresponding to Grignard rearrangements to **31** and **46** in Eqs. (41) and (42) were observed, despite the thermodynamic and kinetic preference expected *(31)* for cleavage to a secondary radical. It is significant that these cases, in which radical rearrangement is stereoelectronically inhibited, appear to rearrange fairly readily in the Grignard process.

The concept of a preferred perpendicular attack of the carbon–magnesium bond on the double bond may also help to rationalize the stereoselectivity in the Grignard cyclization of Eq. (39). Examination of models shows that conformations exist in which the carbon chain may be coiled so as to locate the carbon–magnesium bond over and approximately perpendicular to the double bond. Conformations are drawn in Fig. 5a and 5b that would lead, on addition, to the trans and cis isomers,

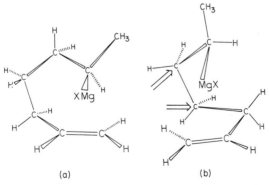

(a)   (b)

FIG. 5. Perpendicular transition-state models for ''heptenyl Grignard'' cyclization leading to (a) trans and (b) cis adducts.

respectively. In both conformations, the methyl group is rotated anti to the rest of the carbon chain. It appears that chain conformation a, leading to the trans isomer, can readily be attained with approximate staggering of methylene groups, while conformation b may experience near eclipsing between the methylene groups indicated by arrows, or may have excessive congestion in the plane of the double bond. The distinction is not entirely clear-cut, since there is substantial uncertainty in the choice of a "best" positioning of the reacting bonds, leading to some considerable variation in the chain conformation. Nevertheless, the "perpendicular" model appears to offer the potential of a rationale for the stereoselectivity, where the planar model makes a clear prediction that is contrary to experiment.

It might be noted also that either model is capable of explaining the irregular order $1° < 2° > 3°$ for reactivity in cyclization (see Section II,D,6). In either case, one methyl group on C_α may occupy a conformation in which it is fairly well out of the way, allowing the electronic destabilization of the organometallic to produce a reactivity increase. However, if there are two methyls on C_α, one must be in a position of substantial steric congestion.

Brief reference should also be made to a study of cyclization of hexenylaluminum compounds (66). In the cyclization analogous to Eq. (39) with R = hydrogen, the cis isomer was the major product, in contrast to the Grignard cyclization. The results were rationalized in terms of a chairlike π-complex transition state, in which substituent groups prefer equatorial positions.

III

REARRANGEMENT OF ALLYLIC ORGANOMAGNESIUM COMPOUNDS

On the basis of considerable spectroscopic and chemical evidence, allylic organomagnesium compounds appear to be best represented as a mixture of rapidly equilibrating allylic isomers [Eq. (44)]. As such, by

$$\tag{44}$$

their very nature they *incorporate* a rearrangement. Furthermore, their reaction with carbonyl compounds and other electrophilic reactants appears to occur with allylic rearrangement. For example, the Grignard reagent from crotyl chloride, which is largely the primary isomer, reacts mainly at the secondary carbon, as shown in Eq. (45).

$$
\begin{array}{ccc}
\underset{\text{CH}_3\text{CHCH}=\text{CH}_2}{\overset{\text{MgX}}{|}} \rightleftharpoons \underset{\text{CH}_3\text{CH}=\text{CHCH}_2}{\overset{\text{MgX}}{|}} & \xrightarrow[\text{2. H}_2\text{O}]{\text{1. R}_2\text{CO}} & \underset{\text{CH}_3\text{CH}-\text{CH}=\text{CH}_2}{\overset{\overset{\displaystyle\text{R}}{|}}{\text{R}-\text{C}-\text{OH}}} \end{array} \qquad (45)
$$

It was recently reported that reaction of this Grignard reagent with trimethylchlorosilane gave about 50% of reaction at each end of the allylic system, but there is no indication as to which allylic Grignard isomer was the reacting species (61).

The fairly recent literature has included a review of the preparation and reactions of allyl and "butenyl" Grignard reagents (5), a review of the reactions of allylic organometallic compounds (11), and a review of organomagnesium rearrangements, in which allylic derivatives were discussed in some detail (29). Since more recent publications do not appear to include new information that would significantly modify earlier conclusions, the previous discussions should suffice.

One relatively recent development might be treated briefly. It was reported by Benkeser in 1969 (7) and by Miginiac in 1970 (51) that the products of some allylic Grignard addictions to hindered ketones are isomerized after longer reaction times to products with "inversion" of the allylic group [e.g., Eq. (46)].

$$
\text{CH}_3\text{CH}=\text{CHCH}_2\text{MgX} \; + \; \underset{\diagdown}{\overset{\diagup}{\text{C}}}=\text{O} \longrightarrow \underset{\text{CH}_3\text{CH}-\text{CH}=\text{CH}_2}{\overset{\overset{\displaystyle +}{|}}{\rangle-\text{C}-\text{OMgX}}}
$$

$$
\Big\downarrow \qquad\qquad\qquad (46)
$$

$$
\underset{\diagup\diagdown}{\overset{+}{\text{CH}_3\text{CH}=\text{CHCH}_2\text{COMgX}}}
$$

In the case of allylic zinc derivatives, which isomerize much more readily in a similar fashion, the rearrangement takes place by reversal of the addition, and readdition of the allylic organometallic to give the thermodynamically favored, less hindered product (51). More recently, Benkeser has demonstrated the reversibility of the addition of the allyl Grignard itself (6).

Another newly discovered rearrangement process with allylic Grignard compounds is their photochemical cyclization (10). The rearrangements illustrated in Eqs. (47)–(51) are found to occur on irradiation of the Grignard solutions with a mercury arc lamp through quartz. These rearrangements presumably correspond to the photochemical electro-

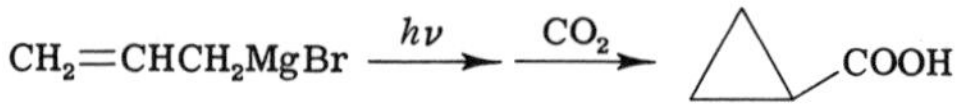

$$CH_2{=}CHCH_2MgBr \xrightarrow{h\nu} \xrightarrow{CO_2} \text{△}COOH \qquad (47)$$

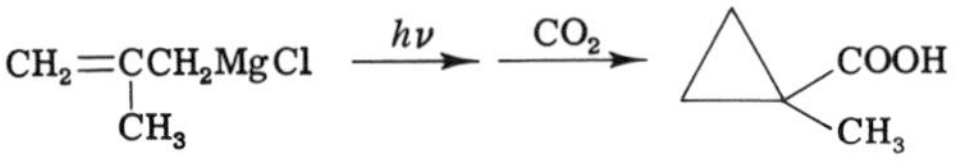

$$CHD{=}CDCH_2MgBr \xrightarrow{h\nu} \xrightarrow{CO_2} \underset{D}{\text{△}}\underset{D}{COOH} + \underset{D}{\text{△}}\underset{COOH}{D} \qquad (48)$$

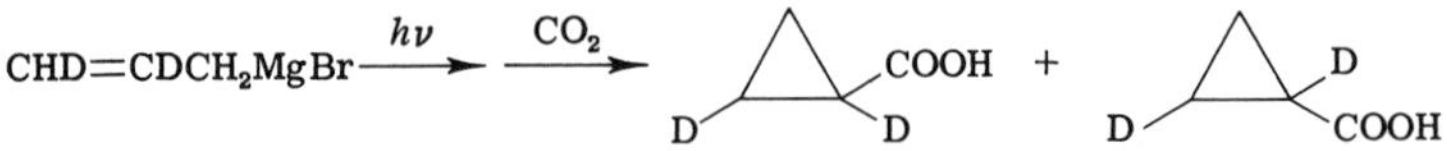

(1:1 ratio)

$$\underset{\underset{CH_3}{|}}{CH_2{=}C}CH_2MgCl \xrightarrow{h\nu} \xrightarrow{CO_2} \text{△}\underset{CH_3}{COOH} \qquad (49)$$

$$C_6H_5CH{=}CHCH_2MgBr \xrightarrow{h\nu} \xrightarrow{CO_2} \underset{H_5C_6}{\text{△}}COOH + \underset{H_5C_6}{\text{△}}COOH$$

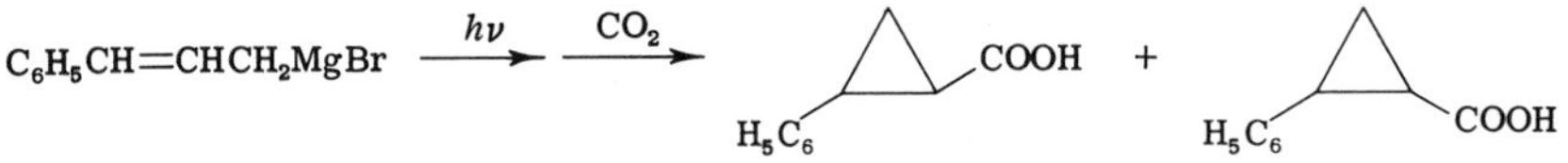

(trans:cis = 49:1)

(50)

$$CH_2{=}CHCH{=}CHCH_2MgCl \xrightarrow{h\nu} \xrightarrow{CO_2} \text{△}COOH + \text{△}COOH$$

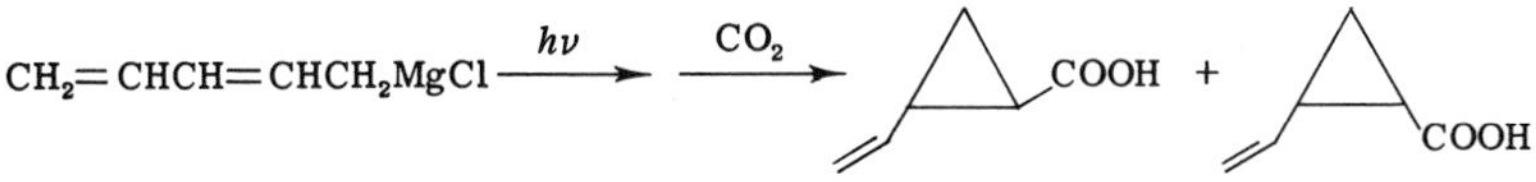

(trans : cis = 6:1)

(51)

cyclic reaction of an allyl carbanion, which is expected to be allowed in the disrotatory sense (68, pp. 38–47). It was not possible to determine the stereochemical sense of the reaction because of the configurational lability of allylic Grignards.

IV

1,2-SHIFTS

The least-represented class of rearrangement is the 1,2-shift. The carbanionic 1,2-shift of alkyl or hydrogen is a forbidden process on orbital symmetry grounds. Wittig and Stevens rearrangements, which are formally 1,2-anionic shifts, appear most probably to follow radical cleavage mechanisms [see Hill (29) for leading references]. Carbanionic aryl migration suffers no such problems, since it corresponds essentially to a nucleophilic aromatic substitution process. However, there are still few instances in organomagnesium chemistry, compared with the alkali metals, because of the greater degree of covalency of the carbon–

magnesium bond. Two examples that have been observed (20, 24) are shown in Eqs. (52) and (53).

$$C_6H_5O—C(C_6H_5)_2 \xrightarrow{\text{HMPT}} \overset{+}{X}Mg\overset{-}{O}—C(C_6H_5)_3 \qquad (52)$$
$$\underset{\mathrm{MgX}}{|}$$

$$\qquad (53)$$

As shown in Eq. (15) in Section II,B, a formal 1,2-rearrangement of an allyl group has been reported in the product of addition of allyl- or methallylmagnesium chloride to styrene (45). The rearrangement would appear to involve reversal of the addition step, followed by readdition.

Aryl migration, probably radical in nature, has also been observed during the formation and reaction of Grignard reagents. For example, rearranged hydrocarbon by-products were found in the reaction of neophyl chloride with magnesium, but little or none of the Grignard formed was rearranged (22, 59). The formation of some rearranged Grignard reagent has been confirmed in the reaction of magnesium with 2,2,2-triphenylethyl chloride in ether (28). However, once formed, the Grignard reagent does not rearrange further.

REFERENCES

1. Ashby, E. C., Lopp, I. G., and Buhler, J. D., *J. Am. Chem. Soc.* **97**, 1964 (1975).
1a. Ashby, E. C., and Smith, M. B., *J. Am. Chem. Soc.* **86**, 4363 (1964).
2. Beckwith, A. L. J., *Chem. Soc., Spec. Publ.* **24**, 239 (1970).
3. Beckwith, A. L. J., Blair, I., and Phillipou, G., *J. Am. Chem. Soc.* **96**, 1613 (1974).
4. Beckwith, A. L. J., and Phillipou, G., *Aust. J. Chem.* **29**, 123 (1976).
5. Benkeser, R. A., *Synthesis* 347 (1971).
6. Benkeser, R. A., and Siklosi, M. P., *J. Org. Chem.* **41**, 3212 (1976).
7. Benkeser, R. A., and Broxterman, W. E., *J. Am. Chem. Soc.* **91**, 5162 (1969).
8. Benson, S. W., Cruickshank, F. R., Golden, D. M., Haugen, G. R., O'Neal, H. E., Rodgers, A. S., Shaw, R., and Walsh, R., *Chem. Rev.* **69**, 279 (1969).
9. Bodewitz, H. W. H. J., Blomberg, C., and Bickelhaupt, F., *Tetrahedron* **31**, 1053 (1975).
10. Cohen, S., and Yogev, A., *J. Am. Chem. Soc.* **98**, 2013 (1976).
11. Courtois, G., and Miginiac, L., *J. Organometal. Chem.* **69**, 1 (1974).
12. Cowan, D. O., Krieghoff, N. G., Nordlander, J. E., and Roberts, J. D., *J. Org. Chem.* **32**, 2639 (1967).
13. Crandall, J. K., and Michaely, W. J., personal communication.
14. Dagonneau, M., and Vialle, J., *Tetrahedron* **30**, 3119 (1974).
15. Demole, H., *Helv. Chim. Acta* **47**, 319 (1964).
16. Derocque, J.-L., and Sundermann, F.-B., *J. Org. Chem.* **39**, 1411 (1974).

17. Derocque, J.-L., Beisswenger, U., and Hanack, M., *Tetrahedron Lett.* 2149 (1969).
18. Dolzine, T. W., and Oliver, J. P., *J. Am. Chem. Soc.* **96**, 1737 (1974).
19. Dolzine, T. W., and Oliver, J. P., *J. Organometal. Chem.* **78**, 165 (1974).
20. Ebel, H. F., Dörr, V., and Wagner, B. O., *Angew. Chem., Int. Ed. Engl.* **9**, 163 (1970).
21. Epiotis, N. D., *J. Am. Chem. Soc.* **95**, 191 (1973).
22. Fainberg, A. H., and Winstein, S., *J. Am. Chem. Soc.* **78**, 2763 (1956).
23. Felkin, H., Kwart, L. D., Swierczewski, G., and Umpleby, J. D., *J. Chem. Soc., Chem. Commun.* 242 (1975).
24. Fraenkel, G., and Cooper, J. W., *J. Am. Chem. Soc.* **93**, 7228 (1971).
25. Freeman, P. K., George, D. E., and Rao, V. N. M., *J. Org. Chem.* **29**, 1682 (1964).
26. Freeman, P. K., George, D. E., Rao, V. N. M., and Fenwick, G. L., *J. Org. Chem.* **32**, 3958 (1967).
27. Gerard, F., and Miginiac, P., *J. Organometal. Chem.* **111**, 17 (1976).
28. Grovenstein, E., Jr., personal communication.
29. Hill, E. A., *J. Organometal. Chem.* **91**, 123 (1975).
30. Hill, E. A., and Chen, A. T., unpublished observations.
31. Hill, E. A., Chen, A. T., and Doughty, A., *J. Am. Chem. Soc.* **98**, 167 (1976).
32. Hill, E. A., and Hsieh, K., unpublished work.
33. Hill, E. A., Hsieh, K., and Elgas, D., unpublished work.
34. Hill, E. A., and Myers, M. M., unpublished work.
35. Hill, E. A., and Ni, H.-R., *J. Org. Chem.* **36**, 4133 (1971).
36. Hill, E. A., and Shih, G. E.-M., *J. Am. Chem. Soc.* **95**, 7764 (1973).
37. Hill, E. A., Theissen, R. J., Cannon, C. E., Miller, R., Guthrie, R. B., and Chen, A. T., *J. Org. Chem.* **41**, 1191 (1976).
38. Hill, E. A., Theissen, R. J., Doughty, A., and Miller, R., *J. Org. Chem.* **34**, 3681 (1969).
39. Hill, E. A., Theissen, R. J., and Taucher, K., *J. Org. Chem.* **34**, 3061 (1969).
40. Howden, M. E. H., Maercker, A., Burdon, J., and Roberts, J. D., *J. Am. Chem. Soc.* **88**, 1732 (1966).
41. Jackson, R. A., *J. Chem. Soc. B* 58 (1970).
42. Jones, P. R., *J. Org. Chem.* **37**, 1886 (1972).
43. Julia, M., *Acc. Chem. Res.* **4**, 386 (1971).
44. Kossa, W. C., Jr., Rees, T. C., and Richey, H. G., Jr., *Tetrahedron Lett.* 3455 (1971).
45. Lehmkuhl, H., Bergstein, W., Henneberg, D., Janssen, E., Olbrysch, O., Reinehr, D., and Schomburg, G., *Liebigs Ann. Chem.* 1176 (1975).
46. Lehmkuhl, H., Reinehr, D., Henneberg, D., Schomburg, G., and Schroth, G., *Liebigs Ann. Chem.* 119 (1975).
47. Lehmkuhl, H., Reinehr, D., Schomburg, G., Henneberg, D., Damen, H., and Schroth, G., *Liebigs Ann. Chem.* 108 (1975).
48. Maercker, A., and Geuss, R., *Angew. Chem., Int. Ed. Engl.* **9**, 909 (1970); *Chem. Ber.* **106**, 773 (1973).
49. Maercker, A., Guthlein, P., and Wittmayr, H., *Angew. Chem., Int. Ed. Engl.* **12**, 774 (1973).
50. Maercker, A., and Streit, W., *Angew. Chem., Int. Ed. Engl.* **11**, 542 (1972).
50a. Maercker, A., and Streit, W., *Chem. Ber.* **109**, 2064 (1976).
51. Miginiac, P., *Bull. Soc. Chim. Fr.* 1077 (1970).
52. Patel, D. J., Hamilton, C. L., and Roberts, J. D., *J. Am. Chem. Soc.* **87**, 5144 (1965).
53. Richey, H. G., Jr., and Kossa, W. C., Jr., *Tetrahedron Lett.* 2313 (1969).
54. Richey, H. G., Jr., and Rothman, A. M., personal communication; Rothman, A. M., Ph.D. Thesis, Pennsylvania State University, 1969.

55. Richey, H. G., Jr., and Veale, H. S., *J. Am. Chem. Soc.* **96**, 2641 (1974).
56. Richey, H. G., Jr., and Veale, H. S., *Tetrahedron Lett.* 615 (1975).
57. Roberts, J. D., and Mazur, R. H., *J. Am. Chem. Soc.* **73**, 2509 (1951).
58. Roberts, J. D., Trumbull, E. R., Jr., Bennett, W., and Armstrong, R., *J. Am. Chem. Soc.* **72**, 3116 (1950).
59. Rüchardt, C., and Trautwein, H., *Chem. Ber.* **95**, 1197 (1962).
60. St. Denis, J., Oliver, J. P., Dolzine, T. W., and Smart, J. B., *J. Organometal. Chem.* **71**, 315 (1974).
61. Sakurai, H., Kudo, Y., and Miyoshi, H., *Bull. Chem. Soc. Jpn.* **49**, 1433 (1976).
62. Santelli, M., and Bertrand, M., *C. R. Acad. Sci., Paris, Ser. C* 757 (1970).
63. Silver, M. S., Shafer, P. R., Nordlander, J. E., Rüchardt, C., and Roberts, J. D., *J. Am. Chem. Soc.* **82**, 2646 (1960).
64. Smart, J. B., Hogan, R., Scherr, P. A., Emerson, M. T., and Oliver, J. P., *J. Organometal. Chem.* **64**, 1 (1974).
65. Smith, L. I., and McKenzie, S., Jr., *J. Org. Chem.* **15**, 74 (1950).
66. Stefani, A., *Helv. Chim. Acta* **57**, 1346 (1974).
67. Walker, F. W., and Ashby, E. C., *J. Am. Chem. Soc.* **91**, 3845 (1969).
68. Woodward, R. B., and Hoffmann, R., "The Conservation of Orbital Symmetry," pp. 114–132. Verlag Chemie, Weinheim, Germany, 1971.
69. Zimmerman, H. E., and Zweig, A., *J. Am. Chem. Soc.* **83**, 1196 (1961).

Aryl Migrations in Organometallic Compounds of the Alkali Metals

ERLING GROVENSTEIN, JR.

School of Chemistry
Georgia Institute of Technology
Atlanta, Georgia

I

INTRODUCTION

A. Scope

Rearrangements of aryl groups the type $1 \to 2$, where M is an alkali metal, will be surveyed with respect to likely reaction mechanism(s), the

$$-\overset{|}{\underset{|}{C}}-(\overset{|}{\underset{|}{C}})_n-\overset{|}{\underset{|}{\overset{*}{C}}}M \longrightarrow M\overset{|}{\underset{|}{C}}-(\overset{|}{\underset{|}{C}})_n-\overset{*|}{\underset{|}{C}}-$$
$$\quad\,\, Ar \qquad\qquad\qquad\qquad\qquad Ar$$

$$(1) \qquad\qquad\qquad (2)$$

migratory aptitude of groups, the distance over which migrations occurs ($n = 0, 1, 2, 3$), and variation of the alkali metal cation (M^+), the solvent or ligands of the alkali metal cation. The case where aryl migration occurs in competition with cleavage (and sometimes migration) of benzyl will also be considered. In general this review will concentrate upon organic moieties having only carbon and hydrogen; however, systems containing heteroatoms will be briefly mentioned. A number of recent reviews (*60, 61, 67, 76, 80*) have appeared on rearrangements in systems containing heteroatoms—sometimes referred to as Stevens rearrangements, Sommelet–Hauser rearrangements, Wittig ether rearrangements, Wawzonek rearrangements, silyl migrations, etc. Zimmerman (*98*) and Cram (*10*) have given early reviews of rearrangements of carbanions including migrations from carbon to carbon. Buncel (*5*) and Hunter (*54*) have recently surveyed proton transfer reactions, geometrical isomerism, electrocyclic reactions, homoconjugation with resultant rearrangements, and sigmatropic rearrangements of carbanions. Hill (*51*) has recently reviewed the rearrangement of organomagnesium compounds and some related organoalkali compounds including cyclopropylcarbinyl and homoallyl rearrangements. Nevertheless, no in-depth review has previously been made of much of the material in the present survey.

B. *Historical Aspects*

In 1948, when the author began work in the area, it appeared that while rearrangements of carbonium ions were common (Wagner–Merwein rearrangements), and rearrangements of free radicals were beginning to become well known (*85, 93*), skeletal rearrangements of carbanions were unknown[1] save in systems involving heteroatoms—the Stevens and Sommelet rearrangements of quaternary ammonium ions and the Wittig rearrangement of ethers. A more detailed survey of the literature revealed that certain highly arylated olefins when treated with sodium were reported to give products of rearranged structure (*58, 75, 97*) and that a compound that Cone and Robinson (*9*) had provisionally assigned the structure of 2-chloro-1,1,1-triphenylethane was reported by Wooster and Mitchell (*94*) to give an anion of rearranged structure. Since the rearrangement claimed by Wooster and Mitchell appeared to resemble migration of phenyl groups in free radicals or carbonium ions, the author began a detailed investigation of this system. It soon became evident to the author (*23*) and, independently at about the same time, to Hughes and co-workers (*7, 53*) that the rearrangement which Wooster and Mitchell had reported actually took place during the preparation of the chloride of Cone and Robinson. Moreover, reaction (*7*) of authentic 2-chloro-1,1,1-triphenylethane with sodium in liquid ammonia under the conditions of Wooster and Mitchell gave essentially unrearranged 1,1,1-triphenylethane.

---

[1] The author wishes to thank Professor C. Gardner Swain for pointing out to him in 1947 the dearth of examples of rearrangements of carbanions.

Since 2,2,2-triphenylethylsodium might be expected to undergo protonation by liquid ammonia faster than it rearranged to 1,1,2-triphenyl-
ethylsodium, reaction of authentic 2-chloro-1,1,1-triphenylethane with
sodium metal in more inert solvents was investigated. Hughes and co-
workers (7) found that reaction with sodium was tardy and complicated
by thermal decomposition of the chloride; at 120°C in petroleum
solution, triphenylethylene was isolated from the reaction products. We
recovered largely unreacted chloride from attempted reaction with
sodium sand in *n*-pentane. Eventually, however, the author (24) found
that the chloride reacted readily with molten sodium in dioxane to give
the rearrangement product 1,2,2-triphenylethylsodium. Independently,
at essentially the same time, Zimmerman and Smentowski (99) discovered that this reaction took place also with a sodium dispersion in
isooctane–diethyl ether solution.

C. Experimental Techniques

The rearrangements described in this review have nearly all been run
in a Morton high-speed stirring apparatus (65a) under an atmosphere of
purified nitrogen. As noted above, difficulty may be encountered in
reaction of finely divided solid alkali metals with bulky arylalkyl halides.
Although this difficulty is decreased by use of ethereal solvents (99), we
have often found it to be highly advantageous to employ finely dispersed
liquid alkali metals rather than solid alkali metals. With sodium (m.p.
97.8°C) the solvent dioxane at its boiling point (101°C) has proved to be
an effective medium for conducting reactions with alkyl halides (24) or
quaternary ammonium salts (28); however, this solvent readily proton-
ates carbanions having a localized negative charge. The less acidic
solvent tetrahydrofuran (b.p. 65°–66°C) can be effectively employed (42)
with molten potassium (m.p. 63.7°C). This solvent has also been our
favorite solvent rather than diethyl ether for reactions with cesium (m.p.
28.6°C) or with the minimum melting alloys (27) of cesium with potassium (m.p. −45°C) and of cesium with sodium and potassium (m.p.
−79°C). Indeed some reactions that proceed readily in tetrahydrofuran
proceed at a negligible rate in diethyl ether, evidently because the
products of the reaction are insoluble in diethyl ether and coat the
surface of the molten metal; also cesium *dissolves* in tetrahydrofuran at
low temperatures, as is evident from the blue color of the solution
("solvated electrons") with the pure metal or its alloys. Hence we
recommend Cs–K–Na alloy in tetrahydrofuran at −40° to −75°C as a
reaction medium par excellence for the production of organocesium

compounds (*35*). This medium permits the complete reaction of gram quantities of alkyl halides in reaction times of less than 30 seconds at $-75°C$ for a well-stirred reaction mixture (*34*).

II

MIGRATIONS OF ARYL GROUPS IN ANIONS

A. [1,2] Migrations

The most common and thoroughly studied rearrangements of organoalkali compounds are [1,2] sigmatropic shifts of aryl groups. That the reaction of 2-chloro-1,1,1-triphenylethane (**3**) with sodium in dioxane (*24*) or isooctane–diethyl ether (*99*) gives 1,1,2-triphenylethylsodium (**5**)

$$Ph_3CCH_2Cl \xrightarrow{\text{Na}} [Ph_3CCH_2Na] \rightarrow NaCPh_2CH_2Ph$$
$$\quad\quad (3) \quad\quad\quad\quad (4) \quad\quad\quad\quad (5)$$

does not prove that rearrangement of intermediate 2,2,2-triphenylethylsodium **4** has occurred. The chloride **3** was shown to be thermally stable in refluxing dioxane under the general conditions for the reaction with sodium. With amylsodium in isooctane–ether at 35°C, the chloride underwent α-elimination (*99*) with rearrangement to give triphenylethylene. While triphenylethylene is reduced to 1,1,2-triphenylethylsodium (**5**) with sodium in liquid ammonia (*96*), with excess molten sodium in dioxane this olefin yields (*24*) a disodium adduct; presumably, therefore, this olefin is not a direct intermediate leading to **5** under the conditions for reaction of the chloride with sodium.

More direct information that 2,2,2-triphenylethylsodium (**4**) is an intermediate leading to the rearranged anion is the observation (*24*) that reaction of the chloride with sodium in dioxane containing *t*-amyl alcohol gave a product consisting of 94% of 1,1,1-triphenylethane and 6% of 1,1,2-triphenylethane, whereas, if *t*-amyl alcohol was added after the reaction with sodium in dioxane had been completed, the product was primarily the rearranged hydrocarbon. These reactions were taken to indicate that in presence of *t*-amyl alcohol most of the 2,2,2-triphenylethylsodium (**4**) is protonated before it has had time to rearrange to **5**.

Similarly, cleavage (*37*) of the chloride **3** or of the corresponding quaternary ammonium salt, 2,2,2-triphenylethyltrimethylammonium iodide (**6**), with sodium in liquid ammonia gave 1,1,1-triphenylethane containing $11 \pm 3\%$ of 1,1,2-triphenylethane (or its cleavage products); the percentage rearrangement of the hydrocarbon product from the quaternary ammonium salt was not appreciably affected by addition of enough ammonium chloride to destroy all the sodium amide formed in the cleavage. According to a theory (38) for cleavage of quaternary ammonium salts by sodium in liquid ammonia (the theory was advanced to explain the relative ease of cleavage of primary, secondary, and tertiary alkyl groups from quaternary nitrogen), primary alkyl groups such as methyl and ethyl cleave from quaternary nitrogen as carbanions rather than as free radicals. Application of this theory to 2,2,2-triphenyl-ethyltrimethylammonium ion suggests that cleavage occurs as shown in Scheme 1. The 2,2,2-triphenylethyl group cleaves from quaternary nitrogen 10^4 times more readily than methyl (after correction for statistical factors that arise because of unequal numbers of the groups being cleaved). In contrast, ethyl cleaves 0.008 times as fast as methyl; hence the 2,2,2-triphenylethyl group cleaves 1.2×10^6 times faster than ethyl. Evidently the much more facile cleavage of triphenylethyl than ethyl is due to the electrical effect of phenyls to stabilize the 2,2,2-triphenylethyl anion; the effect cannot be due to a direct neighboring group participation of phenyl [cf. the 4×10^6 estimated anchimeric acceleration in solvolysis with carbonium ion rearrangement of 2,2,2-triphenylethyl brosylate (92)] because most of the final hydrocarbon product is of unrearranged structure.

Additional more-direct evidence that a normal unrearranged organoal-kali compound is formed prior to rearrangement was the observation (42) that lithium reacts with 2-chloro-1,1,1-triphenylethane at temperatures of $-65°$ to $-30°C$ in tetrahydrofuran (THF) to give 2,2,2-triphenyl-

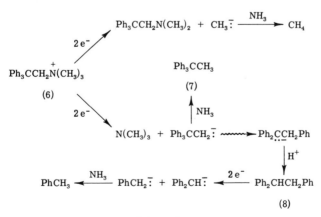

Scheme 1

ethyllithium (9), which is comparatively stable at these temperatures but rearranges rather readily at temperatures of 0°C or higher to give 1,1,2-triphenylethyllithium (10).

$$Ph_3CCH_2Cl \xrightarrow[-65° \text{ to } -30°]{2 \text{ Li}} Ph_3CCH_2Li \xrightarrow[0°]{} LiCPh_2CH_2Ph$$

$$(9) \qquad\qquad (10)$$

In contrast potassium reacts (42) with 2-chloro-1,1,1-triphenylethane to give 1,1,2-triphenylethylpotassium as the only detectable monopotassium derivative even at −50°C in 1,2-dimethoxyethane. Similarly Cs–K–Na alloy reacts with the chloride in THF at −65°C to give chiefly 1,1,2-triphenylethylcesium (27, 35). The dependence of the extent of rearrangement upon the alkali metal used constitutes part of the general evidence in favor of these rearrangements being rearrangements of organometallic compounds rather than of free radicals.

In all the above examples 2,2,2-triphenylethyl alkali metal compounds were generated by reactions of alkali metals upon halides or quaternary ammonium salts. Similar results have been obtained by metal–metal exchange reactions not involving elementary metals. Thus bis(2,2,2-triphenylethyl)mercury (11) upon reaction with butyllithium in THF below 0°C is reported (17) to give 2,2,2-triphenylethyllithium, which rearranges (evidently at higher temperatures) to 1,1,2-triphenylethyllithium.

$$(Ph_3CCH_2)_2Hg \xrightarrow[<0°]{BuLi} Ph_3CCH_2Li \rightarrow LiCPh_2CH_2Ph$$

$$(11)$$

Also 2,2,2-triphenylethyllithium reacts with excess potassium tert-butoxide in THF at −75°C to give complete rearrangement of carbon skeleton in less than 10 minutes (45); this reaction is believed to occur by way of intermediate 2,2,2-triphenylethylpotassium (44).

$$Ph_3CCH_2Li \xrightarrow[THF, -75°]{KO-t-Bu} Ph_3CCH_2K \rightarrow KCPh_2CH_2Ph$$

An acid–base reaction has also been used to generate a derivative of 2,2,2-triphenylethylsodium with resultant rearrangement. Thus Hauser and co-workers (12) allowed 1,1,1,2-tetraphenylethane (12) to react with phenylsodium in decane-heptane at 90°–100°C to give after protonation 1,1,2,2-tetraphenylethane (13). This rearrangement was reasonably interpreted to occur by the process of Scheme 2.

The above work provides compelling evidence that 2,2,2-triphenylethylalkali metal compounds are rearrangement precursors. However, the circumstance that a rearrangement starts and ends with an organoalkali metal compound does not prove that the rearrangement itself directly

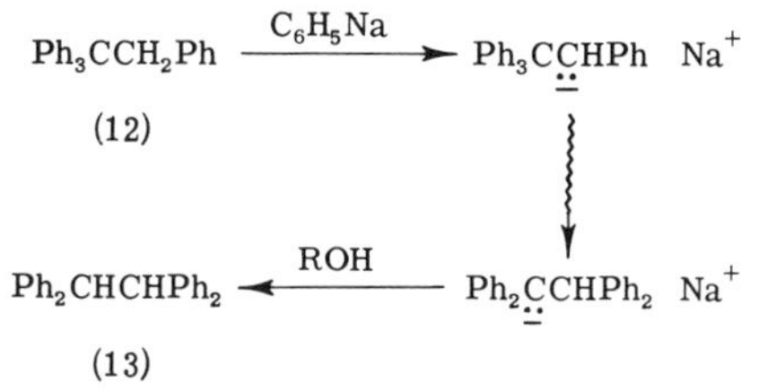

$$Ph_3CCH_2Ph \xrightarrow{C_6H_5Na} Ph_3C\underset{\cdot\cdot}{C}HPh \quad Na^+$$

(12)

$$Ph_2CHCHPh_2 \xleftarrow{ROH} Ph_2\underset{\cdot\cdot}{C}CHPh_2 \quad Na^+$$

(13)

Scheme 2

occurred in the organometallic compound or even in an anion derived from the organometallic compound (65). Thus the rearrangement might occur instead in an intermediate free radical or methylene; it might occur by an inter- rather than by an intramolecular process. What evidence is there concerning the detailed mechanism of rearrangement?

One possible mechanism for phenyl migration would be by an elimination–readdition process, thus 2,2,2-triphenylethyllithium could eliminate phenyllithium, which could readd to the resulting 1,1-diphenylethene to give 1,1,2-triphenylethyllithium.

$$Ph_3CCH_2Li \rightarrow Ph_2C{=\!=}CH_2 + PhLi \rightarrow LiCPh_2CH_2Ph$$

As a test for this mechanism, 2,2,2-triphenylethyllithium was allowed to undergo rearrangement in the presence of ^{14}C-labeled phenyllithium (39, 40). The 2,2,3-triphenylpropanoic acid formed upon carbonation of the final reaction mixture was indistinguishable in radioactivity from background radioactivity; accordingly, less than 0.05% (if any) of the activity of the starting phenyllithium was incorporated in the 2,2,3-triphenylpropanoic acid. Repetition of the rearrangement of 2,2,2-triphenylethyllithium in the presence of benzyllithium rather than phenyllithium [since benzyllithium adds some 10^4 times more readily to 1,1-diphenylethene than does phenyllithium at 0.1 M concentration of organolithium reagents (86)] gave less than 0.1% (if any) incorporation of benzyl group in the rearranged organolithium reagent. While these experiments establish that rearrangement of 2,2,2-triphenylethyllithium occurs as an *intramolecular* process, they do not eliminate some subtle modifications of the elimination–readdition process. Thus 2,2,2-triphenylethyllithium may eliminate phenyl anion to give 1,1-diphenylethene, but the phenyl anions and the 1,1-diphenylethene may exist in a solvent "cage" and recombine before radioactive phenyllithium or benzyllithium can diffuse into the cage. An attractive variation (10) of this idea is that rearrangement of 2,2,2-triphenylethyllithium occurs with elimination of phenyllithium as an ion pair to which 1,1-diphenylethene is bound as a ligand of the lithium ion; the ion pair–olefin complex then collapses to final product before an external organolithium compound can attack the 1,1-diphenyl-ethene in the complex.

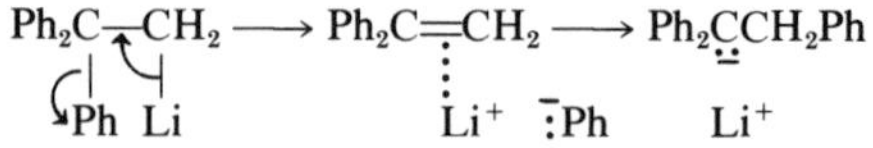

$$Ph_2C\!-\!CH_2 \longrightarrow Ph_2C\!=\!CH_2 \longrightarrow Ph_2\underset{\cdot\cdot}{C}CH_2Ph$$
$$\underset{Ph\ \ \ Li}{} \qquad\qquad Li^+\ \ \ \ \ \ \ :Ph \qquad Li^+$$

Important contributions toward understanding the mechanism of phenyl migration come from work on 2,2-diphenylpropylalkali metal compounds. 2,2-Diphenylpropyllithium **14** in ethereal solution was obtained (*100*) either by reaction of lithium with 1-chloro-2,2-diphenylpropane at 0°C or by reaction of lithium with bis(2,2-diphenylpropyl)mercury. The solution upon heating at reflux for 3 hours gave 1-methyl-1,2-diphenylethyllithium **15** as the major organolithium product.

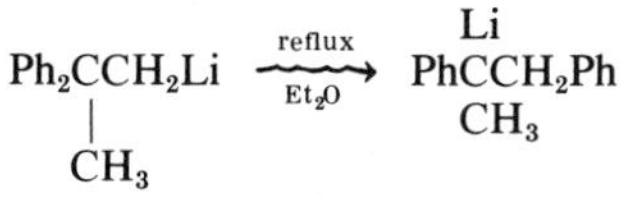

$$Ph_2CCH_2Li \xrightarrow[Et_2O]{reflux} PhCCH_2Ph$$
$$\underset{CH_3}{|} \qquad\qquad\qquad \overset{Li}{\underset{CH_3}{|}}$$

$$(14) \qquad\qquad\qquad (15)$$

In contrast, when a solution of **14** in benzene was warmed at 35°C for 15 hours, little, if any, rearrangement occurred. Reaction (*100*) of bis(2,2-diphenylpropyl)mercury with potassium in THF at room temperature gave 1-methyl-1,2-diphenylethylpotassium, presumably by way of intermediate 2,2-diphenylpropylpotassium.

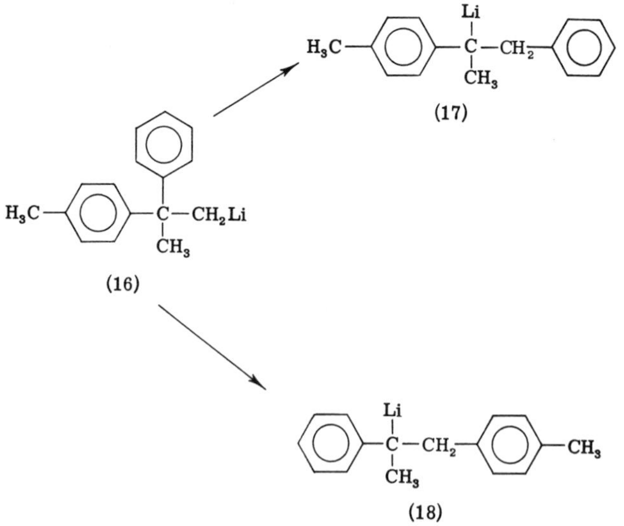

Investigation (*100*) of 2-phenyl-2-*p*-tolylpropyllithium (**16**) showed that this organolithium compound rearranged in ethereal solution at reflux temperature with formation of about a 11:1 ratio of the product of phenyl migration (**17**) to the product of *p*-tolyl migration (**18**). Zimmerman and Zweig (*100*) argued that if rearrangement of **16** occurred in the intermediate free radical **19** there should be no appreciable selectivity in

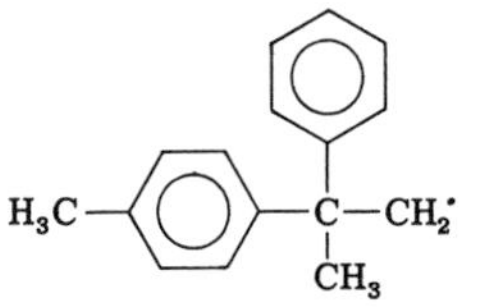

(19)

migration of phenyl versus *p*-tolyl [in what is perhaps the most accurate measurement of the migratory aptitude (*72*) of *p*-tolyl versus phenyl in the rearrangement of neophyl radicals, the *p*-tolyl group migrated at about 0.7 times as fast as phenyl]. On the other hand, if rearrangement occurred in the anion **20**, *p*-tolyl would be expected to migrate appreciably more slowly than phenyl, since the expected intermediate or transition state (**21**) of *p*-tolyl migration should be less stable than that (**22**) for phenyl migration. Hence from the relative migratory aptitudes of

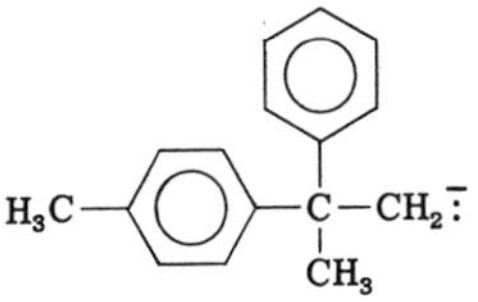

(20)

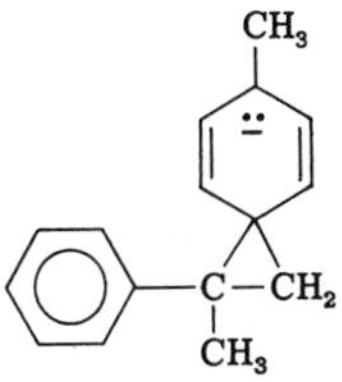

(21)

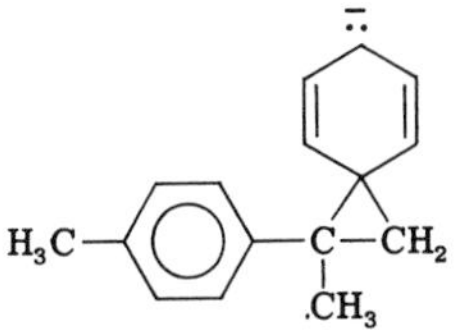

(22)

p-tolyl versus phenyl, Zimmerman and Zweig argued in favor of their rearrangement being a carbanion rearrangement rather than a free-radical rearrangement. They also noted that the rearrangement was not likely to proceed by the elimination of phenyllithium (or *p*-tolyllithium) and readdition to the resulting olefin, since *p*-tolyllithium was found not to add to α-methylstyrene under their reaction conditions.

As we noted earlier in the discussion of triphenylethylalkali metal compounds, some subtle modifications of the elimination–readdition mechanism are hard to disprove. Fortunately, therefore, the interpretation of Zimmerman and Zweig is supported by recent work of Eisch and Kovacs (*17*) upon the halide **23**. This halide is very similar to the

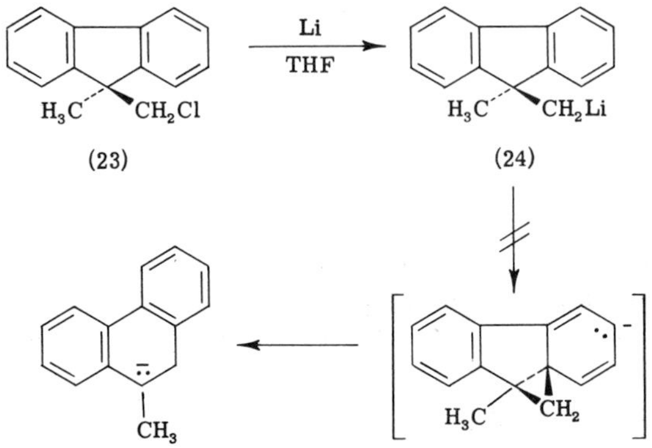

1-chloro-2,2-diphenylpropane used in the rearrangements of Zimmerman and Zweig, save for the presence of the extra carbon–carbon bond of the five-membered ring. Hence it is significant that a skeletal rearrangement did *not* occur in the reaction of **23** with lithium. The interpretation is that rearrangement requires the production of the highly strained spiro intermediate or transition state **25** and therefore fails. Had aryl migration proceeded by an elimination readdition process, the organolithium compound **24** would have been expected to rearrange to **26**.

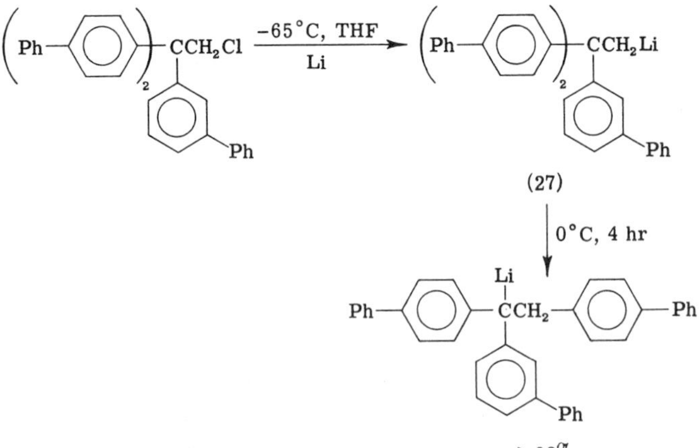

Further evidence in favor of spiro intermediates or transition states during aryl migrations in organoalkali compounds is found in the

preferential migration of p-biphenylyl in 2-m-biphenylyl-2,2-bis(p-biphenylyl)ethyllithium **27** (*41*). The p-biphenylyl group of **27** may be calculated after statistical correction for unequal numbers of biphenylyl groups, to migrate more than 24.5 times more readily than the m-biphenylyl group. This result can be explained on the basis that the rearrangement should proceed much more readily via the transition state or intermediate **28** than via **29** because of the more extensive charge

(28)   (29)

delocalization into the p-phenyl group of **28** than into the m-phenyl of **29**. Had the organolithium compound **27** rearranged by elimination of a biphenylyl anion or free radical little difference in migratory aptitude of m- and p-biphenylyl groups would have been expected, in part since the unshared electron pair of the biphenylyl anion or the unpaired electron in the biphenylyl radical is in an orbital orthogonal to the π-orbitals of the biphenylyl group and hence insensitive to structural changes within this group. Indeed, for elimination of a biphenylyl anion the m-biphenylyl anion would have been expected to be somewhat more readily eliminated than the p-biphenylyl anion, since biphenyl undergoes deuterium exchange with lithium cyclohexylamide in cyclohexylamine 1.6 times more readily at a single meta position than at a para position (*82*).

To what extent are [1,2] migrations of aryl groups in anions general phenomena? In the examples cited so far, two or three aryl groups have been present on the carbon next to the anionic center, such that the product of rearrangement was more stable than the starting anion. Are aryl migrations limited to such cases? Staley and Erdman (*79*) have observed the rearrangement of 6-methyl-6-phenylcyclohexadienyl anion (**30**) produced by reaction of alkali metal amides with 5-methyl-5-phenyl-1,3-cyclohexadiene in liquid ammonia. While this anion underwent a complex series of reactions (see Scheme 3), o-methylbiphenyl (**32**) was reasonably interpreted as having arisen by rearrangement of phenyl to give anion **31**, which aromatizes to give **32** by loss of hydride ion. In this example the rearranged anion **31** is expected to be somewhat *less* stable than the starting anion **30** because of the destabilizing effect of the methyl substituent. Perhaps this is why rearrangement competes so

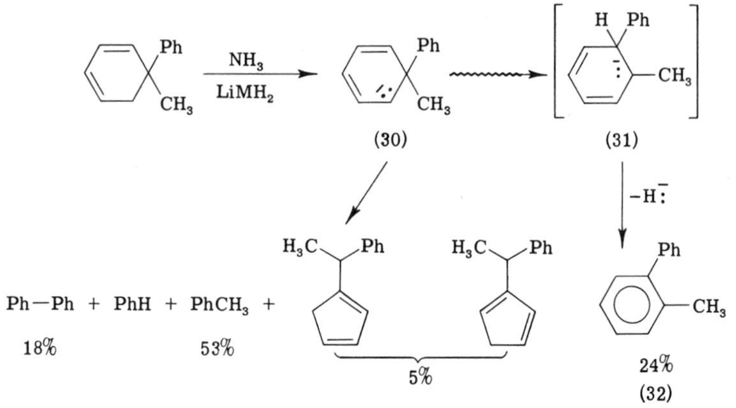

Scheme 3

poorly with other processes, especially when larger alkali metal counterions are present (with $NaNH_2$, 6% of **32**; with $CsNH_2$, 3% of **32**).

Cram and Dalton (*11*) have observed that reaction of *threo*-2-phenyl-3-pentyl methanesulfonate (**33**) with potassium in 1,2-dimethoxyethane gave a hydrocarbon product containing 94% of 2-phenylpentane and 6% of 3-phenylpentane (product of phenyl migration).

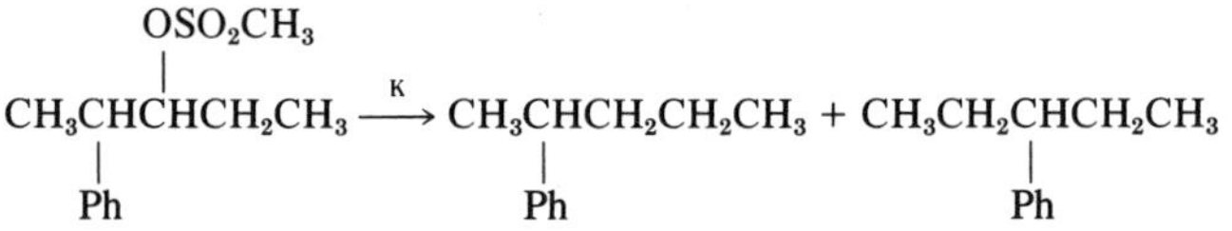

Similarly, treatment of *threo*-3-phenyl-2-pentyl methanesulfonate with sodium in liquid ammonia gave hydrocarbon products containing 3% of 2-phenylpentane (product of phenyl migration). Treatment of 1,1-dideuterio-1-methoxy-2-(1-naphthyl)ethane (**34a**) or 1,1-dideuterio-2-(1-naphthyl)ethyl methanesulfonate (**34b**) with potassium in 1,2-dimethoxy-

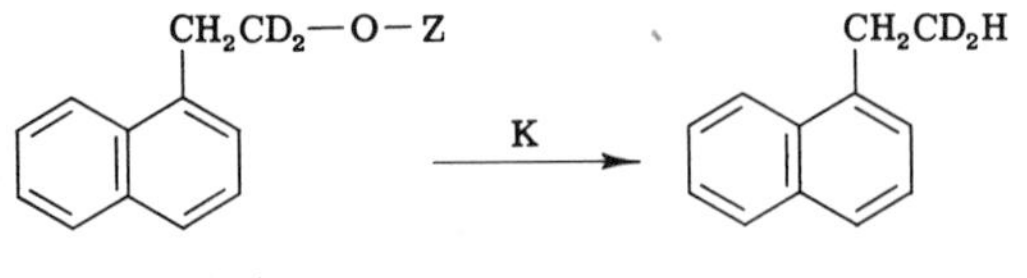

(34)

a: Z = CH_3
b: Z = SO_2CH_3

ethane gave 1-ethylnaphthalene deuterated only in the methyl group. The minor to negligible amount of rearrangement observed in these reactions is perhaps to be attributed to the solvents used, which may have largely protonated the expected intermediate anions before rearrangement of aryl occurred. To be sure, these rearrangements are of unestablished mechanism.

In another study (*30*) 1-chloro-2-*p*-biphenylylethane-1,1-d_2 was allowed to react with various alkali metals in tetrahydrofuran (organoalkali compounds appear to have longer lifetimes in tetrahydrofuran than in 1,2-dimethoxyethane). Reaction with lithium at −70°C gave a good yield of 2-*p*-biphenylyllithium-1,1-d_2 (**35**); however on warming to 0°C this

Ph—⟨○⟩—CH₂CD₂Cl $\xrightarrow[\substack{THF \\ -70°C}]{Li}$ Ph—⟨○⟩—CH₂CD₂Li

(35)

$\Big\downarrow$ 0°C

Ph—⟨○⟩—CD₂CH₂Li

< 3%

organolithium compound failed to undergo appreciable rearrangement, as judged by the position of the deuterium label in the products of carbonation and protonation even though the conditions were so severe that most of the reagent had decomposed to *p*-biphenylylethane. The failure of the lithium compound **35** to rearrange appreciably under conditions that effect the complete rearrangement of 2,2,2-triphenylethyllithium shows that the extra phenyl groups at the reaction center in the latter compound play an important role in the migration of phenyl [probably by stabilizing the anion toward protonation by solvent, by helping to delocalize the negative charge of the anion in the transition state for rearrangement (*100*), and/or by providing steric acceleration (*43*)]. In contrast, reaction (*30*) of 1-chloro-2-*p*-biphenylylethane-1,1-d_2 with cesium or potassium metal in THF at 65°C gave an essentially 50/50 mixture of 1-*p*-biphenylylethane-2,2-d_2 and 1-*p*-biphenylylethane-1,1-d_2 while at −65°C with Cs–K–Na alloy scrambling of the label in the *p*-biphenylylethane was only partial (9–22% migration of *p*-biphenylyl). Repetition of the reaction with potassium in presence of a small amount of *tert*-butyl alcohol resulted in a greatly reduced amount of rearrangement in the product *p*-biphenylylethane. These results indicate that, unlike the case where lithium is the counterion, 2-*p*-biphenylylethylpotassium and -cesium readily undergo migration of the *p*-biphenylyl group

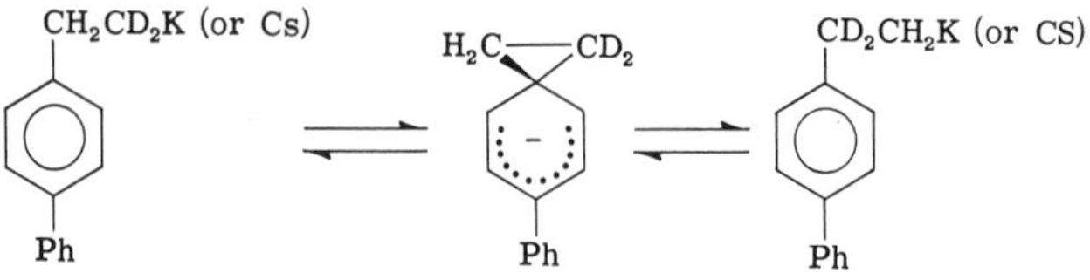

(36)

provided that protonation by solvent or solvent component does not intervene.

Various heteroatom analogs of [1,2] aryl migration are known. Thus benzyldiphenylamine is rearranged (*19*) by treatment with *n*-butyllithium, doubtless by the process outlined below:

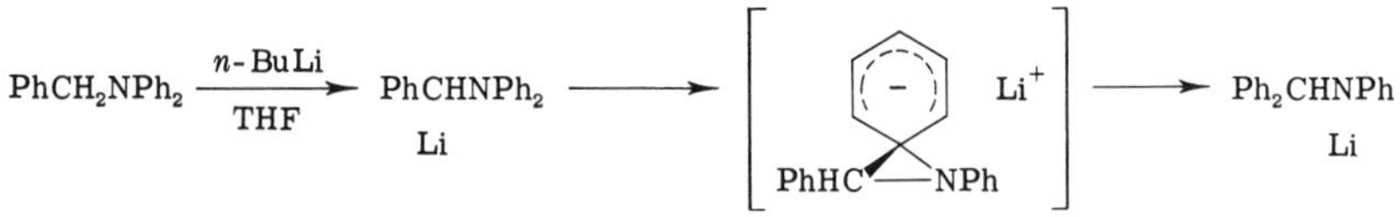

Evidence for the above mechanism comes from the failure of the closely related *N*-benzylcarbazole to undergo an analogous rearrangement, probably because of the large amount of steric strain in the necessary spiro intermediate:

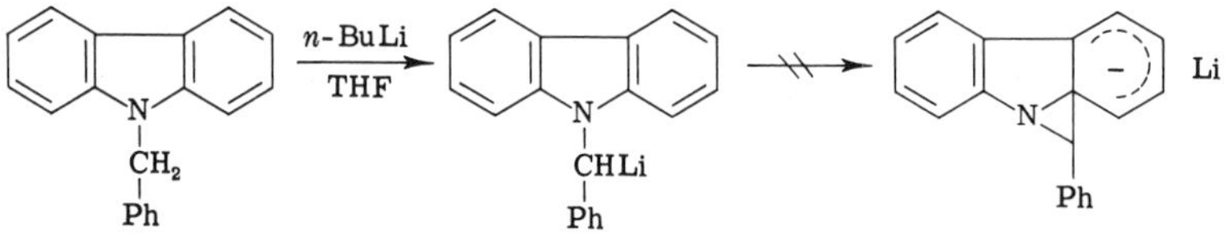

The same criterion applied to the Wittig ether [1,2] rearrangement of aryl groups suggests that this rearrangement occurs by an alternative elimination–readdition pathway, presumably involving radicals (*19*). While monoanions of 1,1-diphenylhydrazine (as lithium salt) do not undergo [1,2] aryl migration even at 100°C in bis(2-methoxyethyl) ether, the dianion (*91*) rearranges slowly but quantitatively in diethyl ether at 30°C:

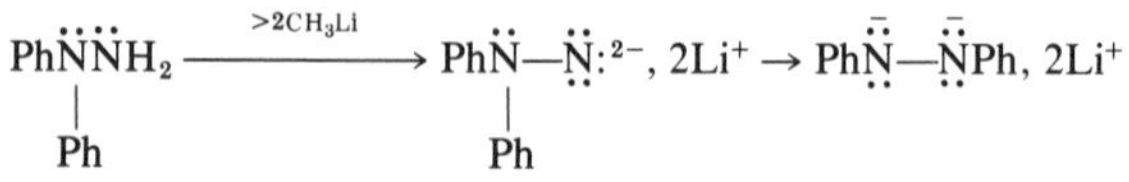

Aryl migrations also occur in Stevens rearrangement of quaternary ammonium salt salts (*50, 84*).

B. [1,3] Migrations and Related Reactions

Sigmatropic shifts of aryl in organoalkali compounds of order [1,3] seem yet not to have been observed because of the occurrence of side reactions. Thus, Wooster and Morse (95) found that 3-iodo-1,1,1-triphenylpropane reacts with sodium in liquid ammonia to give a red solution, which after decomposition with ammonium nitrate afforded triphenylmethane. The triphenylmethane was shown not to have been formed by cleavage of intermediate 1,1,1-triphenylpropane, since this hydrocarbon was not appreciably cleaved under the reaction conditions. The cleavage reaction was interpreted as involving fission of intermediate 3,3,3-triphenylpropyl anion into ethylene and triphenylmethyl anion. In a more quantitative study (37), 3,3,3-triphenylpropyltrimethylammonium iodide was cleaved by sodium in liquid ammonia to give the products shown in Scheme 4. The 3,3,3-triphenylpropyl group cleaves 92-fold faster (after statistical correction) from quaternary nitrogen than does methyl or 11,000-fold more readily than ethyl. The facile cleavage of the 3,3,3-triphenylpropyl group is evidently attributable to an electrical effect of the triphenylmethyl group and yet there is little, if any, direct fragmentation into the triphenylmethyl anion and ethylene during the cleavage from quaternary nitrogen. The main product resulted from protonation of the 3,3,3-triphenylpropyl anion by the liquid ammonia solvent. In the less acidic solvent THF, 3,3,3-triphenylpropyllithium undergoes rapid fragmentation into ethylene and triphenylmethyllithium even at −40°C; similar fragmentations occur with 3,3-diphenylbutyllithium and 2-(9-methyl-9-fluorenyl)-ethyllithium (20).

That yet other processes may intervene is indicated by the reaction (70) of 3-methyl-3-phenyl-1-chlorobutane (37) with potassium in cyclo-

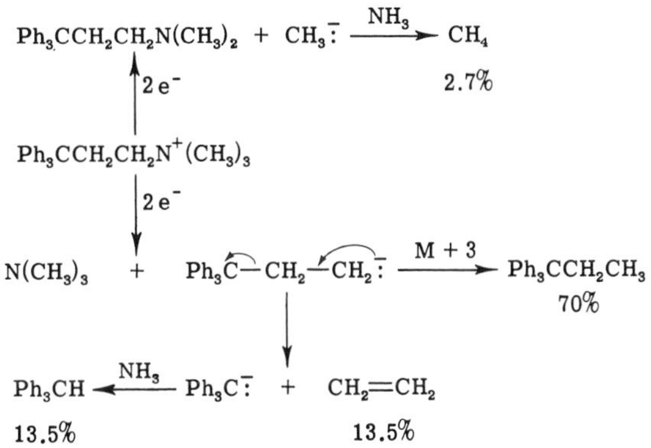

$$Ph_3CCH_2CH_2N(CH_3)_2 \ + \ CH_3^{\overline{\cdot}} \ \xrightarrow{\ NH_3\ } \ CH_4$$

$$\uparrow 2\,e^- \qquad\qquad 2.7\%$$

$$Ph_3CCH_2CH_2N^+(CH_3)_3$$

$$\downarrow 2\,e^-$$

$$N(CH_3)_3 \ + \ Ph_3\overset{\frown}{C}\!-\!CH_2\overset{\frown}{-}CH_2^{\overline{\cdot}} \ \xrightarrow{\ M\,+\,3\ } \ Ph_3CCH_2CH_3$$

$$70\%$$

$$Ph_3CH \xleftarrow{\ NH_3\ } Ph_3C^{\overline{\cdot}} \ + \ CH_2{=}CH_2$$

$$13.5\% \qquad\qquad\qquad 13.5\%$$

Scheme 4

hexane at reflux temperature. The products are as shown in Scheme 5. The most interesting novel product is the 1,1-dimethylindane **40**, which is reasonably interpreted as coming via ortho-cyclization of the intermediate organopotassium compound **38**. A better relative yield of the indane **40** (ca. 50% yield) along with about an equal amount of *t*-pentylbenzene may be obtained by reaction of isopropylbenzene with ethylene (at a pressure of about 30 atmospheres) at 190°C in the presence of a catalyst consisting of potassium and anthracene *(74)*; this reaction is interpreted as proceeding by metallation of isopropylbenzene to give the anion **41** which reacts with ethylene as shown in the scheme. According to Pines and Mark *(69)*, even triphenylmethylsodium will combine with ethylene under pressure to give 1,1,1-triphenylpropane (reversal of the cleavage reaction shown in Scheme 4). The formation of indanes from various 3-phenylpropylpotassium compounds is a general process occurring during the potassium-catalyzed reaction of alkylbenzenes with ethylene, but toluene gives only 2% yield of indane (and 98% of *n*-propylbenzene), whereas ethylbenzene gives 14% of 1-methylindane. α-Alkylation therefore increases the rate of cyclization of the

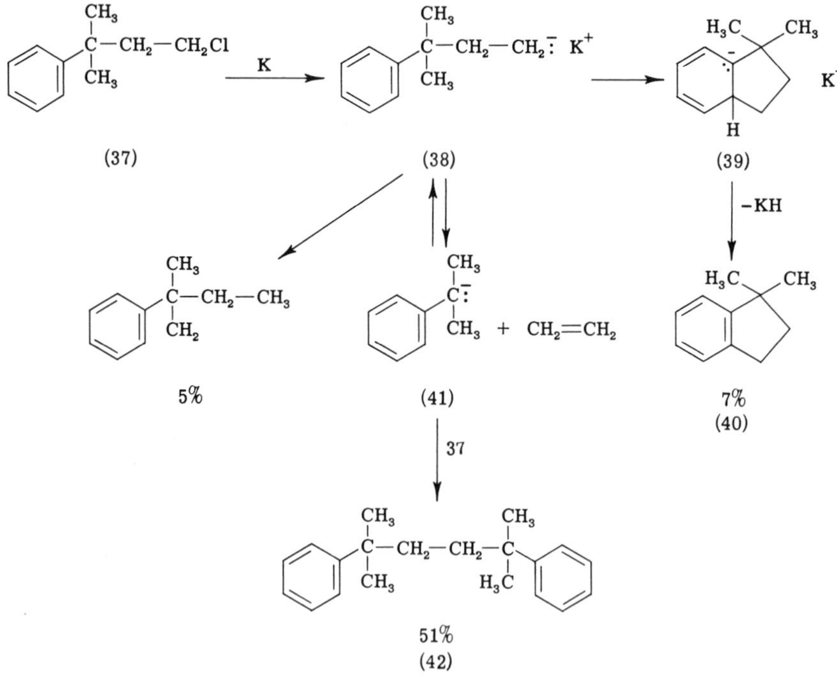

Scheme 5

intermediate 3-phenylpropyl anion relative to the rate of protonation of this anion. Unlike intermediate 3-phenylpropylpotassium compounds, 3-phenylpropylsodium compounds formed during sodium-catalyzed reaction (71) of alkylbenzenes with ethylene do *not* cyclize to indanes, but only suffer protonation to give 3-phenylpropanes. The potassium-catalyzed reaction (68, 81) of 1-methylnaphthalene at 105°C with ethylene gave initially **43** which yielded, among other products, dihydrophenalene **45** (19% yield) while 2-methylnaphthalene at 90°C similarly gave the indane **48** (42% yield). Evidently no appreciable amount of cyclization occurs at β-positions to give **49** and **50** because these anions have less

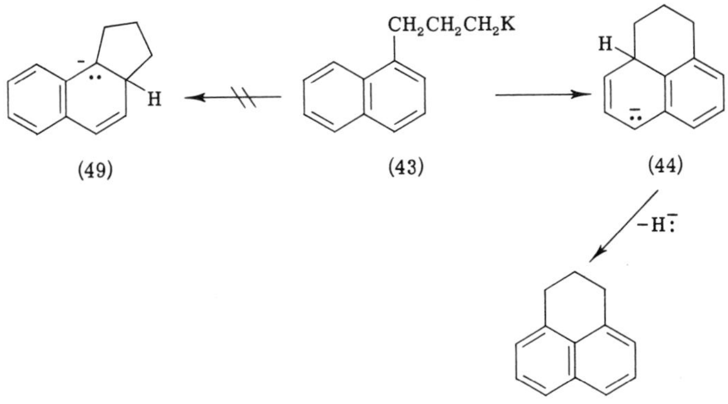

(49)   (43)   (44)

(45)

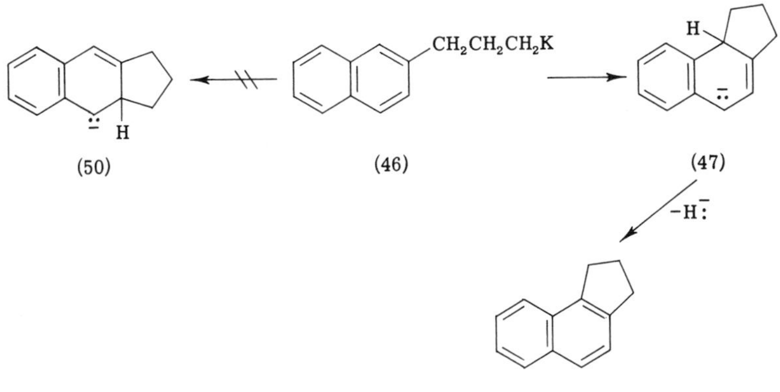

(50)   (46)   (47)

(48)

delocalization energy than **44** and **47**, respectively. While the cyclic anions **39, 44,** and **47** are speculative intermediates, these appear reasonable in view of observations to be discussed later in this review.

3-Methyl-3-(4-pyridyl)butyl chloride upon reaction with lithium in THF undergoes fragmentation (21) into ethylene and 2-(4-pyridyl)-2-lithiopropane in a manner analogous to its carbon analog 37.

C. [1,4] Migrations and Related Reactions

In view of the ready [1,2] migration of phenyl that has been observed in 2,2,2-triphenylethylalkali compounds, [1,4] migration of phenyl was sought in reactions of 4-chloro-1,1,1-triphenylbutane with alkali metals (27). Reaction of this chloride with lithium in THF at −50° to −60°C gave, as expected, 4,4,4-triphenylbutyllithium; however, when this compound was warmed to 25°C, primarily 1,1,1-triphenylbutane was obtained, but no product of [1,4] rearrangement of phenyl was found.

$$Ph_3CCH_2CH_2CH_2Cl \xrightarrow[\substack{THF \\ -55°}]{Li} Ph_3CCH_2CH_2CH_2Li \xrightarrow[THF]{25°} Ph_3CCH_2CH_2CH_3$$

Reaction of the same chloride with molten potassium in THF at 65°C gave, according to the results of carbonation, primarily hydrocarbons and only some 0.5% of the organopotassium compound (52) corresponding to [1,4] migration of phenyl. The volatile hydrocarbons consisted of 84% of 1,1-diphenyl-1,2,3,4-tetrahydronaphthalene (54) and 14% of 1,-1,1-triphenylbutane (55). These products may be rationalized according to Scheme 6. With cesium rather than potassium in THF at 33° to 40°C, the chloride yielded only about 1% of 52 and gave mainly hydrocarbons containing 8% of 54 and 80% of 55 (or products derived therefrom). At lower temperature (−40°C) with Cs–K alloy the product derived from 52 rose to about 10% of the total volatile products, 80% of the remainder being 55 (and products derived therefrom). The interpretation of these reactions is complicated by the likelihood that the formation of 53 from 4,4,4-triphenylbutyl anion is a reversible reaction. Also the primary products from the reaction, especially 55, frequently underwent secondary reactions, and not all the reaction products were volatile under the conditions of gas chromatographic analysis.

Since [1,4] migration of phenyl was a minor process in all the above reactions of 4-chloro-1,1,1-triphenylbutane with alkali metals, reactions of 4-chloro-1-p-biphenylyl-1,1-diphenylbutane were next studied (36) because of the expected superior migratory aptitude of p-biphenylyl over phenyl in anionic rearrangements. Reaction in THF with potassium or cesium at 65°C or with Cs–K–Na alloy at −75°C gave a high yield of 4-p-biphenylyl-1,1-diphenylbutyl anion (58) as deduced from the NMR spectrum of the anion and products of carbonation and protonation.

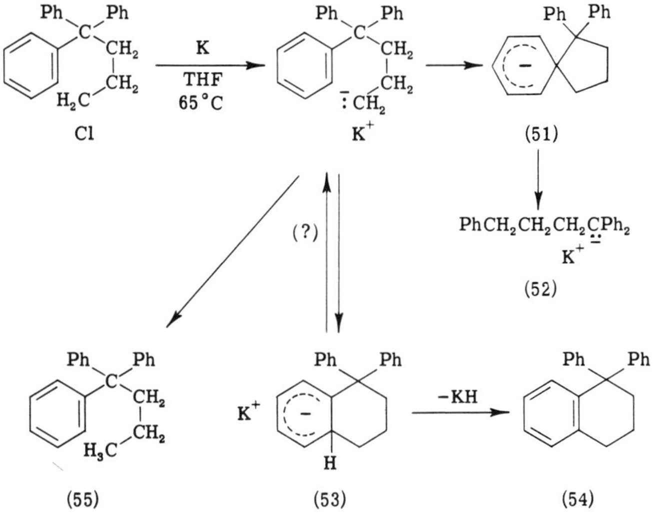

Scheme 6

When the reaction with potassium was run in presence of a small quantity of a sufficiently active proton donor, such as *tert*-butyl alcohol, the product was unrearranged 1-*p*-biphenylyl-1,1-diphenylbutane or products derived from the latter by subsequent reduction of the *p*-biphenylyl group. Likewise, with sodium in refluxing dioxane the chloride gave 1-*p*-biphenylyl-1,1-diphenylbutane containing only some 7% of the rearranged hydrocarbon 4-*p*-biphenylyl-1,1-diphenylbutane. In contrast to sodium, with potassium or cesium in dioxane chiefly the rearranged hydrocarbon was obtained. From these results it is concluded that the chloride reacts with sodium, potassium, and cesium as shown in Scheme 7. The initial product is the anion **56** which is either rapidly protonated to give 1-*p*-biphenylyl-1,1-diphenylbutane or undergoes rearrangement to **58** evidently via the spiro anion **57**. Even in a reaction time of 2 minutes at −75°C it was not possible to detect **57** which evidently suffers ready cleavage to **58**.

4-*p*-Biphenylyl-4,4-diphenylbutyllithium (**56**, M = Li), unlike compounds of the other alkali metals, gave little if any rearrangement even upon standing at 0°C in THF but was largely protonated to give 1-*p*-biphenylyl-1,1-diphenylbutane. Yet addition of potassium *tert*-butoxide to the solution of this organolithium compound in THF at −75°C brought about ready [1,4] migration of *p*-biphenylyl to give **58**. This catalysis by potassium *tert*-butoxide is doubtless brought about by conversion of the

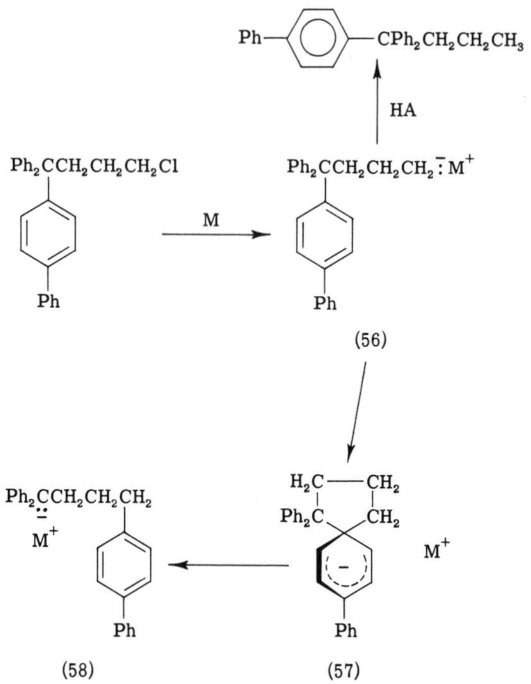

$$Ph-\underset{Ph}{\bigcirc}-CPh_2CH_2CH_2CH_3$$

$$\uparrow \quad HA$$

$$Ph_2CCH_2CH_2CH_2Cl \qquad Ph_2CCH_2CH_2CH_2\overset{-}{:}M^+$$

$$\underset{Ph}{\bigcirc} \qquad \xrightarrow{\quad M \quad} \qquad \underset{Ph}{\bigcirc}$$

$$(56)$$

$$Ph_2\underset{\cdot\cdot}{C}CH_2CH_2CH_2 \qquad \overset{H_2C-CH_2}{\underset{Ph_2C\diagdown CH_2}{|\quad\quad|}}$$

$$M^+ \underset{Ph}{\bigcirc} \quad \longleftarrow \quad \underset{Ph}{\bigcirc(-)} \quad M^+$$

$$(58) \qquad\qquad (57)$$

Scheme 7

organolithium compound into the corresponding organopotassium compound, which readily rearranges (36).

From the data cited, it is evident that the relative rate of [1,4] migration of p-biphenylyl versus the rate of protonation of the organoalkali compound by solvent increases along the series: Li ≪ Na ≪ K or Cs. Also [1,4] migration of aryl competes less favorably in competition with protonation by solvent than does [1,2] migration.

D. [1,5] Migrations and Related Reactions

The reaction of alkali metals with 5-chloro-1,1,1-triphenylpentane might be expected to proceed according to Scheme 8. In fact, the reaction of potassium with this chloride in THF at 65°C gave predominantly unrearranged hydrocarbon **63** along with cleavage products; the yield of rearranged products **62** and **65** did not exceed more than a few percent. With Cs–K–Na alloy in THF at −68°C, the chloride gave after

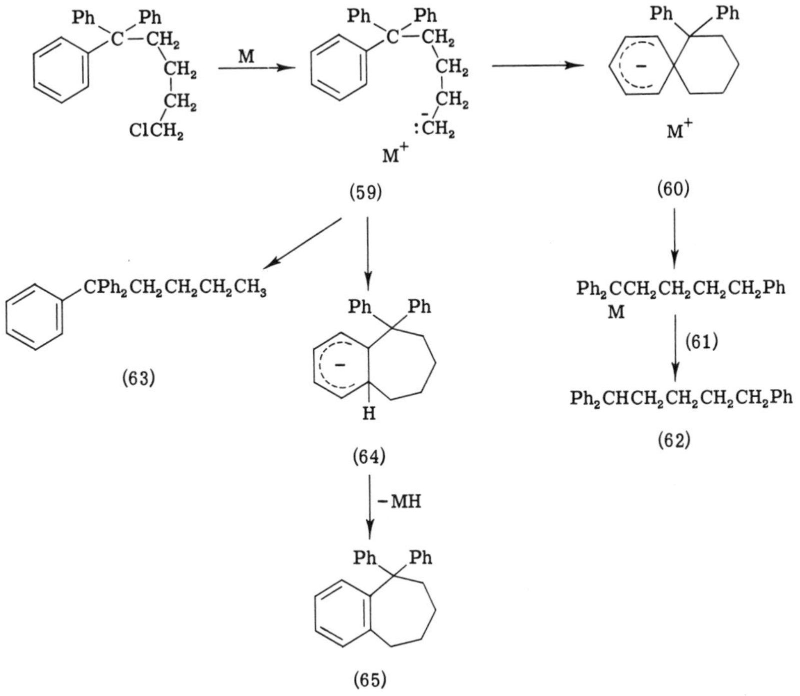

Scheme 8

protonation 6 ± 3% of rearranged hydrocarbon **62**; the remainder of the product was unrearranged hydrocarbon **63** or products derived therefrom by further reaction with the alloy (27). It thus appears that [1,5] migration of phenyl is somewhat less favorable than [1,4] migration. Also ortho-cyclization of 5,5,5-triphenylpentylpotassium to form the seven-membered ring in **64** is considerably less favorable than forming the six-membered ring in **53** from 4,4,4-triphenylbutylpotassium. These observations are in accord with empirical evidence that five- and six-membered rings are generally formed more readily than seven-membered rings in organic reactions.

E. Direct Observation of Intermediate Spiro Anions

The preceding sections have reviewed the evidence for [1,n] migrations of aryl groups in organoalkali compounds. These rearrangements were all postulated to proceed via intermediates or transition states (see

21, 22, 25, 28, 29, 36, 57, 60) of spiro anionic structure of general formula **66** for migration of phenyl. Are these structures transition states or reaction intermediates? We were encouraged to think that **66** may sometimes be a stable intermediate on the basis that pentadienyllithiums can be prepared (*1, 2*), including 6,6-dimethylcyclohexadienyllithium **67**,

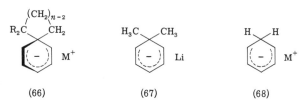

(66)   (67)   (68)

and that these are reasonably stable at room temperature according to NMR spectral studies. This is in sharp contrast to cyclohexadienyllithium itself, which under the general conditions for preparation of **67** (by reaction of 1,4-cyclohexadiene with *n*-butyllithium in THF at about −30°C) gives quantitatively only benzene; similarly, reaction of 1,4-cyclohexadiene with amylsodium in petroleum ether gives (*66*) benzene and sodium hydride. Since cyclohexadienyl anion is a proposed intermediate in the Birch reduction of benzene, it is gratifying that this anion has now also been prepared (*57*) by reaction of 1,4-cyclohexadiene with potassium anide in liquid ammonia and has been found to be quite stable at −60°C but to decompose to benzene fairly readily at −20°C. The observed behavior of cyclohexadienyl anion is therefore in accord with the proposed chemical behavior of the related intermediates (**31, 39, 44, 47, 49, 53, 64**) which have been discussed earlier.

In our previous discussion it was pointed out that 4-chloro-1-*p*-biphenylyl-1,1-diphenylbutane reacts with Cs–K–Na alloy in THF at −75°C to give only rearranged product **58** (see Scheme 7) in a reaction time as short as 2 minutes. In view of the known stability of species such as **67** and **68** is it reasonable to write the spiro anion **57** as an intermediate in this reaction? Probably this question should be answered in the affirmative on the basis that we have been able to prepare (*25, 26*) the closely related anion **70** in 76% yield by reaction between 4-chloro-1-*p*-biphenylylbutane and Cs–K–Na alloy in THF at −70°C (see Scheme 9). The same spiro anion may also be prepared, but only in some 22% yield, by reaction between the chloride and cesium or potassium metal in THF at 65°C. Experiments with sodium and potassium, in which dioxane or tertiary alcohols in THF were used as a proton source to "trap" intermediate anions, indicate that 4-*p*-biphenylylbutyl anion (**69**) is a precursor of **70** (see Scheme 9). The organosodium intermediate **69** (M$^+$ = Na$^+$) in dioxane at 101°C gave primarily 1-*p*-biphenylylbutane

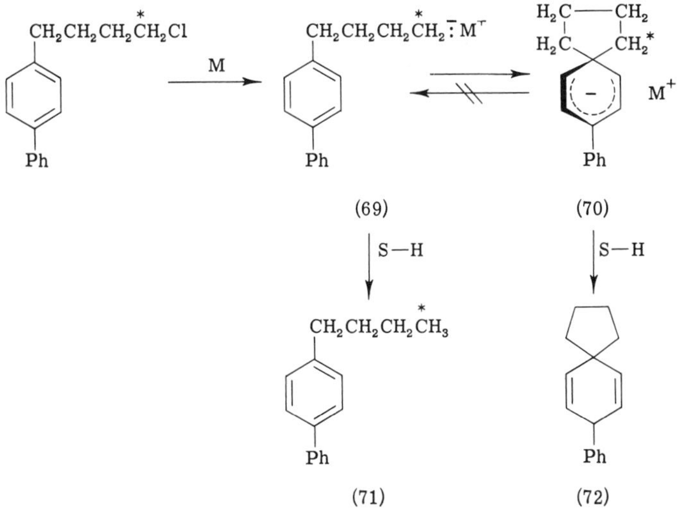

$CH_2CH_2CH_2\overset{*}{C}H_2Cl$

$\xrightarrow{M}$

$CH_2CH_2CH_2\overset{*}{C}H_2\bar{:}M^r$

$\underset{\xleftarrow{}}{\xrightarrow{\cancel{}}}$

H_2C——CH_2
H_2C $\overset{*}{C}H_2$

M^+

Ph   Ph   Ph

(69)   (70)

S—H   S—H

$CH_2CH_2CH_2\overset{*}{C}H_3$

Ph   Ph

(71)   (72)

Scheme 9

(71) whereas the potassium compound under these conditions gave about a 3:1 mixture of **71** and **72**. Spiro cyclization obviously occurred more readily with the potassium compound than the sodium compound, just as [1,4] migration of p-biphenylyl occurred more readily in potassium than sodium compounds (Scheme 7). From labeling experiments conducted with 4-p-biphenylyl-1-chlorobutane-1,1-d_2 no evidence was ever found for scrambling of the position of the label between the 1- and 4-positions of the butane system in the hydrocarbon **71** from reaction of the chloride with potassium in THF at 65°C or in dioxane at 101°C. The spiro anion **70** undergoes other reactions, such as protonation and polymerization, faster than ring opening to regenerate **69**. Evidently the 2,2-diphenyl groups of the spiro anion **57** greatly accelerate ring opening of the spiro anion to give **58**; these groups also accelerate ring closure of **56** more than 200-fold with K^+ to give **57** (see Scheme 7). A pyridine analog of the spiro anion **70** has been prepared by reaction of 1-carbethoxy-4,4-tetramethylene-1,4-dihydropyridine with n-butyllithium and other organometallic compounds (22).

Are spiro anions intermediates in other [1,n] rearrangements of aryl groups, especially where $n = 2$? Staley and co-workers (78) have attempted to generate the spiro[2,5]octa-4,6-dienyl anion (73) by reaction of the spirodiene (72) with potassium amide in liquid ammonia at temperatures as low as −65°C or with n-butyllithium in THF-hexane at room temperature (see Scheme 10). In either case, however, only

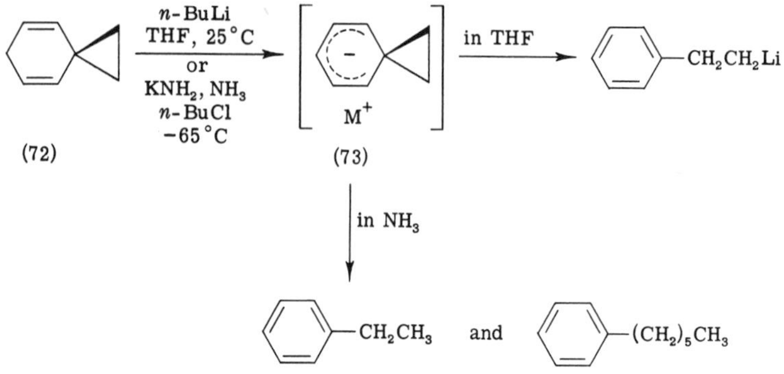

Scheme 10

products in which the cyclopropane ring was opened were obtained. It is not clear from these experiments whether **73** was an intermediate or a transition state; however, the related species **74** could be prepared (*78*) in liquid ammonia and was stable in this solvent at −30°C. The comparative stability of **74** toward ring opening to **75** is attributable to

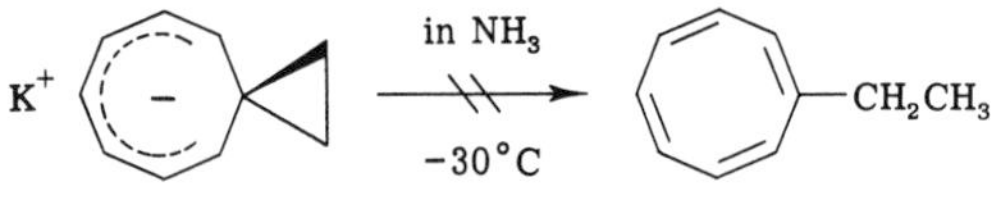

the fact that the product would be a nonaromatic cyclooctatetraene derivative whereas the product from **73** is aromatic.

In view of the foregoing observations, we (*34*) first examined the reaction between 1-*p*-biphenylyl-3-chloropropane and Cs–K–Na alloy. That the hoped for spiro anion **76** was formed was indicated by the products of carbonation, which included, after a reaction time of 1 minute, the corresponding spiro acid **77** (36% yield); the cleavage

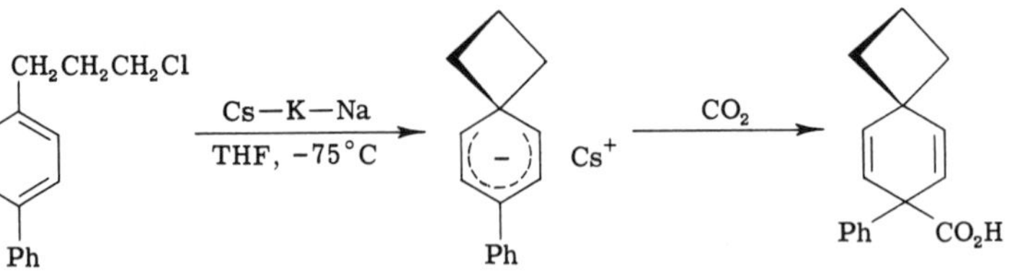

product, *p*-biphenylylacetic acid (14%); and the α-metallation product, 2-*p*-biphenylylbutanoic acid (3%). The half-life of **76** in THF at −75°C was about 13 minutes. Since a repetition of this experiment with the lower homolog 1-*p*-biphenylyl-2-chloroethane gave a complex mixture of products containing evidently only about 1% (if any) of spiro acid and much *p*-biphenylylethane, a method was sought to increase the stability of the hoped for spiro anion.

A possible method seemed to be to make use of the Thorpe–Ingold, or "*gem*-dimethyl," effect which, in one statement (49), is that "the accumulation of substituents along the chain of a bifunctional system appears to selectively stabilize the corresponding cyclic forms." Maercker and co-workers (63) have recently used this effect to shift the equilibrium between a 3-butenyl Grignard reagent and a cyclopropylmethyl Grignard reagent in favor of the cyclopropylmethyl compound.

Initial experiments (34) upon 1-chloro-2-methyl-2-*p*-biphenylylpropane **78** with Cs–K–Na alloy in THF at −75°C were complicated by α-elimination and metallation of the resulting olefin. Reaction of the chloride with lithium metal gave the expected organolithium compound **79** accompanied by some 10–15% of the rearrangement product **80**.

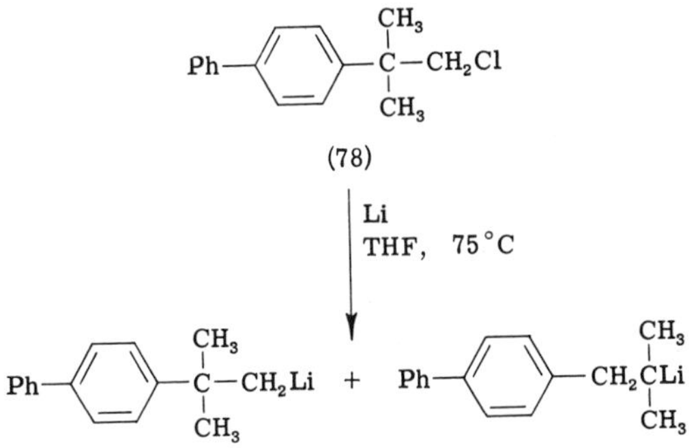

$$Ph-\bigcirc-\underset{\underset{CH_3}{|}}{\overset{\overset{CH_3}{|}}{C}}-CH_2Cl$$

(78)

Li
THF, 75 °C

$$Ph-\bigcirc-\underset{\underset{CH_3}{|}}{\overset{\overset{CH_3}{|}}{C}}-CH_2Li \quad + \quad Ph-\bigcirc-CH_2\underset{\underset{CH_3}{|}}{\overset{\overset{CH_3}{|}}{C}}Li$$

(79)   (80)

Treatment of the resulting mixture of organolithium reagents with potassium *t*-butoxide at −75°C gave upon carbonation after 10 minutes a complex mixture of products containing evidently acids resulting from ring metallation. These negative experiments were interpreted to mean that either the spiro anion **81** was not formed at a satisfactory rate

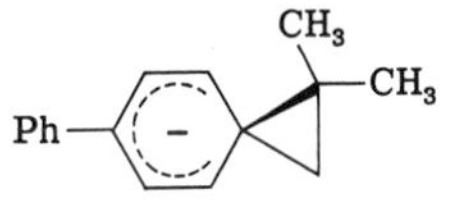

$$Ph-\bigcirc-\triangleleft\overset{CH_3}{\underset{}{}}$$

(81)

relative to competing reactions or else lacked suitable stability to survive the reaction conditions employed.

Since the Thorpe–Ingold effect is expected to be larger for four methyl groups than for two (63), the tetramethyl chloride **82** was next studied (34). Reaction of this chloride with excess of Cs–K–Na alloy in THF at −75°C gave within 35 seconds a deep-red solution, which, upon carbonation after a total reaction time of 1 minute, afforded a 28% yield of the spiro acid **83** and 7% of **84**. The structure of **83** has been fully confirmed

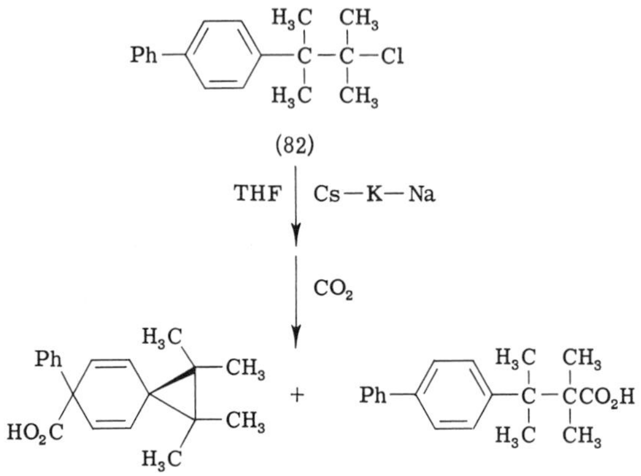

by spectroscopic methods and by X-ray diffraction (3) upon a single crystal. This experiment establishes beyond reasonable doubt that the compound **85** is formed during reaction of the chloride **82** with Cs–K–Na alloy. What remains to be established is the mechanism of formation of **85** and whether or not **85** is in mobile equilibrium with the open isomer **86a** at −75°C. The anion **85** has a half-life of about 22 minutes in THF at −75°C. With lithium in place of cesium alloy, the chloride **82** gives the lithium analog of **86a** without detectable spiro anion. Pyridine analogs (**86b**) of the spiro anion **85** have been prepared by reaction of the *N*-methiodide of 2-methyl-2-(4-pyridyl)propyl chloride with lithium in THF

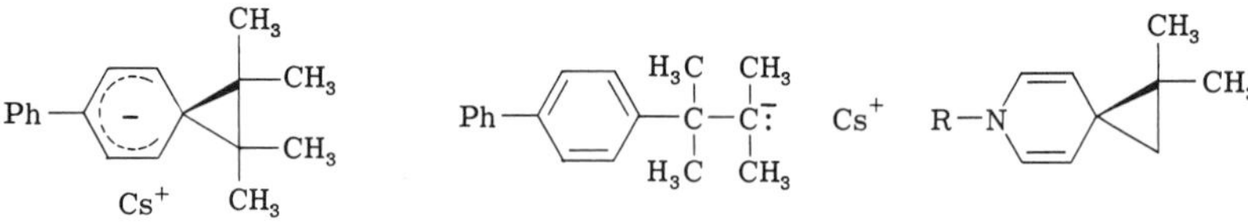

(18) or by reaction of 2-(4-pyridyl)-2-methylpropyl metal derivatives (Mg, Li, and Hg) with acyl chlorides or trimethylsilyl chloride (21).

The work cited here on direct observation of spiro anions from reaction of arylalkyl chlorides with alkali metals, along with the work described on heteroatom analogs, establishes spiro anions as fully plausible intermediates during aryl migrations in organoalkali compounds.

III

MIGRATIONS OF ARYL IN INTERMEDIATE RADICALS

The foregoing reactions have been considered to be rearrangements of carbanions or organoalkali compounds on the basis that evidence was at hand that an unrearranged organoalkali compound was first formed and subsequently rearranged, the migratory aptitudes of groups agreed with those expected for anionic rearrangements, and, in favorable cases, the direct observation of intermediate spiro anions. It must not be believed, however, that all reactions between metals and alkyl halides that give rise to rearrangements proceed by carbanionic mechanisms. Thus there has been evidence for some time that radicals are formed during reactions of alkyl halides with magnesium (56). Rüchardt and Trautwein (73) reported formation of 1–6% of rearranged hydrocarbons during reaction of neophyl chloride with magnesium even though the Grignard reagent itself was said to be immaterially contaminated with isomers. Similarly, Eberson (16) observed some 6% isobutylbenzene in the tert-butylbenzene from cathodic reduction of neophyl chloride. Reaction (29) of neophyl chloride with lithium in THF at −65°C for 30 minutes gives an organolithium reagent containing 6% of 1,1-dimethyl-2-phenylethyllithium (88) in addition to the expected product 2-methyl-2-phenylpropyllithium (87). Moreover, the normal product 87 can be shown not to rearrange to 88 under the reaction conditions for formation of the organolithium reagent. These rearrangements and the related racemization (87–89) during reaction of 1-halogeno-1-methyl-2,2-diphenylcyclopropane with magnesium and lithium seem to occur by way of intermediate radicals that in part are free enough from the metal surface to racemize or rearrange (see Scheme 11). Also the unusual migratory aptitude of groups, α-pyridyl > phenyl > γ-pyridyl, observed during reaction (18) of 2-(α or γ-pyridyl)-2,2-diphenylethyl chloride with lithium in THF has been explained on the basis that rearrangement takes place in an intermediate radical rather than an anion (17).

In another example, 4-chloro-1-p-biphenylyl-1,1-diphenylbutane (36) was found to react with lithium metal in THF at −75°C to give 4-p-biphenylyl-4,4-diphenylbutyllithium (89) containing up to 47% of the rearranged compound 4-p-biphenylyl-1,1-diphenylbutyllithium (90). Attempts to induce thermal rearrangement of 89 to 90 were ineffective and

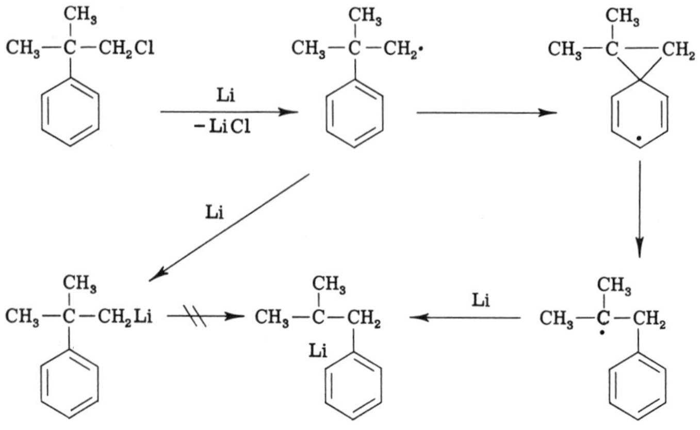

(87) (88)

Scheme 11

resulted mostly in protonation of **89**. These results can be explained by a mechanism analogous to that shown in Scheme 11 for neophyl chloride. Some confirmation of this mechanism was provided by the results of reaction of 4-chloro-1-*p*-biphenylyl-1,1-diphenylbutane with lithium bi-

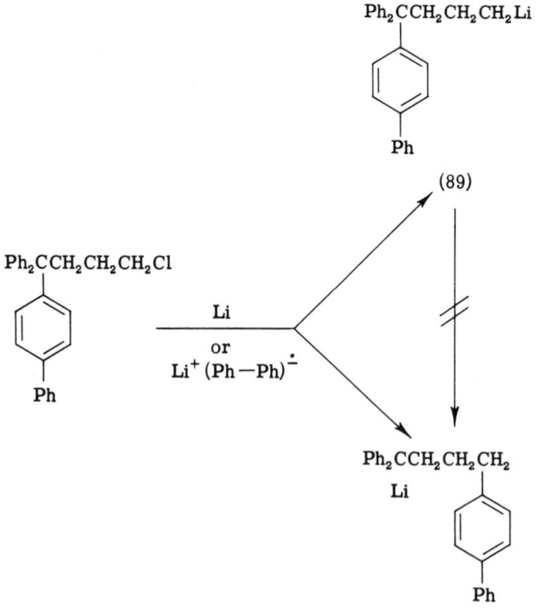

(90)

phenylide. Addition of the chloride to an excess of lithium biphenylide in THF at $-75°C$ gave **89** containing 12% of **90** at an initial lithium biphenylide concentration of 0.011 M or 2% of **90** at a lithium biphenylide concentration of 0.24 M. Evidently lithium biphenylide is able to trap an intermediate (presumably the proposed free radical) prior to rearrangement—the higher the concentration of lithium biphenylide, the more effective the trapping of the intermediate.

According to the general mechanism of Scheme 11, an intermediate leading to the product of biphenylyl migration is the bridged radical **91**.

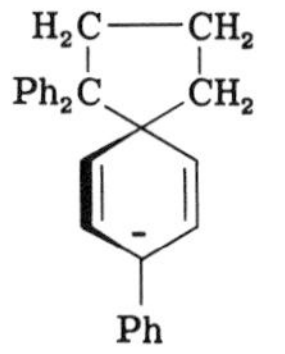

$$H_2C—CH_2$$
$$Ph_2C\quad CH_2$$

(91)

This interpretation is strengthened by the observation (*25*) that 4-chloro-1-*p*-biphenylylbutane reacts with lithium in THF at $-70°C$ to give 4-*p*-biphenylylbutyllithium **92** containing 37% of 8-phenylspiro[4.5]-6,8-deca-dienyllithium **93**. Repetition of this experiment with 4-*p*-biphenylyl-1-chlorobutane-1,1-d_2 gave similar proportions of 4-*p*-biphenylylbutylli-thium-1,1-d_2 (**92-d**) and spiro anion (**93-d**). Since deuterium was distrib-uted at C-1 in **92-d** but not at C-4, it follows that **92-d** cannot be formed

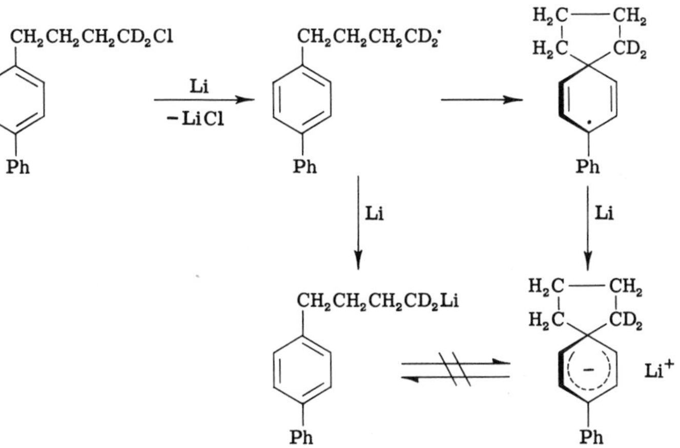

(92-d)   (93-d)

Scheme 12

from a spirocyclic intermediate analogous to **91** (whether an anion or a radical). Also, since in all known cases organolithium compounds do not undergo [1,*n*] aryl migrations in THF at −70°C, **93-d** must be formed by cyclization of some other species, reasonably an intermediate radical (see Scheme 12). Julia and Malassiné (*55*) have obtained a related spiro compound **94** from reaction of *p*-(4-halobutyl)benzoic acids with alkali

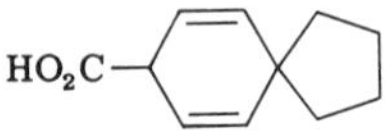

(94)

metals in liquid ammonia but have proposed a different detailed mechanism that accounts for the variation in yield of **94** versus competing products as the halogen is varied.

IV

MIGRATIONS OF ARYL IN RADICAL ANIONS OF AROMATIC HYDROCARBONS AND RELATED REACTIONS

The reductive rearrangement of 1,1,3-triphenylindene by sodium was discovered, but misinterpreted, by Schlenk and Bergmann (*75*). According to Ziegler and Crössman (*97*), the reaction proceeds as follows:

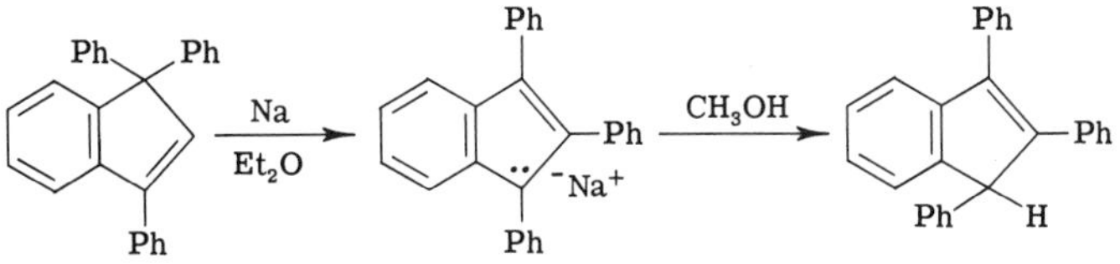

A closely related example was discovered by Koelsch (*58*)[2] (see Scheme 13).

In a reinvestigation of the reaction of Schlenk and Bergmann, Miller and Boyer (*64*) found that reaction of 1,1,3-triphenylindene with sodium or sodium–potassium alloy in THF gave, after addition of water, 1,2,3-triphenylindane (rather than the indene). These workers suggest that the reaction likely proceeds via the radical anion **95** which by [1,2] migration of phenyl gives a more stable *o*-quinodimethane anion radical **96** which upon reduction leads to the dianion **97**. Evidently in THF hydride loss from **97** is slower than in diethyl ether, where **99** is formed prior to

---

[2] That the compound of Koelsch rearranged while the related compound **24** of Eisch did not rearrange is of interest. The explanation may be that Eisch studied a lithium compound whereas Koelsch worked with a more reactive sodium compound. Obviously also the detailed structures and mechanisms of the reactions are different.

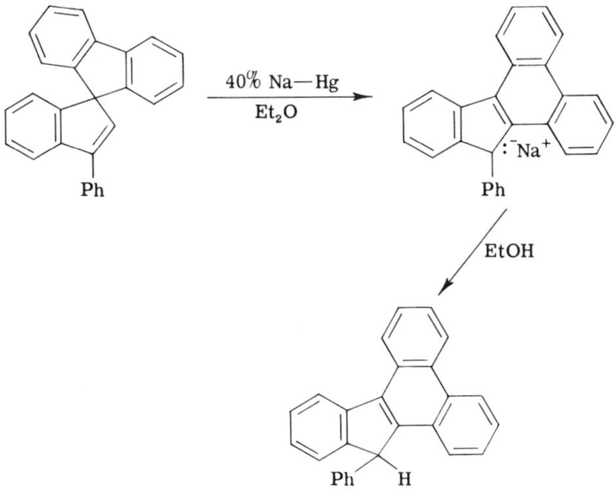

Scheme 13

protonation. Deuterium tracer experiments showed that the proton on C-2 of **95** remained on C-2 in the final product **98**. The work to date leaves unanswered the question of whether phenyl migration proceeds via an

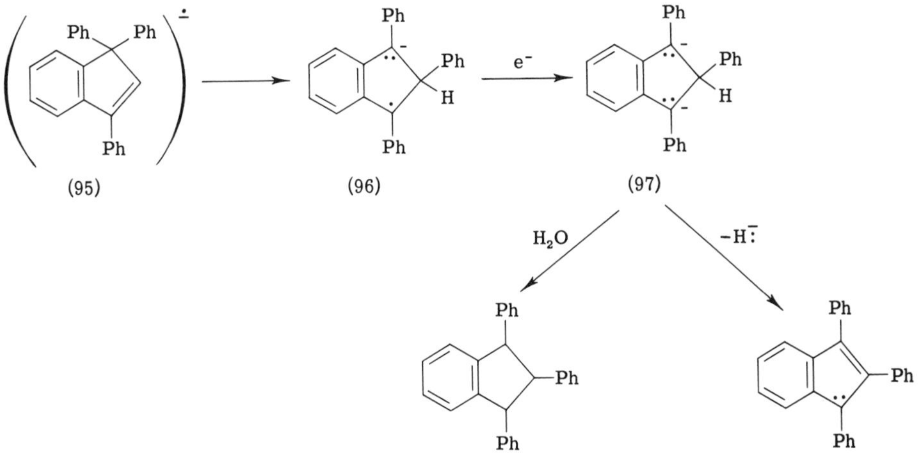

intermediate like structure **100** or like structure **101** (or whether the rearrangement is more like that of a carbanion or that of a radical), if indeed **100** and **101** are not mere canonical structures of the intermediate. As in both anionic and radical rearrangements, a phenyl group migrates much more readily than an alkyl group, then 1-methyl-1-

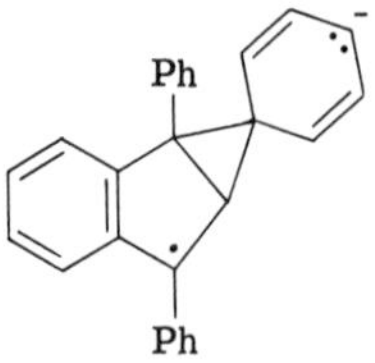

(100) 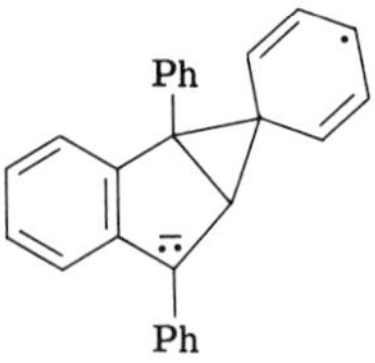   (101)

phenylindene **102** gives 3-methyl-2-phenylindene **104**. As Miller and Boyer pointed out, it is also not certain at what stage addition of the second electron occurs in the mechanism **95** → **97** or **102** → **103**; thus it might occur prior to the migration of phenyl.

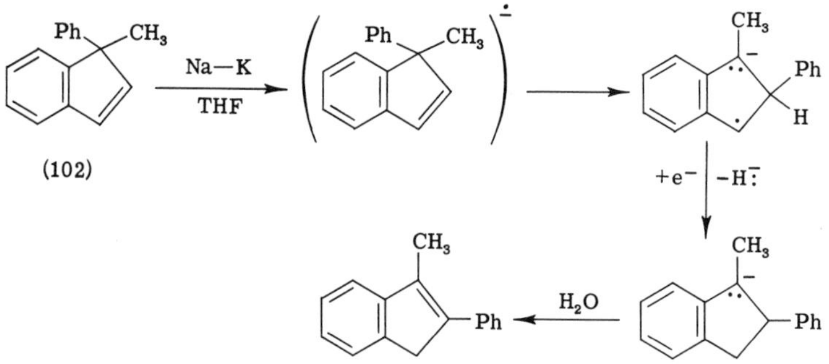

(102)

(104)   (103)

That alkyl groups may migrate in radical anions under favorable circumstances is borne out by the rearrangement (*14*) of naphthobicyclobutane radical anion **105** to the radical anion of pleiadiene **106**. This

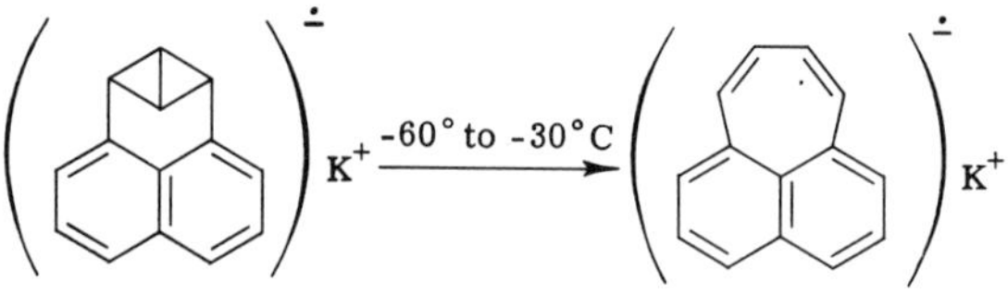

(105)   (106)

rearrangement is formally a cycloreversion of the type $_{\sigma}2_a + {_{\sigma}}2_a$, which is forbidden for a simple thermal process but allowed as a photochemical process. The rearrangement of radical anions, therefore, may in general more nearly parallel the chemistry of the excited state than of the ground state. This possibility would appear to be worthy of further exploration—photochemical rearrangements being myriad.

We have indeed in the course of some of our reactions (*27*) of aryl-substituted alkyl halides **107** with cesium and cesium alloys found final products unlike what we had anticipated. We have already noted that the corresponding hydrocarbons **108** are frequently major initial products in some of these reactions.

$$\text{Ph}_3\text{C}(\text{CH}_2)_n\text{Cl} \xrightarrow[\text{THF}]{\substack{\text{Cs}-\text{K} \\ \text{or} \\ \text{Cs}-\text{K}-\text{Na}}} \text{Ph}_3\text{C}(\text{CH}_2)_n\text{H}$$

(107)   (108)

Further study revealed that **108** will react with cesium alloys in THF to give novel red dianions. Thus 1,1,1-triphenylethane with excess Cs–K–Na alloy in THF at −70°C gives an essentially quantitative yield of **109** (evidently as a single stereoisomer), which was characterized by the

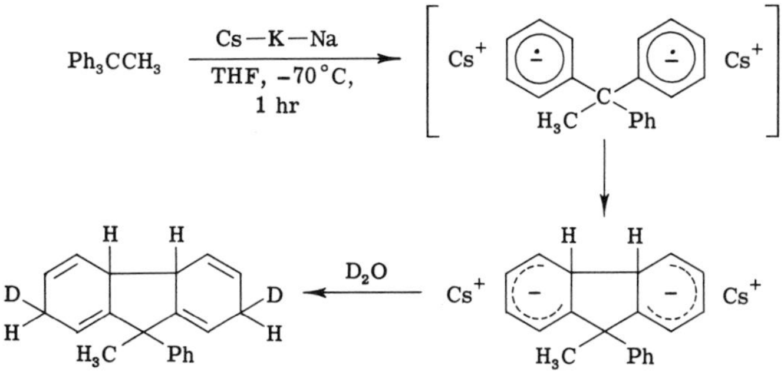

product of decomposition with H_2O or D_2O (**110**). Similarly 2,2-diphenylpropane (*32*) gives **111** and the hydrocarbon **112** whose geometry may be assigned as shown based on its NMR spectrum.

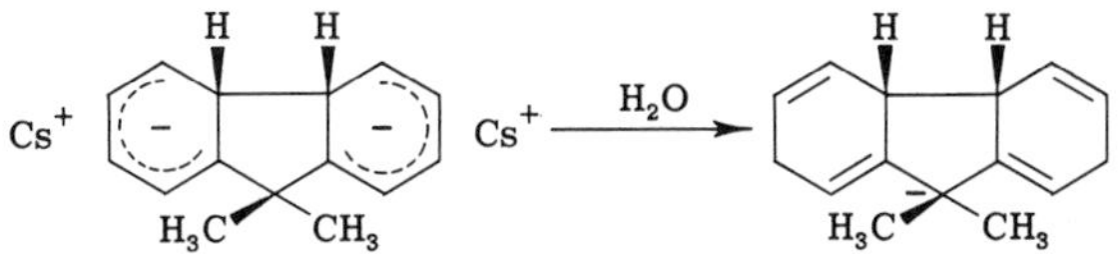

The reactions of the above di- and triphenylalkanes with cesium alloys caused us to wonder whether benzene would give a similar reaction. The literature revealed that Hackspill (*46*) had found that benzene reacted with cesium at 28°C to give a black solid of cesium content near that of C_6H_5Cs so that the compound was initially supposed to be phenylcesium; however, since treatment of the black precipitate with water yielded hydrogen and biphenyl, the assignment of structure was later questioned by Hackspill (8, 47). The black solid was reanalyzed by de Postis (13, 48), who claimed it had the empirical formula $C_6H_6Cs_6$ and suggested that it was a loose addition compound of benzene with cesium.

We (33) find that benzene reacts with cesium or Cs–K–Na alloy in THF at −70°C to give a good yield of black precipitate of formula C_6H_6Cs whose ESR spectrum indicates formation of a radical in a doublet state. Hydrolysis of the black precipitate gives primarily 1,4-dihydrobenzene (and benzene); hence the black precipitate is evidently cesium benzenide (113). Warming the black precipitate to −20°C, or,

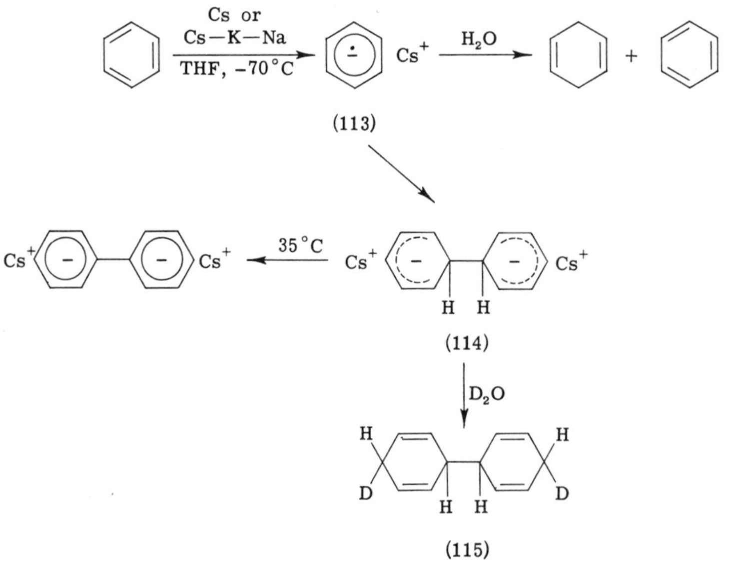

(113)

(114)

(115)

better, carrying out the reaction in THF with Cs–K–Na at about −45°C (76% yield), gives a yellow precipitate 114 (of greatly reduced radical concentration by ESR) which upon decomposition with D_2O gives primarily 115. Thus the reaction of benzene with cesium closely parallels that of 1,1,1-triphenylethane. Warming of 114 to 35°C gives dicesium biphenylide, which is of the same empirical composition as phenylcesium and presumably was the compound prepared by Hackspill.

Similar reactions occur with toluene (35), which with Cs–K–Na alloy in THF at about −45°C gives 116 which with water yields 46% of 117. Dehydrogenation of 117 with 2,3-dichloro-4,5-dicyanobenzoquinone (DDQ) gives dimethylbiphenyl, which contains more than 99% of 3,3'-dimethylbiphenyl 118.

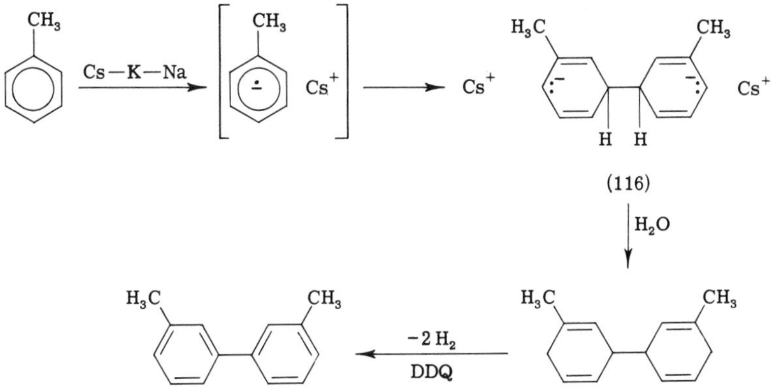

It is notable that in all the above reactions of Cs–K and Cs–K–Na alloys it is cesium which is the metal which reacts, not sodium or potassium. This fact has been established (35) by starting with known amounts of alloy of known composition, carrying out reaction, and then separating some of the unreacted alloy and measuring its composition. The final composition of the alloy agreed with that calculated on the assumption that cesium alone supplied the alkali metal to give the amount and type of reduction products observed.

The reaction of benzene with cesium and cesium alloys to form cesium benzenide is remarkable. In contrast benzene in 0.01 M solution in 2:1 by volume of THF and 1,2-dimethoxyethane with Na–K alloy according to ESR analysis gave (59) concentrations of radical anion at equilibrium of 10^{-6} to $10^{-5} M$ as the temperature decreased from $-20°$ to $-83°C$. The superior reducing power of cesium and its alloys was perhaps to be anticipated in view of the superior reducing power of cesium over potassium in aqueous solution and the appreciably lower ionization potential of cesium compared to potassium in the gas phase. These properties will be influenced by differential solvation of potassium and cesium ions by tetrahydrofuran and by the nature of the ion pairs produced. For 9-fluorenyl salts the fraction of solvent-separated ion pairs has been shown (52) to decrease rapidly in the order Li > Na > K > Cs and is a sensitive function of the solvating power of the medium. The cesium salt of fluorene in THF at $-70°C$ has been shown to exist essentially entirely as contact ion pairs whereas the sodium and lithium salts were completely solvent-separated. The reluctance of cesium cations to become solvent-separated from counteranions means that cesium ions are available for strong electrostatic interaction with anions.

The ionic radius of cesium (1.69 Å) is near that of benzene (1.39 Å from center to the carbon atoms or 2.47 Å to the hydrogen atoms) and the cyclohexa-1,3-dienyl radical; hence extensive electrostatic interaction is anticipated between the corresponding anions on intimate contact with cesium cations. Such similarity in size may facilitate close packing in the crystal lattice; cesium benzenide appears to be very insoluble in tetrahydrofuran.

V

MIGRATION OF PHENYL IN COMPETITION WITH CLEAVAGE OF BENZYL

The observations on aryl migrations in organometallic compounds of the alkali metals which have so far been discussed have largely been of a qualitative nature. What is required for a more detailed description of reaction mechanism is good quantitative data, especially a study of the kinetics of rearrangement. This data is difficult to obtain. Organoalkali compounds combine readily with oxygen, water, various impurities which are apt to be in solvents, and, most important of all, with ethereal solvents themselves. There are no known solvents for n-alkylalkali compounds of sodium, potassium, and cesium with which the organoalkali compounds do not react; organolithium compounds are much less reactive and are commercially available in hydrocarbon solvents and, for more stable lithium compounds, even in diethyl ether. Indeed, it is not known whether or not alkyl compounds of sodium through cesium are soluble in ethers because of the rapidity of combination of these compounds with ethers and the difficulty of distinguishing very fine dispersion of such compounds as produced in usual preparations from true solutions. In contrast, the products of most of the carbanion rearrangements discussed previously have a delocalized negative charge, are much more stable, and are clearly soluble in solvents such as tetrahydrofuran. Another problem is that the rearrangements of organoalkali compounds of sodium, potassium, and cesium are so rapid that the usual simple kinetic techniques cannot be employed. It follows from what has been said that only the kinetics of rearrangement of organolithium compounds appears to be amenable to direct study.

A less arduous approach is to determine relative rate constants. This is possible for the rearrangement of 2,2,3-triphenylpropylalkali compounds because these compounds have been found to undergo two competing reactions—[1,2] migration of the phenyl group and cleavage of benzyl, generally followed by readdition of benzyl for an overall [1,2] migration of benzyl (see Scheme 14). That migration of benzyl occurs by an elimination–readdition mechanism was established by carrying out the rearrangement of 2,2,3-triphenylpropyllithium in presence of [α-^{14}C]benzyllithium; the resulting 2,2,3-triphenylpropyllithium had a radioactivity which agreed with that calculated for addition of benzyllithium of an activity corresponding exactly to that in the prevailing solution (*39, 40*). The addition of benzyllithium to 1,1-diphenylethene was shown to be irreversible under the conditions of the rearrangement. The intermediate 1,1-diphenylethene can be intercepted by other reagents; rearrangement of **119** in the presence of isopropyllithium gave 3-methyl-1,1-diphenylbutyllithium. Reaction of 2,2,3-triphenylpropyl chloride at −75°C with cesium or potassium in THF in the presence of 18-crown-6 gave high yields of 1,1-diphenylethyl anion (**123**) doubtless because of reduction of the intermediate 1,1-diphenylethene by the powerful reducing medium (*15, 44, 62*).

The results of rearrangement of various 2,2,3-triphenylpropyl alkali metal compounds (*44*) are summarized in Table I, in which "phenyl migration" refers to the relative mole percentage of compound **122** and "benzyl cleavage" to the combined relative mole percentages of **120** and **123**. Whereas 2,2,3-triphenylpropyllithium does not rearrange at an

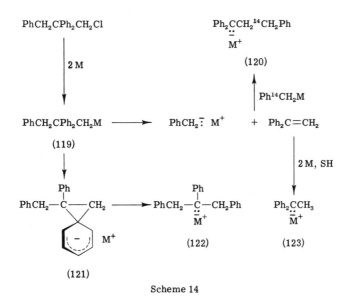

Scheme 14

TABLE I

REARRANGEMENTS OF 2,2,3-TRIPHENYLPROPYL ALKALI METAL COMPOUNDS

Conditions	Temperature (°C)	Phenyl migration (%)	Benzyl cleavage (%)
$PhCH_2CPh_2CH_2Li$			
7 Hr, THF	−75	0	0
2(18-Crown-6), 3.3 hr, THF	−75	0	0
30 Min, THF	0	0	100
3 Hr, Et_2O	+35	100	0
2 NaO-*t*-Bu, THF	−75	0	100
2 KO-*t*-Bu, THF	−75	37	63
2 CsO-*t*-Bu, THF	−75	72	28
2 KO-*t*-Bu·18-crown-6, THF	−75	0	100
2 CsO-*t*-Bu·18-crown-6, THF	−75	0	100
$PhCH_2CPh_2CH_2Cl$			
K, THF	+65	90	10
Cs, THF	+65	96	4
Cs, THF	−75	67	33
18-Crown-6, K, THF	−75	0	100
18-Crown-6, Cs, THF	−75	<8	>92

appreciable rate in THF at −75°C even in presence of 18-crown-6, rearrangement is nearly complete at this temperature in 30 minutes in the presence of the alkali metal alkoxides listed in Table I, doubtless because of the following metathetical reaction:

$$PhCH_2CPh_2CH_2Li + MO\text{-}t\text{-}Bu \rightarrow PhCH_2CPh_2CH_2M + LiO\text{-}t\text{-}Bu$$

Alternatively, the potassium and cesium compounds were prepared by direct reaction of 2,2,3-triphenylpropyl chloride with potassium or cesium.

Examination of Table I reveals that the following systems undergo reaction almost exclusively by benzyl cleavage: 2,2,3-triphenylpropyllithium and -sodium at 0° and −75°C, respectively; 2,2,3-triphenylpropyl potassium and -cesium 18-crown-6 complexes in THF at −75°C. The following undergo rearrangement almost exclusively by phenyl migration: 2,2,3-triphenylpropyllithium in ethyl ether at 35°C; 2,2,3-triphenylpropylpotassium and -cesium in THF at 65°C. The other systems react in comparable amounts by the two processes. Stated otherwise benzyl cleavage is favored by low temperatures, good solvents for solvating cations, coordination of cations by 18-crown-6, and Li^+ and Na^+ as opposed to K^+ and Cs^+; phenyl migration is favored by high temperatures, poorly solvating solvents, absence of good ligands such as 18-crown-6, and large rather than small alkali metal cations. The conditions that favor benzyl cleavage are those that favor loose or separated ion pairs, whereas those that favor phenyl migration are those that favor tight or contact ion pairs (*77, 83*). It appears, therefore, that benzyl cleavage occurs in a transition state resembling a loose ion pair whereas phenyl migration occurs in a transition state resembling a tight ion pair.

These observations may be rationalized (*44*) on the basis that benzyl cleavage occurs in the usual *antiperiplanar* conformation **124** which gives rise to a delocalized benzyl anion separated from the countercation by a molecule of 1,1-diphenylethene; clearly the negative charge is being greatly separated from the cation in this transition state, and the electrostatic work required for such charge separation is greatly reduced by solvation of the cation as in a loose ion pair. In contrast for the transition state leading to the spiro anion **121**, the alkali metal cation in the tight ion pair may accompany the anionic center during bridging (see sketch **125**) so that no great charge separation is necessary at any time

(124)   (125)

for the cyclization. Hence phenyl migration does not benefit as much from solvation of the cation as does benzyl cleavage. During phenyl migration the alkali metal cation exchanges an alkyl anion for a cyclohexadienyl anion as a ligand. Potassium and cesium have ionic radii (1.33 and 1.69 Å, respectively) of a size that appears to be about ideal for strong symmetrical interaction with the π-electron cloud of benzene (1.39 Å from the center to the carbon atoms) and the derived cyclohexadienyl anion.

In a related study, 2,2-diphenyl-4-pentenyl alkali metal compounds (*31*) have been found to undergo phenyl migration in competition with allyl migration. While the mechanistic details of allyl migration have subtleties that need not concern us here, it is of interest that the experimental conditions that favor allyl versus phenyl migration are entirely similar to those that favor benzyl versus phenyl migration in 2,2,3-triphenylpropyl alkali metal compounds. Hence allyl migration occurs in a transition state solvated as in a loose ion pair whereas phenyl migration occurs in a transition state resembling a tight ion pair.

VI

CONCLUSIONS

Some progress has been made in understanding aryl migrations in organometallic compounds of the alkali metals, but much remains to be clarified. Thus organoalkali compounds normally exist as associated compounds; ethyllithium and *n*-butyllithium are hexamers, and *tert*-butyllithium is a tetramer in benzene solution (4). In ethyl ether *n*-butyllithium exists as a tetramer (90). Although reaction with solvent precludes colligative measurements upon *n*-butyllithium in THF at room temperature, rate studies (6) upon *n*-butyllithium with triphenylmethane and with triphenylethylene in THF at −30°C give the order of reaction in *n*-butyllithium as 0.5. From such studies it has been concluded that the dominant form of *n*-butyllithium is the dimer [however, see ref. (90)] while only the small amount of monomer in equilibrium with the dimer reacted with triphenylmethane or with triphenylethylene.

Because of the complications of association and various degrees of solvation, all the equilibria below need to be considered in discussion of the mechanism of rearrangement of organoalkali compounds:

$$\frac{1}{n}(RM)_n \rightleftharpoons R:^-, M^+ \rightleftharpoons R:^- \| M^+ \rightleftharpoons R:^- + M^+$$

where $(RM)_n$ refers to the associated form(s); $R:^-, M^+$ to the tight ion pair; $R:^- \| M^+$ to the loose ion pair; and $R:^- + M^+$ to free ions. In absence of information to the contrary and for simplicity, we have assumed that rearrangements of organometallic compounds of the alkali metals occur in monomeric forms, just as has been proposed for reaction of *n*-butyllithium with triphenylmethane or triphenylethylene. Since free ions and loose ion pairs have similar degrees of solvation, the studies upon competitive reactions that were mentioned in the preceding section are also in accord with benzyl cleavage occurring in a free anion; however, because the concentration of such anions must be very small, it is reasonable to assume that it is predominantly the loose ion pair that undergoes benzyl cleavage.

A generalization worthy of note is that loose ion pairs of arylalkylalkali compounds, which cannot undergo a ready side reaction such as β-elimination of benzyl or allyl anion, most commonly undergo reaction with the ethereal solvent (protonation or alkylation) rather than [1,*n*] migration of aryl. This generalization may be rationalized on the basis that in the loose ion pair the negatively charged end of the arylalkyl anion remains in close proximity to the cation, separated from the cation by an intervening solvent molecule, which serves as one of the ligands of the cation. What then occurs is that the "naked" anion rapidly combines with the intervening solvent molecule whose reactivity is enhanced by coordination with the cation. Only in the case of anions stabilized by substituents, e.g., 2,2,2-triphenylethyl anion from 2,2,2-triphenylethyllithium in THF, does the naked anion in the loose ion pair survive long enough to undergo [1,*n*] migration of aryl. It therefore follows that frequently the most interesting chemistry, such as [1,*n*] migrations of aryl or related spiro cyclizations, can only be observed in tight ion pairs wherein the countercation shields the anion from interaction with the solvent. It is for this reason, among many others, that the organometallic chemistry of potassium, rubidium, and cesium needs to be extended from its current state of infancy and holds promise of providing many new useful reactions.

ACKNOWLEDGMENTS

The author wishes to acknowledge the many contributions of current and former students and research associates who did so much of the work described herein. Support of the more recent work, especially the unpublished experiments, by the National Science Foundation is gratefully acknowledged. The author is indebted to the American Chemical Society for permission to adapt the following reaction schemes to this article: Schemes 1, 4, 6–8, 14.

REFERENCES

1. Bates, R. B., Brenner, S., Cole, C. M., Davidson, E. W., Forsythe, G. C., McCombs, D. A., and Roth, A. S., *J. Am. Chem. Soc.* **95**, 926 (1973).
2. Bates, R. B., Gosselink, D. W., and Kaczynski, J. A., *Tetrahedron Lett.* 199, 205 (1967).
3. Bertrand, J. A., Grovenstein, E., Jr., Lu, P.-C., and VanDerveer, D., *J. Am. Chem. Soc.* **98**, 7835 (1976).
4. Brown, T. L., *Advan. Organomet. Chem.* **3**, 365 (1965).
5. Buncel, E., "Carbanions. Mechanistic and Isotopic Aspects," see especially Chapters 4–6. Elsevier, New York, 1975.
6. Carpenter, J. G., Evans, A. G., and Rees, N. H., *J. Chem. Soc., Perkin Trans. 2* 1598 (1972).
7. Charlton, J. C., Dostrovsky, I., and Hughes, E. D., *Nature (London)* **167**, 986 (1961).
8. Clusius, K., and Mollet, H., *Helv. Chim. Acta* **39**, 363 (1956).
9. Cone, L. H., and Robinson, C. S., *Chem. Ber.* **40**, 2160 (1907).
10. Cram, D. J., "Fundamentals of Carbanion Chemistry," pp. 176–256. Academic Press, New York, 1965.
11. Cram, D. J., and Dalton, C. K., *J. Am. Chem. Soc.* **85**, 1268 (1963).
12. Crimmins, T. F., Murphy, W. S., and Hauser, C. R., *J. Org. Chem.* **31**, 4273 (1966).
13. de Postis, J., *Proc. Int. Congr. Pure Appl. Chem., 11th* **5**, 867 (1947).
14. Dodd, J. R., Winton, R. F., Pagni, R. M., Watson, C. R., Jr., and Bloor, J., *J. Am. Chem. Soc.* **96**, 7846 (1974).

15. Dye, J. L., DeBacker, M. G., and Nicely, V. A., *J. Am. Chem. Soc.* **92**, 5226 (1970).
16. Eberson, L., *Acta Chem. Scand.* **22**, 3045 (1968).
17. Eisch, J. J., *Ind. Eng. Chem., Prod. Res. Develop.* **14**, 11 (1975).
18. Eisch, J. J., and Kovacs, C. A., *J. Organomet. Chem.* **25**, C33 (1970).
19. Eisch, J. J., and Kovacs., *J. Organomet. Chem.* **30**, C97 (1971).
20. Fischer, H. P., Kaplan, E., and Neuenschwander, P., *Chimia* **22**, 338 (1968).
21. Fraenkel, G., and Cooper, J. W., *J. Am. Chem. Soc.* **93**, 7228 (1971).
22. Fraenkel, G., Ho, C. C., Liang, Y., and Yu, S., *J. Am. Chem. Soc.* **94**, 4732 (1972).
23. Grovenstein, E., Jr., Southwide Chem. Conf., Atlanta, Georgia, Oct. 16, 1950; The Filter Press (Georgia Section of the Am. Chem. Soc.), V. No. 8, 12 (1950).
24. Grovenstein, E., Jr., *J. Am. Chem. Soc.* **79**, 4985 (1957).
25. Grovenstein, E., Jr., and Akabori, S., *J. Am. Chem. Soc.* **97**, 4620 (1975).
26. Grovenstein, E., Jr., Akabori, S., and Rhee, J.-U., *J. Am. Chem. Soc.* **94**, 4734 (1972).
27. Grovenstein, E., Jr., Beres, J. A., Cheng, Y.-M., and Pegolotti, J. A., *J. Org. Chem.* **37**, 1281 (1972).
28. Grovenstein, E., Jr., Blanchard, E. P., Jr., Gordon, D. A., and Stevenson, R. W., *J. Am. Chem. Soc.* **81**, 4842 (1959).
29. Grovenstein, E., Jr., and Cheng, Y.-M., *Chem. Commun.* 101 (1970).
30. Grovenstein, E., Jr., and Cheng, Y.-M., *J. Am. Chem. Soc.* **94**, 4971 (1972).
31. Grovenstein, E., Jr., and Cottingham, A. B., unpublished results (1975).
32. Grovenstein, E., Jr., and Longfield, T. H., unpublished research (1973).
33. Grovenstein, E., Jr., Longfield, T. H., and Quest, D. E., unpublished results (1976).
34. Grovenstein, E., Jr., and Lu, P.-C., unpublished results (1976).
35. Grovenstein, E., Jr., and Quest, D. E., unpublished results (1976).
36. Grovenstein, E., Jr., and Rhee, J.-U., *J. Am. Chem. Soc.* **97**, 769 (1975).
37. Grovenstein, E., Jr., and Rogers, L. C., *J. Am. Chem. Soc.* **86**, 854 (1964).
38. Grovenstein, E., Jr., and Stevenson, R. W., *J. Am. Chem. Soc.* **81**, 4850 (1959).
39. Grovenstein, E., Jr., and Wentworth, G., *J. Am. Chem. Soc.* **85**, 3305 (1963).
40. Grovenstein, E., Jr., and Wentworth, G., *J. Am. Chem. Soc.* **89**, 1852 (1967).
41. Grovenstein, E., Jr., and Wentworth, G., *J. Am. Chem. Soc.* **89**, 2348 (1967).
42. Grovenstein, E., Jr., and Williams, L. P., Jr., *J. Am. Chem. Soc.* **83**, 412 (1961).
43. Grovenstein, E., Jr., and Williams, L. P., Jr., *J. Am. Chem. Soc.* **83**, 2537 (1961).
44. Grovenstein, E., Jr., and Williamson, R. E., *J. Am. Chem. Soc.* **97**, 646 (1975).
45. Grovenstein, E., Jr., and Williamson, R. E., unpublished experiments (1975).
46. Hackspill, L., *Proc. Int. Congr. Appl. Chem., 8th* **2**, 113 (1912); *Ann. Chim Phys.* **28**, 653 (1913).
47. Hackspill, L., *Helv. Chim. Acta* **11**, 1026 (1928).
48. Hackspill, L., in "Nouveau Traite de Chemie Minérale" (P. Pascal, ed.), Vol. 3, p. 125. Masson, Paris, 1956.
49. Hammond, G. S., in "Steric Effects in Organic Chemistry" (M. S. Newman, ed.), pp. 460–469. Wiley, New York, 1956.
50. Heaney, H., and Ward, T. J., *Chem. Commun.* 810 (1969).
51. Hill, E. A., *J. Organomet. Chem.* **91**, 123–271 (1975).
52. Hogen-Esch, T. E., and Smid, J., *J. Am. Chem. Soc.* **88**, 307, 318 (1966).
53. Hughes, E. D., *Bull. Soc. Chim. Fr.* [5] **18**, C.41 (1951).
54. Hunter, D. H., in "Isotopes in Organic Chemistry" (E. Buncel and C. C. Lee, eds.), pp. 135–229. Elsevier, New York, 1975.
55. Julia, M., and Malassiné, B., *Tetrahedron Lett.* 2495 (1972).
56. Kharasch, M. S., and Reinmuth, O., "Grignard Reactions of Nonmetallic Substances," pp. 58–68. Prentice-Hall, New York, 1954.

57. Kloosterziel, H., and van Drunen, J. A. A., *Rec. Trav. Chim. Pays-Bas* **89**, 368 (1970).
58. Koelsch, C. F., *J. Am. Chem. Soc.* **55**, 3394 (1933); **56**, 480, 1605 (1934).
59. Kooser, R. G., Volland, W. V., and Freed, J. H., *J. Chem. Phys.* **50**, 5243 (1969).
60. Lepley, A. R., *in* "Chemically Induced Magnetic Polarization" (A. R. Lepley and G. L. Closs, eds.), pp. 323-384. Wiley (Interscience), New York, 1973.
61. Lepley, A. R., and Giumanini, A. G., *in* "Mechanisms of Molecular Migrations," (B. S. Thyagarajan, ed.), Vol. 3, pp. 297-440. Wiley (Interscience), New York, 1971.
62. Loc, M. T., Tehan, F. J., and Dye, J. L., *J. Phys. Chem.* **76**, 2975 (1972).
63. Maercker, A., Güthlein, P., and Wittmayr, H., *Angew. Chem., Int. Ed. Engl.* **12**, 774 (1973).
64. Miller, L. M., and Boyer, R. F., *J. Am. Chem. Soc.* **93**, 646 (1971).
65. Morton, A. A., and Lanpher, E. J., *J. Org. Chem.* **21**, 93 (1956).
65a. Morton, A. A., and Redman, L. S., *Ind. Eng. Chem.* **40**, 1190 (1948).
66. Paul, R., and Tchelitcheff, S., *C. R. Acad. Sci. Paris* **239**, 1222 (1954).
67. Pine, S. H., *Org. React.* **18**, 403-464 (1970).
68. Pines, H., *Acc. Chem. Res.* **7**, 155 (1974).
69. Pines, H., and Mark, V., *J. Am. Chem. Soc.* **78**, 4316 (1956).
70. Pines, H., and Schaap, L., *J. Am. Chem. Soc.* **80**, 4378 (1958).
71. Pines, H., Vesely, J. A., and Ipatieff, V. N., *J. Am. Chem. Soc.* **77**, 554 (1955).
72. Rüchardt, C., and Hecht, R., *Tetrahedron Lett.* 961 (1962).
73. Rüchardt, C., and Trautwein, H., *Chem. Ber.* **95**, 1197 (1962).
74. Schaap, L., and Pines, H., *J. Am. Chem. Soc.* **79**, 4967 (1957).
75. Schlenk, W., and Bergmann, E., *Justus Liebigs Ann. Chem.* **463**, 125 (1928).
76. Schöllkopf, U., *Angew. Chem., Int. Ed. Eng.* **9**, 763 (1970).
77. Smid, J., *in* "Ions and Ion Pairs in Organic Reactions" (M. Szwarc, ed.), Vol. 1, pp. 85-151. Wiley(Interscience), New York, 1972.
78. Staley, S. W., Cramer, G. M., and Kingsley, W. G., *J. Am. Chem. Soc.* **95**, 5052 (1973).
79. Staley, S. W., and Erdman, J. P., *J. Am. Chem. Soc.* **92**, 3832 (1970).
80. Stevens, T. S., and Watts, W. E., "Selected Molecular Rearrangements," pp. 81-118. Van Nostrand-Reinhold, London, 1973.
81. Stipanovic, B., and Pines, H., *J. Org. Chem.* **34**, 2106 (1969).
82. Streitwieser, A., Jr., and Lawler, R. G., *J. Am. Chem. Soc.* **85**, 2854 (1963); **87**, 5388 (1965).
83. Szwarc, M., *in* "Ions and Ion Pairs in Organic Reactions" (M. Szwarc, ed.), Wiley (Interscience), Vol. 1, pp. 1-26. New York, 1972.
84. Truce, W. E., and Heuring, D. L., *Chem. Commun.* 1499 (1969).
85. Urry, W. H., and Kharasch, M. S., *J. Am. Chem. Soc.* **66**, 1438 (1944).
86. Waack, R., and Doran, M. A., *J. Am. Chem. Soc.* **91**, 2456 (1969).
87. Walborsky, H. M., and Aronoff, M. S., *J. Organometal. Chem.* **4**, 418 (1965).
88. Walborsky, H. M., and Aronoff, M. S., *J. Organometal. Chem.* **51**, 55 (1973).
89. Walborsky, H. M., and Young, A. E., *J. Am. Chem. Soc.* **86**, 3288 (1964).
90. West, P., and Waack, R., *J. Am. Chem. Soc.* **89**, 4395 (1967).
91. West, R., Stewart, H. F., and Husk, G. R., *J. Am. Chem. Soc.* **89**, 5050 (1967).
92. Winstein, S., Lindegren, C. R., Marshall, H., and Ingraham, L. I., *J. Am. Chem. Soc.* **75**, 147 (1953).
93. Winstein, S., and Seubold, F. H., Jr., *J. Am. Chem. Soc.* **69**, 2916 (1947).
94. Wooster, C. B., and Mitchell, N. W., *J. Am. Chem. Soc.* **52**, 1042 (1930).
95. Wooster, C. B., and Morse, R. A., *J. Am. Chem. Soc.* **56**, 1735 (1934).
96. Wooster, C. B., and Ryan, J. F., *J. Am. Chem. Soc.* **54**, 2419 (1932).

97. Ziegler, K., and Crössman, F., *Chem. Ber.* **62**, 1768 (1929).
98. Zimmerman, H. E., *in* "Molecular Rearrangements," (P. de Mayo, ed.), Vol. 1, pp. 345–406. Interscience, New York, 1963.
99. Zimmerman, H. E., and Smentowski, F. J., *J. Am. Chem. Soc.* **79**, 5455 (1957).
100. Zimmerman, H. E., and Zweig, A., *J. Am. Chem. Soc.* **83**, 1196 (1961).

SUPPLEMENTARY REFERENCES

Since this article was written, the following papers have appeared.

Dagan, A., and Rabinovitz, M., *J. Am. Chem. Soc.* **98**, 8268 (1976). [1,4]Migration of *o*-biphenylyl; Formation of dibenzo[*cd,gh*]pentalenide dianion by a novel base-catalyzed rearrangement.

Grovenstein, E., Jr., and Cottingham, A. B., *J. Am. Chem. Soc.* **99**, 1881 (1977). Rearrangements of 2,2-diphenyl-4-pentenyl alkali metal compounds.

Grovenstein, E., Jr., Longfield, T. H., and Quest, D. E., *J. Am. Chem. Soc.* **99**, 2800 (1977). Reactions of cesium or cesium-potassium-sodium alloy with benzene and toluene.

Fluxional and Nonrigid Behavior of Transition Metal Organometallic π-Complexes

J. W. FALLER

Department of Chemistry
Yale University
New Haven, Connecticut

I

INTRODUCTION

The stereochemistry of products derived from reactions of coordinated olefins and the stereochemistry of polymers formed in reactions catalyzed by transition metals are ultimately determined by the conformational stability of π-complexed intermediates. For example, the cis:trans ratios and the relative amounts of 1,2- versus 1,4-polymer units obtained in diene polymerization are determined by stabilities of syn and anti isomers of π-allyls and the relative stabilities of various orientations of substituted olefins bound to metals (105). The interconversion rates of these isomers and the thermodynamic preferences of olefin–metal conformations should explain observed product distributions and provide a rational basis for catalyst design.

Molecules that undergo intramolecular rearrangements at rates that influence nuclear magnetic resonance (NMR) line shapes in accessible temperature ranges are termed *stereochemically nonrigid* (31). Nonrigid molecules, of which all observable interconverting species are chemically and structurally equivalent, are designated as *fluxional*.

Fluxional and other stereochemically nonrigid organometallic molecules were the subject of extensive reviews by Cotton *et al.* (24, 31) and Vrieze (106); these reviews cover the literature through mid-1973 (70).

Cotton's personal account of the development of this area appeared in the golden issue (Vol. 100) of the *Journal of Organometallic Chemistry* (*32*).

The study of nonrigid systems has closely followed the technical advance of NMR instrumentation. In the last 5 years, the rapid advances in ^{13}C spectroscopy have made previously undetectable carbonyl rearrangements, conformational interconversions, and fluxionality accessible. Carbonyl rearrangements will be reviewed by J. Takats in a future volume of this series; major developments in the understanding of nonrigidity in π-complexes are discussed here.

II

DEVELOPMENTS IN INSTRUMENTATION AND TECHNIQUE

The rapid developments in this field make a comprehensive review impractical; hence, this report will concentrate on recent developments in technique that are likely to be used more extensively, potential problems in interpretation of dynamic NMR spectra, and observations that establish general patterns of rearrangement or conflict with currently accepted mechanisms.

The measurement of line broadening and coalescence of resonances has been the dominant mode of analyzing kinetic data in dynamic NMR spectra. In addition, detailed analyses of relative broadenings of resonances pioneered at the Massachusetts Institute of Technology and Yale (*9, 112, 102*) have allowed the elucidation of mechanistic pathways. Calculation of line shapes has allowed the determination of exchange rates as a function of temperature and, therefore, of the activation parameters associated with the rearrangements. Since these rearrangements are intramolecular, significant activation entropies are not observed, and thus in carefully executed experiments log A generally lies between 12.3 and 13.5. A value outside of this range can generally be attributed to poor data rather than to an intrinsic anomaly in entropy factors. Since the determination of accurate activation parameters from line shapes as a function of temperature is a tedious procedure (see below), it is often far more practical to measure one rate accurately at a known temperature and to utilize absolute rate theory to evaluate ΔG^*. As the primary goal in most studies is the determination of mechanism, the variations in rates reflected by differences in ΔG^* are adequate for most purposes.

Although computer calculations of line shapes allow detailed matching of spectra (see Fig. 1), many systems can be evaluated utilizing simple two-site exchange formulas. Thus, for two sites with equal population,

the first-order rate constant, k, for leaving a given site may be determined from the initial broadening in excess of the natural linewidth, W,
by

$$k = \pi W \tag{1}$$

and after coalescence for a shift difference, $\delta\nu$, by

$$k = \pi(\delta\nu)^2/2W \tag{2}$$

Since the chemical shift difference at the temperature of the experiment

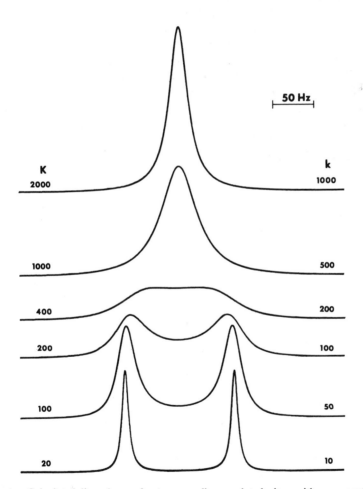

FIG. 1.   Calculated line shapes for two equally populated sites with a separation in
resonance frequency of 100 Hz. Exchange probabilities are $p_{12} = p_{21} = 0.5$. The rate
constant (sec^{-1}) for leaving each site is $k_{12} = k_{21} = 0.5\ K$, where $K =$ the overall rate
constant or τ^{-1}. Natural line widths of 4 Hz were assumed.

is required in Eq. (2), large errors are often incurred owing to variations of chemical shift with temperature. This is particularly the case when the chemical shift differences are small (<25 Hz). Hence, to be accurate, measurements must extrapolate chemical shift variations into the region of interest. Sources of error in line-shape calculations have been reviewed by Allerhand *et al.* (2).

These equations provide an insight into many problems, but quantitative experiments require that the equations not be applied beyond the limit of the conditions for which they were derived.

For example, Eq. (1) is viable before coalescence for unequal populations, but Eq. (2) is valid only for equal populations. Neglect of spin-spin coupling between exchanging sites can lead to significant errors. Another common error results from the use of τ as defined in the original Holm–Gutowsky papers, in which the rate constant $1/\tau$ differs by a factor of 2 from the rate constant for leaving a given site.

Reference to Eq. (2) indicates that a greater separation of resonances will require a higher temperature for complete averaging. Thus, for the same E_a, closely spaced resonances will average at lower temperatures, even though all broaden equally at the initial stages (see Fig. 2).

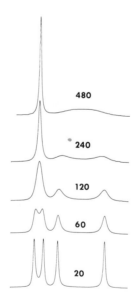

FIG. 2. A comparison of coalescence and approach to limiting spectra for equal exchange rates between two closely spaced resonances ($\delta\nu = 20$ Hz) and two more widely spaced resonances ($\delta\nu = 100$ Hz). The overall rate constant or τ^{-1} is given in sec^{-1}, and natural line widths of 2 Hz were assumed.

It is also informative to examine the effect of unequal populations on the line shape as shown in Fig. 3. The maximum amount of broadening observed for the dominant resonance is determined by the relative populations (the greater the minor, the broader) and the separations in Hertz (the greater $\delta\nu$, the broader).

Although analytical solutions exist for many simple systems, a serious venture into dynamic NMR usually utilizes computer simulation of NMR data (*13, 14, 95, 101, 102, 112*) and, for the most part, algebraic expressions in terms of coalescence temperatures are no longer used. Nevertheless, the users of packaged programs should recognize the significant distinction between the reciprocal lifetime, τ^{-1}, and the rate constant for leaving each site, k, which most often corresponds to a chemist's intuitive notion of the rate for the process. Therefore, the user must relate the "rate," "preexchange lifetime," or τ, in the input to the program to the rate for an elementary process, such as the migration of some group, which corresponds to the conventional rate for the reaction. Neglect of these relationships produces errors of factors of 2 in equal-population cases. The relationships between the transfer probabilities and rates have been outlined most clearly by Johnson and Moreland (*72*).

Most modern programs are based on the Kubo–Sack matrix approach as modified by Gordon and McGinnis (*63, 71, 72, 100*). This method casts the problem in the form of matrices representing transition

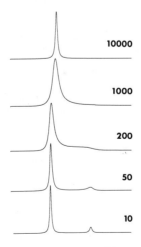

FIG. 3. Exchange between sites ($\delta\nu = 100$ Hz) with a population ratio of 85:15 and exchange probabilities of $p_{12} = 0.15$ and $p_{21} = 0.85$. The overall rate constant or τ^{-1} is given in sec^{-1}, and natural line widths of 4 Hz were assumed.

probabilities and relative populations. Thus, for populations in the ratio 85:15, as shown in Fig. 3, the relative probability of migrating from site 1 to 2, p_{12} is 0.15; whereas that for the reverse direction p_{21}, is 0.85. To preserve the equilibrium populations, the rate from 1 to 2 must be equal to that from 2 to 1. The rate constant, k, is generally given as $\tau^{-1} \cdot p_{ij}$ or $K \cdot p_{ij}$. Hence, the mass transfer balance can be checked by $K \cdot p_{ij}P_i = K \cdot p_{ji}P_j$. In this case $K \cdot 0.15 \cdot 0.85 = K \cdot 0.85 \cdot 0.15$. These problems are invariably written in terms of lifetimes or overall rates, and it is up to the user to deconvolute these values into the rates for leaving a particular site and the rate of the reaction. Thus, in the initial phases of resonance broadening on raising the temperature, the relative amount of broadening in site 1 is 5.7 times greater than that in site 2, as one may determine by use of Eq. (1).

The discussion above is indicative of the most common sources of error in computation, but reference to specific literature will provide a more extensive discussion of errors (2, 11–13, 71, 72). It is essential, however, that published studies of rates be accompanied by a description of the relationship between τ and the probabilities in the transfer matrix.

If one assumes that there is a single isomer in solution, a minor component can produce anomalous broadening due to averaging. A minor component might escape detection because it will be broader when one is near the "limiting low-temperature spectrum." This effect can be observed in the $K = 50$ sec^{-1} resonances in Fig. 3, but is much more pronounced if the minor component is less than 3%. In these situations the minor component may not appear until the temperature is lowered another 15°C, even though the "limiting spectrum" has apparently been reached from the point of view of the major component.

Furthermore, there are situations where high concentrations of a second isomer can be detected by infrared (IR) at room temperature, but the equilibrium constant changes as the temperature is lowered such that the minor component can no longer be detected in the limiting low-temperature spectrum (46).

The principal problems arise from underestimating the complexity of the systems under consideration. Generally one is tempted to assume that only one process or one rearrangement pathway is responsible for a dynamic NMR spectrum. Multiple pathways were found early in the study of organometallic systems (25), but often were separated by significant differences in activation energy. Furthermore, studies of fluxional molecules have generally assumed only one structure or conformation to be present. These simplifications have often sufficed; nevertheless, they have proved inadequate often enough that one should take care to avoid the pitfalls of oversimplification. The best rule of thumb is that calculated spectra should accurately reproduce the observed spectra, and deviations, particularly if they are different for each resonance, indicate a more complex situation. That is, the correct interpretation will explain all the observations, and neglect of apparently minor deviations may lead to misinterpretation.

The simple notion that in the initial stages of broadening, the *line widths will be broader in resonances from which nuclei leave faster* allows one to qualitatively predict mechanisms of reaction. This is illustrated for the rearrangements of $C_8H_8Ru(CO)_3$, where it is clear that a 1,2-shift interchanges environments in the following manner (*16*).

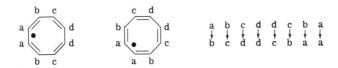

This leads to the expectation that the b and c resonances should broaden more than the a and d resonances in the slow-exchange region of the proton spectrum. This was observed qualitatively in the proton spectra of $C_8H_8Ru(CO)_3$ (*16*) and $C_8H_8Os(CO)_3$ (*22*), and thus by analogy a 1,2-shift was presumed in the Fe case (*16*). Owing to the coupling of protons, the broadenings did not lend themselves to quantitative determination of rates. Proton decoupled ^{13}C spectra provided an opportunity to examine line shapes free from the complications of spin–spin coupling, and indeed, the resonances assigned to b and c positions broadened to a greater degree and established the 1,2-shift mechanism (*35*). Thus, the advent of readily obtained ^{13}C spectra provided a ready method for mechanistic studies. The larger chemical shift range of ^{13}C made the differential broadening much more obvious.

Examination of the spectra at high field (67.9 MHz), however, indicates that the two outer resonances do not broaden to the same extent (*61*). This effect, which is shown in Fig. 4, cannot be accounted for by a mixture of processes, i.e., predominant 1,2-shift plus some random, 1,3- or 1,4-shift. It therefore appears to be the result of concurrent averaging with a second conformer that is in low concentration and has resonances slightly displaced from a, b, and c but more greatly displaced from d. The sensitivity of the ^{13}C spectra does not allow ready location of the other resonances although limits can be placed on locations and populations. A further indication of the presence of a second conformer is the failure of carbonyl region line broadening to accurately fit the observed widths (see below). Further experimentation

J. W. FALLER

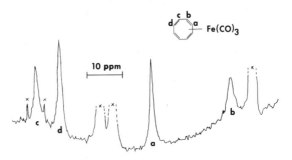

FIG. 4. A 67.9 MHz ^{13}C NMR spectrum of $(C_8H_8)Fe(CO)_3$ at $-115°$ in a 2:3 mixture of CD_2Cl_2–$CHFCl_2$. The region illustrated contains resonances assigned to the carbon nuclei in the cyclooctatetraene ring.

is necessary to establish the nature of the second conformer, but this example demonstrates the advantage of high field spectra.

One can expect the greatest difficulties to arise from low population conformer effects in high-field carbon and phosphorus spectra, where chemical shift differences between conformers may be over 10 ppm (679 Hz for ^{13}C with ^{1}H fields corresponding to 270 MHz). In this instance, averaging with an isomer present to the extent of 1% could produce a broadening of ~12 Hz.

Increased technology allows greater versatility; however, certain difficulties and more pitfalls are encountered. When separations are larger, the coalescence temperature is higher. This can give the impression that two different rates are occurring for the same process (Fig. 2). Further, if two processes occur with significantly different activation free energies, then the effect of going to higher field will be to make the processes overlap to a greater extent. For example, with a 1 ppm shift difference, a ΔG^* of 8 kcal/mole would produce a coalesced line of only 3 Hz in width at $-85°C$ at 60 MHz; but at 270 MHz the resonance would still be broad (57 Hz). Thus, one can anticipate that as higher fields are attained, greater effects from incompletely averaged processes will be observed. Therefore, residual broadenings from hindered rotations about metal–carbon and metal–phosphorus bonds, conformational effects in chelate rings, and rotation of phenyl groups in phosphines may begin to affect low-temperature spectra (55). Situations may therefore arise in which sequential processes with different activation energies may be more readily delineated at lower fields.

Since most modern instrumentation uses the Fourier transform technique, the apparatus is usually available to determine T_1 relaxation times. Initially one might have hoped that in addition to the chemical shift and the spin–spin couplings associated with a particular nucleus, the T_1 might provide a readily interpretable datum that would be useful in characterization. Occasionally T_1 measurements are useful with protons. Quaternary carbons may also be identified by their longer relaxation times; however, the many of effects altering relaxation times do not lend themselves to ready interpretation. Therefore, the introduction of a single dominant relaxation mechanism by the coordination of a paramagnetic metal ion provides a useful method of assigning structure (52, 80).

Conflicting assignments for the predominant isomer in (η^3-allyl)Fe(CO)$_3$X have been given (90, 91, 94), but T_1 measurements provide a straightforward solution to the problem. Coordination to the gadolinium in Gd(fod)$_3$ provides a dominant relaxation path. Thus the nuclei that are closest to the gadolinium ($T_1 \propto r^6$) relax fastest, demonstrating that the endo isomer is most stable (Fig. 5).

Although line-broadening techniques most readily lend themselves to

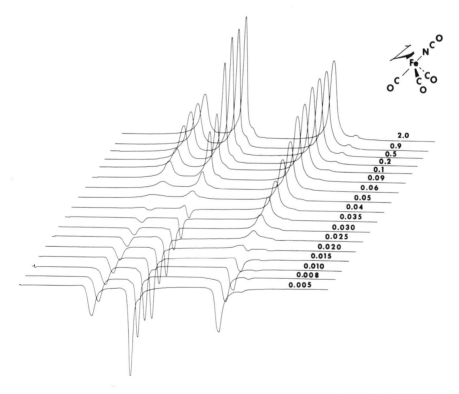

FIG. 5. A 270 MHz ^{1}H 180°-τ-90° inversion recovery determination of T_1 in Gd(fod)$_3$-relaxed (η^3-allyl)Fe(CO)$_3$NCO. The value of τ for each trace is the delay (seconds) between the 180° and 90° pulse. The equilibrium is shifted predominantly toward the endo isomer, so that only the resonances of the central, syn, and anti protons (from left to right) are clearly discernible.

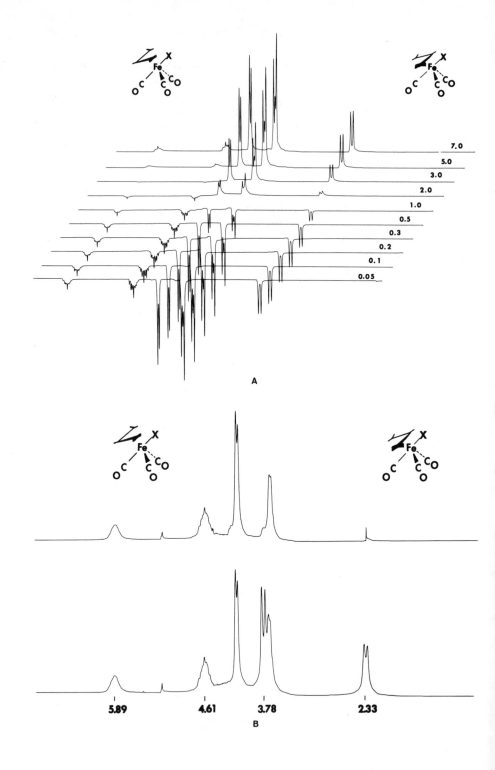

A

B

5.89   4.61   3.78   2.33

mechanism determinations, decomposition at high temperatures some-times thwarts line-shape analysis. Magnetization transfer or spin-saturation transfer techniques, however, often allow site exchanges to be delineated more clearly than line broadening and also may be performed at lower temperatures (50, 68). Pulse spectrometers are particularly well suited to this type of experiment, since they often provide the mechanism for "gated-decoupling." In this experiment nuclei in a given site are saturated, and if they migrate to a new site in a time comparable to that of the new site, saturation will be observed in the new site. If the relaxation times of the sites are known, then the rate of exchange can be determined from the degree of saturation and the T_1 at the site. In general, line-broadening measurements are more accurate, but the saturation transfer technique allows greater delineation of mechanism.

(η^3-Allyl)Fe(CO)$_3$I contains both endo and exo isomers in a ratio of approximately 2:1. These isomers readily interconvert, but raising the temperature above 30°C leads to decomposition and the production of paramagnetic impurities. Relaxation times for the protons in the absence of relaxation reagents show that the T_1's for analogous resonances are comparable for both isomers (Fig. 6A). The slightly shorter time for syn and anti protons can be attributed to the relative proximities of other protons. Saturation of the anti protons of the minor isomer gives rise to saturation in the anti protons in the major isomer (Fig. 6B). This demonstrates that the mechanism for isomer interconversion is a rotation of the allyl group rather than π-σ-π interconversion. Interconversion by the π-σ-π pathway results in interchange of syn and anti protons (see Tsutsui and Courtney, this volume, p. 241).

Magnetization transfer should be quite useful in ^{13}C spectroscopy, where averaging of large chemical shift differences can require extreme temperature variations in order to reach limiting spectra. Gated decoupling capabilities are often not standard items for ^{13}C spectrometers;

Fig. 6. (A) A 270 MHz 180°-τ-90° inversion recovery determination of relaxation times in an endo–exo isomeric mixture of (η^3-allyl)Fe(CO)$_3$I. The value of τ for each trace is the delay (seconds) between the 180° and 90° pulse. The doublets with the greater spacing at 2.33 and 3.78δ are assigned to the anti protons of the exo and endo isomers, respectively. Those at 3.69 and 4.15δ are assigned to the syn resonances. The central proton resonances are at lower field. These traces demonstrate that the relaxation times for the syn and anti protons in both isomers are equal.

(B) The effect of saturating the 2.33δ resonance of the exo isomer at 37°C. Saturation is only transferred from the anti resonance of the exo isomer to the anti resonance of the syn isomer, demonstrating that rotation of the allyl is occurring. The data were collected in a gated decoupling mode; i.e., the 2.33δ resonance was irradiated for 10 seconds, the irradiation was terminated, and after a 0.1-second delay a 90° pulse was applied, followed by accumulation of the FID.

however, rapid multiple pulses of short duration can produce an equivalent effect in saturating a particular resonance. Alternatively, Overhauser enhancement can be transferred by irradiation of specific protons.

At least for the present, proton saturation transfer experiments will be the most tractable. For accurate measurement of the degree of saturation, adequate time must have elapsed for equilibrium to be established. This time is on the order of five times the relaxation time, or k^{-1}. Hence, for the gated decoupling of the protons in $(\eta^3$-allyl$)Fe(CO)_3I$, decoupling must be carried out for 10–20 seconds before the 90° pulse to achieve reproducible results. A second feature, which can often be observed in syn and anti protons, is the signal enhancement arising from the nuclear Overhauser effect (NOE). Thus, the observed intensity is a function of both the NOE and the magnetization transferred into the site. Quantitative determinations must correct for the NOE enhancements.

III

RING WHIZZERS

As summarized by Cotton (31), considerations of line shape have generally led to the conclusion that migration of metals about conjugated rings proceeds via a series of 1,2-shifts. Thus for the $(\eta^1$-C$_5$H$_5)$M complexes listed in Table I, shifts such as those below occur. Similar

TABLE I
DEMONSTRATED 1,2-SHIFTS

Complex	References
$(\eta^1$-C$_5$H$_5)$Fe(CO)$_2(\eta^5$-C$_5$H$_5)$	(9, 29)
$(\eta^1$-C$_5$H$_5)$Ru(CO)$_2(\eta^5$-C$_5$H$_5)$	(26)
$(\eta^1$-C$_5$H$_5)$Ge(CH$_3)_3$	(44)
$(\eta^1$-C$_5$H$_5)$SiCl$_2$CH$_3$	(27, 103)
$(\eta^3$-C$_7$H$_7)$Fe(CO)$(\eta^5$-C$_5$H$_5)$	(20)
$(\eta^3$-C$_7$H$_7)$Mo(CO)$_2(\eta^5$-C$_5$H$_5)$	(46, 8)
$(\eta^3$-C$_7$H$_7)$Co(CO)$_3$	(8)
$(\eta^3$-C$_7$H$_7)$Mo(CO)$_2$[3,5-(CH$_3)_2$C$_3$H$_3$N$_2]_2$BH$_2$	(18, 31)
$(\eta^4$-C$_8$H$_8)$Fe(CO)$_3$	(35)
$(\eta^4$-C$_8$H$_8)$Ru(CO)$_3$	(16)
$[\eta^6$-C$_6$(CO$_2$CH$_3)_6]$Rh$(\eta^5$-C$_5$H$_5)$	(74)
$[\eta^6$-(CH$_3)_4$C$_8$H$_4]$Cr(CO)$_3$	(25)
$[\eta^6$-(CH$_3)_4$C$_8$H$_4]$Mo(CO)$_3$	(25)
$[\eta^6$-(CH$_3)_4$C$_8$H$_4]$W(CO)$_3$	(25)
$(\eta^6$-C$_8$H$_8)[$C$_6$H$_5$CCo$_3$(CO)$_6]$	(97)

rearrangements were noted in η^3-cycloheptatriene, η^4-cyclooctatetraene, and substituted η^6-cyclooctatetraene complexes. Some exceptions might be cited, such as the observations of 1,3-shifts in indenyls; however, these can be attributed to sequential 1,2-shifts in which the o-xylylene intermediate is destabilized (42–44). Following Occam's razor, systems

for which the data are ambiguous, such as $(\eta^1\text{-}C_5H_5)CuL(SO_2)$ (112, 28) or $(\eta^1\text{-}C_5H_5)HgX$ (110), could be assumed to undergo 1,2-shifts until proved otherwise.

The dominance of the 1,2-shift pathway can be attributed to the principle of least motion (67), although orbital symmetry rules (114) have been invoked, particularly for η^1-cyclopentadienyl systems (4, 42, 83, 104). Since the site exchanges are identical for both least-motion and orbital-symmetry control in sigmatropic rearrangements in five-membered rings, a choice of controlling factor cannot be made. Polyhapto systems cannot be treated by conventional Woodward–Hoffmann procedures (114), which consider interaction with a single orbital on the migrating group. There is a single case, however, which has been discussed by Larrabee (82) in which a 1,4-shift inconsistent with least motion was found in $(\eta^1\text{-}C_7H_7)Sn(C_6H_5)_3$. The 1,2-shift would involve a migration of 3.07 Å, whereas the 1,4-molecular long jump involves a distance of 3.62 Å (109). As emphasized by Cotton (31), nontransition elements may satisfy the requirements of orbital symmetry control, whereas the valence d-orbitals of transition metals may provide sufficient flexibility in symmetry types that many paths may be allowed.

Thus up to mid-1974 a proliferation of ring-whizzer papers further augmented the data suggesting that 1,2-shifts were the dominant, if not universal, mode of transition metal migrations about polyene rings.

The complexity of the proton spectra of $(\eta^6\text{-}C_8H_8)Mo(CO)_3$ species was such that line-shape analysis and inference of the rearrangement pathway were not feasible. In 1974 however, Cotton, Hunter, and Lahuerta (30) showed that the four ^{13}C signals of the C_8H_8 ring broadened equally at temperatures about 0°C. This result rules out the predominant occurrence of a 1,2-pathway (see above). A random pathway via a η^8-intermediate appears to account for the observation. A similar path is also observed in the analogous $(\eta^6\text{-}C_8H_8)$Rh (norbornadiene).

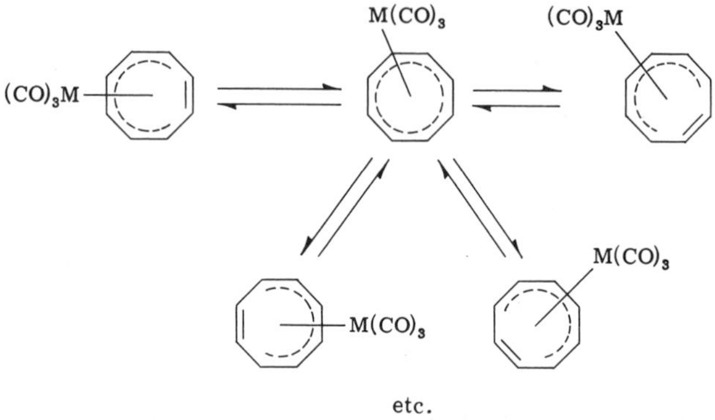

etc.

Thus the useful hypothesis that transition metals migrate via 1,2-shifts is likely to have exceptions in cases that are structurally or electronically similar to those above. The principle of least motion appears to account for most results and it also appears that a determining factor may be the ease with which the ring can adopt a conformation which presents a desirable set of orbitals to the metal. At this stage the capability of predicting random shift preference to 1,2-shift preference is limited, although some speculative attempts to rationalize the existing data have been put forward (35). These will not be reviewed here, but the requirement for a contiguous set of double bonds and appropriate overlap between them is certainly an important consideration. The continuity of the overlap is preserved in migrations in cyclic systems because unstable intermediates are not encountered. The absence of fluxionality in $(\eta^4\text{-}C_7H_8)$Fe(CO)$_3$ (7) can be attributed to the necessity of producing a high-energy intermediate during the migration. This guideline appears to be generally applicable, but a tin complex of cycloheptadiene provides a rare exception (39).

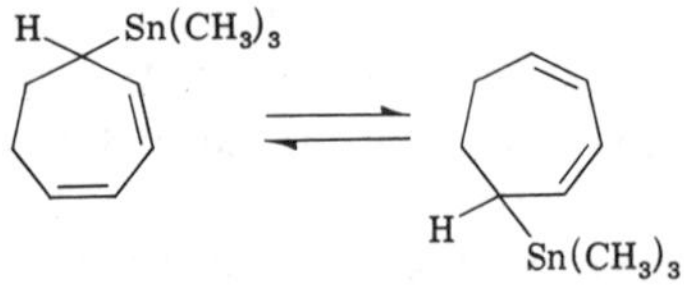

Recently, $[\eta^4\text{-}C_7H_7Ge(CH_3)_3]$Fe(CO)$_3$ has been shown to be fluxional and appears to be a second exception (85a).

IV

ROTATION OF π-COMPLEXES

A. Olefin Rotation

The potential utility of chiral metal complexes in inducing asymmetry in reactions with prochiral olefins has increased interest in conformational equilibria in olefin–metal complexes (15, 84, 88). Stereochemical nonrigidity can be interpreted in these systems in terms of variations in bond strength with orientation of the olefin. Although recent treatments suggest a somewhat more complicated bonding situation (98, 99, 111), the Dewar–Chatt–Duncanson scheme adequately describes the metal-olefin bond for most purposes (99). In Cotton's brief review of olefin rotation (31), it is apparent that rotation about an axis from the metal to the midpoint of the double bond is commonplace (73, 54, 56) and conforms to the original hypothesis of Cramer (37) regarding the fluxional behavior of $(\eta^5\text{-}C_5H_5)Rh(C_2H_4)_2$.

Barriers vary widely with olefin substituent, e.g., 13.6 kcal/mole for C_2H_4 and 20 kcal/mole for C_2F_4 in $(\eta^5\text{-}C_5H_5)Rh(C_2H_4)(C_2F_4)$ (38). Steric factors tend to be important when bulky groups are cis to the olefin in platinum(II) complexes (5, 19). Barriers to rotation tend to increase as one moves down a column in the periodic table; e.g., Ir > Rh (89) or Pt > Pd (69). The barrier is believed to represent the difference in energy between the olefin–metal bonding when the olefin is oriented horizontally and vertically. As emphasized by Cotton (31), electronic contributions to the barrier arise from the nonequivalence of the d-orbitals interacting with the olefin. Thus, in a molecule such as $Cr(CO)_5(C_2H_4)$ there should be no electronic barrier to rotation if one assumes that there are no breathing motions of the carbonyls during rotation; i.e., the $Cr(CO)_5$ moiety retains C_{4v} symmetry.

Conformational equilibria have been treated most extensively in olefin derivatives of "piano stool" cyclopentadienyl compounds. The work of Faller and Johnson (54, 56) and Herberhold, Alt, and Kreiter (66), which is summarized for ethylene complexes in Table II, suggests that decreased backbonding in cationic species accounts for the lower barriers. In this case as well, it is the difference in energy between vertical and horizontal orientations of the olefin rather than the overall bond strength that is important. The difficulty in assessing these energies led to some initial confusion as to the preferred orientation in the two conformations suggested by Green and Nagy (64). The "horizontal" B structure is now generally agreed to be the more stable (40, 41, 54, 56, 66). It has been

TABLE II
ETHYLENE ROTATION BARRIERS

Compound	ΔG^* (kcal/mole)	Reference
$(\eta^5\text{-}C_5H_5)Mn(CO)_2(C_2H_4)$	8.4	(66)
$(\eta^5\text{-}C_5H_5)Cr(CO)(NO)(C_2H_4)$	11.4	(3, 66)
$(\eta^5\text{-}C_5H_5)Fe(CO)[Sn(CH_3)_3](C_2H_4)$	12.8	(56)
$[(\eta^5\text{-}C_5H_5)Fe(CO)_2(C_2H_4)]BF_4$	7.8	(54)
$[(\eta^5\text{-}C_5H_5)Ru(CO)_2(C_2H_4)]BF_4$	7.6	(54)

shown to be the orientation in solid $(\eta^5\text{-}C_5H_5)Fe(CO)_2$(tetramethylallene) by Foxman (62), and the preferred orientation in solution by analysis of ring-current shifts by Faller, Johnson, and Schaeffer (56). In the Mn, Cr,

(A)   (B)

and Fe analogs there is a strong steric interaction between the olefin substituents and the ring, such that the conformation with the substituent oriented away from the ring is more stable. Detailed thermodynamic studies have not been performed, but free-energy differences are on the order of 3 kcal/mole. Increasing alkyl substitution tends to destabilize the metal–olefin bond, and highly substituted olefins can be

displaced readily in $[\eta^5\text{-}C_5H_5Fe(CO)_2(\text{olefin})]^+$ cations in accord with the stability series $C_2H_4 > C_2H_3R > C_2H_2R_2 > C_2H_1R_3 > C_2R_4$. This variation provides for the synthetic utility of the isobutylene complexes (40, 41). Steric interactions are such that deviations from the horizontal plane might be anticipated, particularly with geminal disubstitution (56). Since the NMR technique provides only information concerning the lowest-energy pathway for a given site interchange, we are unable to ascertain whether the olefin executes a full rotation of 360° or merely oscillates ~180°. It is clear that an oscillatory motion should be of lower energy.

Steric interactions also influence the ground-state stability and barrier heights of the (R,R)–(S,S) and (R,S)–(S,R) diastereomers found in (η^5-C_5H_5)Fe(CO)[Sn(CH$_3$)$_3$](CH$_3$CH=CH$_2$) (56). The olefin in one pair of diastereomers is significantly more labile. Diastereomers can also be detected by both IR and NMR in (η^5-C_5H_5)M(CO)(NO)(olefin) where M = Cr, Mo, and W (57, 60, 66). Approximately equal amounts of the diastereomers of the chromium complexes may be prepared readily by irradiation of the nitrosyl dicarbonyl complex in the presence of an olefin, such as propene or styrene (66).

Addition of hydride to [η^5-C_5H_5Mo(CO)(NO)–η^3-C_3H_5]PF$_6$ provides the propene molybdenum analog (6); however, Faller and Rosan (57, 60) have found that *hydride addition via cyanoborohydride is stereospecific and produces only one of the pairs of diastereoisomers*. The asymmetry in charge induced by the difference in electronic demands of CO and NO appear to provide the driving force for this asymmetric induction. This is a particularly important observation because it represents an electronic control of induction, whereas the usual approach has been to utilize steric effects to promote asymmetric induction (88). The importance of the electronic asymmetry is dramatically demonstrated by perturbations of the endo–exo equilibrium in the molybdenum complexes (60). In this instance there are clearly disparate ''trans effects'' of the NO and CO.

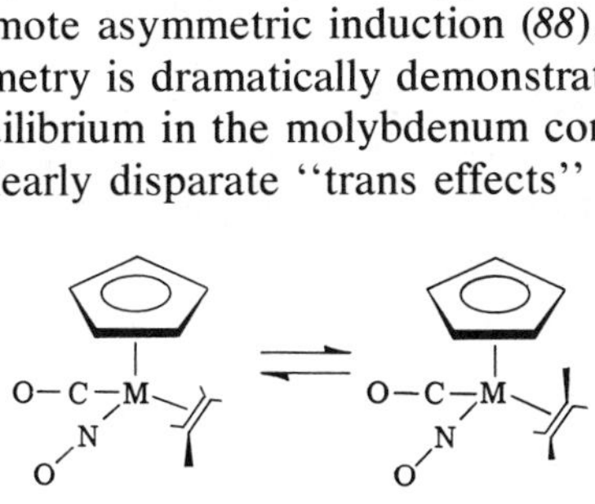

$$K_{exo/endo} = 1.7$$

The donor properties of the methyl group provide a preference for the substituted end of the olefin to be trans to one of the ligands. Thus the inherent instability of the endo isomer and the interaction of the methyl group with the ring are partially overcome. The rotational barriers of substituted olefin derivatives in the molybdenum complexes (ΔG^* ~13–

17 kcal/mole) tend to be higher than those for the complexes listed in Table II. In general it appears that both substituent bulkiness and the presence of NO groups increase the barriers to rotation.

The importance of stereochemistry and fluxionality in these systems is demonstrated in the $Mo(C_2H_4)_2(diphos)_2$ system developed by Osborn (17). The AA'BB' pattern for the olefin averages above 25° to a singlet in the ^{31}P decoupled spectra corresponding to a barrier of ~15 kcal/mole. The preferred conformation was presumed to contain a staggered trans arrangement of olefins (17); however, the complexities of the diphos resonances suggest that a more complicated fluxional behavior is occurring and the 25° spectrum is an average of a lower barrier process.

Addition of acid to the bis(ethylene) complex produces $[HMo(C_2H_4)_2(diphos)_2]^+$, which exhibits averaging between the hydride and only one of the olefins above −85°. This is the first case where the insertion–deinsertion process, fundamental to many catalytic mechanisms, can be observed directly.

A novel fluxional system has been recently investigated in which the interaction of ethylene with metal clusters produces species such as $H_2Os_3(CO)_9(C{=}CH_2)$. Rearrangements in these species, as well as those in benzyne complexes, such as $Os(C_6H_4)P(CH_3)_2(CO)_7$ (45), have been reviewed by Lewis and Johnson (85).

B. Allyl Rotation

The dominant mode of fluxional behavior in η^3-allyl complexes involves a $\eta^3 \rightarrow \eta^1 \rightarrow \eta^3$ interconversion (47, 49). These features have been reviewed by Vrieze (106) and are discussed in this volume by Tsutsui and Courtney. The formation of the η^1-intermediate provides a pathway for interchanging syn and anti substituents and inverting stereochemistry in chiral allyl complexes (49).

Rotation of the allyl group, in which syn–anti interconversion does not occur, is now known in many instances for complexes with coordination numbers greater than four (see Table III). NMR spectra consistent with allyl rotation in most square palladium complexes have been shown to result from ligand exchange (47).

As in olefin systems based on piano stool structures, two conformational isomers exist in $(\eta^5\text{-}C_5H_5)Mo(CO)_2(\eta^3\text{-}C_3H_5)$ and barriers to interconversion are on the order of 15 kcal/mole. Steric interactions with the ring control the equilibria (51, 57); thus when R = H, $K_{exo/endo} = 4.7$, whereas when R = CH_3, $K_{exo/endo} = 0.38$.

Addition of NO^+ inverts the configuration and introduces chirality into

TABLE III
DEMONSTRATED ALLYL ROTATIONS

Complex	Reference
$(\eta^3\text{-}C_3H_5)Mo(CO)_2(\eta^5\text{-}C_5H_5)$	(51)
$(\eta^3\text{-}C_3H_5)Mo(CO)_2[(N_2C_3H_3)_2BH_2]$	(87)
$(\eta^3\text{-}C_3H_5)Mo(CO)_2(diphos)X$	(58)
$(\eta^3\text{-}C_3H_5)W(CO)_4X$	(65)
$(\eta^3\text{-}C_3H_5)Fe(CO)_3X$	(59)
$(\eta^3\text{-}C_3H_5)_2M(CO)_2$; M = Fe, Ru	(23)
$(\eta^3\text{-}C_3H_5)_3Rh$	(93)
$[(\eta^3\text{-}C_3H_5)RhCl]_2$	(23)
$(\eta^3\text{-}C_3H_5)_2Pt$	(86)
$(\eta^3\text{-}C_3H_4R)MoL(CO)(\eta^5\text{-}C_5H_5)$	(21)
$[(\eta^3\text{-}C_3H_5)Mo(CO)(NO)(C_5H_5)]^+$	(57)

the molecule (57). It is unusual that the addition of NO^+ reduces the rate of exo–endo interconversion by a factor of 10^{-6} compared to the dicarbonyl complex. This is consistent with the observation in olefin complexes that barriers are greater in isoelectronic systems containing NO. It contrasts, however, with olefin behavior, for which rotation is generally faster in cations than in neutral species.

The metal center in $(\eta^5\text{-}C_5H_5)M(L)(CO)(allyl)$ can be considered pseudotetrahedral and characterized by the (R) and (S) nomenclature. Neutral chiral complexes of this type have been prepared previously, e.g., phosphine carbonyl derivatives (21, 75) and nitrosyl halide derivatives prepared by carbonyl displacement from nitrosyl cations by iodide (57, 60). Each of the diastereomers produced by the addition of a prochiral allyl moiety to the chiral metal center exhibit endo–exo isomerism. The endo–exo equilibration depends on charge and the neutral nitrosyl iodides rearrange rapidly compared to the cationic nitrosyls.

Great stereoselectivity is shown upon addition of NO^+ to (η^3-crotyl) $(\eta^5\text{-}C_5H_5)Mo(CO)_2$ and one of the possible $(R,R\text{-}S,S)$-exo, $(R,R\text{-}S,S)$-endo, $(R,S\text{-}S,R)$-exo, or $(R,S\text{-}S,R)$-endo isomers is produced in >85% yield. Nucleophilic addition to these cations is regiospecific, and it appears that additions can be forced to occur at either the 1- or 3-position of the allyl moiety by the asymmetry inherent in the $(\eta^5\text{-}C_5H_5)Mo(CO)(NO)$ fragment (60).

The barriers vary widely in these complexes, and it has been suggested that the rotation is facilitated in complexes for which the parent structure is nonrigid (51). Thus, we expect that structures based on $(\eta^5\text{-}C_5H_5)Mo(CO)_2LX$ to rotate more readily than those based on

(η^5-C$_5$H$_5$)Ru(CO)LX (48, 53). Although one cannot expect to be able to describe the detailed path, one should recognize that an allyl (or an olefin for that matter) does not rotate against a rigid framework and some flexing of the other ligands must occur. In the piano stool structures, this bending does not appear to be excessive; however, in some cases it appears that extreme deformations of the other ligands are coupled to the rearrangement of the allyl or olefin (see Section V). In fluxional triazenido complexes related to these allyl complexes, the triazenide retains its stereochemistry even though the carbonyl groups are interchanged (92).

In fluxional (η^3-C$_3$H$_5$)M(CO)$_2$(L-L)X complexes, where L-L is a chelating bisphosphine, the allyl may be capable of rotating, but there is a preferred configuration, so that rotation per se is not observed. The fluxional migration of groups about the metal, however, render the spectrum of the allyl dynamic. This complex is approximately octahedral, and the rearrangement can be conveniently described as a trigonal twist of only one pair of triangular faces. This complex is noteworthy because it demonstrates that a trigonal twist about a single axis in an octahedron inverts certain elements of chirality, but not all of them. Therefore, in certain compounds a trigonal twist mechanism will produce epimerization rather than racemization. Thus, the chirality of the (As-P-X) center is not inverted under the twist mechanism.

Hence, even though the structure is nonrigid, it retains a component of chirality of sufficient asymmetry to produce significant stability differences in the binding of prochiral olefins and allyls. This contrasts with certain guidelines which suggest that the more rigid the structure, the greater will be the asymmetric induction (15, 88).

This system is analogous to the HMo(η^3-C$_3$H$_5$)(diphos)$_2$ system developed by Osborn et al. (17), which provided the first direct observation of the π-allyl–hydride exchange mechanism proposed for the 1,3-hydride shifts found in many metal-catalyzed olefin reactions. Presumably, this hydride undergoes a fluxional rearrangement of the diphos ligands in addition to the metal(allyl)(hydride) ⇌ metal (propene) interconversion.

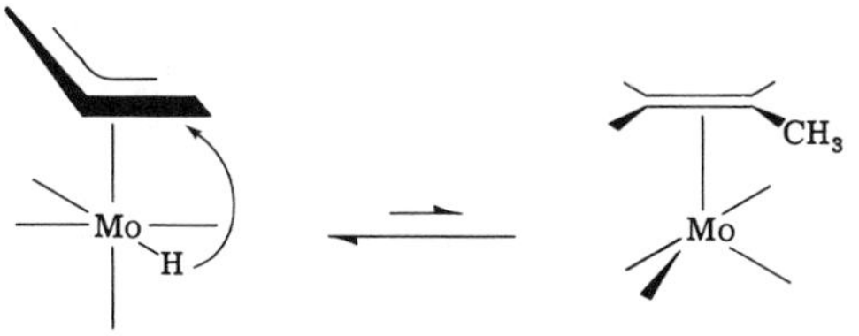

V

COUPLING OF π-LIGAND ROTATION WITH CARBONYL
REARRANGEMENT

A. Olefin Rotation and Carbonyl Scrambling

Olefin rotation in analogs of $[(\eta\text{-}C_5H_5)Fe(CO)_2(\text{olefin})]^+$ is not accompanied by significant rearrangement of the other ligands. Nevertheless, olefin rotation in $Fe(CO)_4(\text{olefin})$ complexes does not occur independently of carbonyl rearrangement (76, 113). The ^{13}C NMR spectrum in the metal carbonyl region of $Fe(CO)_4(\text{diethyl fumarate})$ exhibits two resonances of equal intensity at $-60°$. These resonances coalesce to a single resonance at room temperature, which is indicative of a rearrangement barrier of 14.3 kcal/mole. Barriers increase with the π-acceptor ability of the olefin in the order: (styrene)-<(ethyl acrylate <(benzylideneacetone)<(diethylmaleate)- ~ (diethyl fumarate)-<(trifluorochloroethylene)$Fe(CO)_4$. The olefin occupies an equatorial site in a trigonal bipyramidal (TBP) complex with the olefin lying in the equatorial plane. The reported changes in the ^{13}C NMR spectra are consistent with a synchronous olefin rotation and Berry pseudorotation of the carbonyl groups (76). This implies an intimate coupling in which each rotation of 90° executed by the olefin results in an interchange of

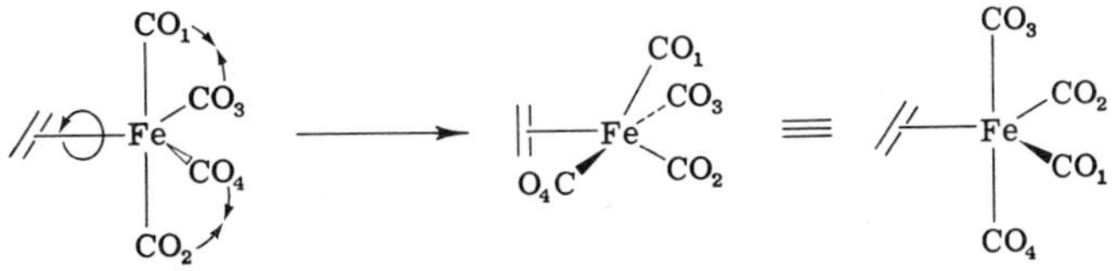

axial and equatorial carbonyls. It would thus appear that olefin rotation would not be a facile process in ground-state TBP structures, in contrast to the apparent rotation found in TBP $\{Os(CO)(NO)[P(C_6H_5)_3]_2(C_2H_4)\}^+$ (73).

The constraints of axial-equatorial preferences of the ligands presumably has a great influence on the apparent coupling of olefin rotation to pseudorotation of the carbonyls. Osborn *et al.* (*113*) investigated this feature by considering the nonrigidity of $Fe(CO)_3(C_6H_5CH_2NC)$(maleic anhydride). Equilibration of the two unequal sets of $2:1$ carbonyl

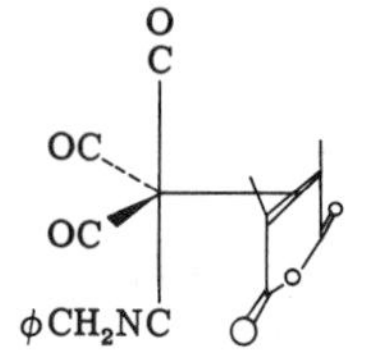

(1a) 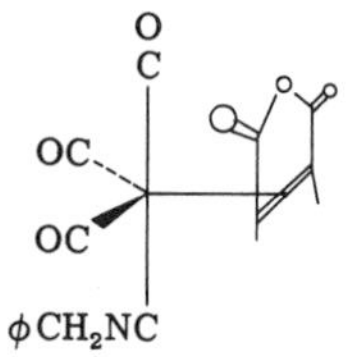   (1b)

resonances observed at $-30°$ produces a single $2:1$ pattern at 25°C. Simultaneous averaging of the olefin resonances suggests that a nonconcerted mechanism may be an attractive alternative to an intimate coupling of olefin rotation and Berry pseudorotation (*113*). In this alternative interpretation, one assumes that once a square pyramidol (SP) intermediate is formed, relatively free rotation of the olefin can occur. Hence, this viewpoint differs in the role played by the SP structure. In the concerted mechanism, the SP structure is considered a transition state; whereas in the other, an intermediate is formed in which the olefin may rotate many times during a single conversion to an SP.

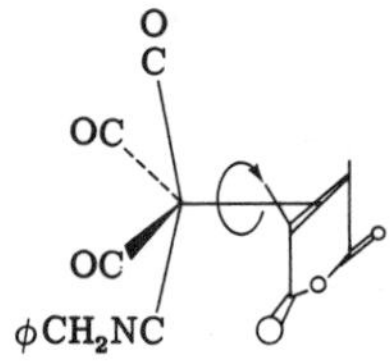

(1c) 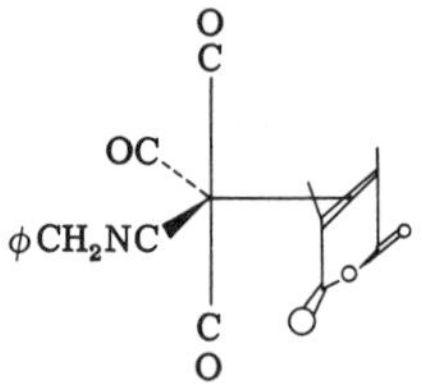   (1d)

It should be recognized, however, that *all* the previously published data are consistent with either interpretation. In each case the equatorial site occupied by the olefin is taken as the pivotal position in forming the square pyramid. As long as only one pivot is used in the pseudorotation, the cyclic order of the ligands as viewed along the pivot does not change. Thus if ligand 1 prefers an axial position, the A $\rightleftarrows$ B equilibrium may be shifted largely toward **A**, and therefore **B** may only be present in undetectable concentrations. In this case axial preferences are $RNC > CO >$ olefin; thus, even the isonitrile results are consistent with a concerted rotation of the olefin because a structure such as **1d** would be destabilized. The preference for the olefin to retain its equatorial position demands that it be the pivot in the lowest-energy pathway. The apparent lack of pseudorotation in $\{Os(CO)(NO)[P(C_6H_5)_3]_2(C_2H_4)\}^+$ may also be attributed to the axial preference of the phosphines. One must remember, however, that these energy surfaces are relatively flat

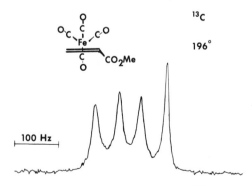

and alternate modes of rearrangement may occur at slightly higher energies if a given pathway is blocked. For example, isocyanide interchange is observed at higher temperatures in $Fe(CO)(C_6H_5CH_2NC)_3$(maleic anhydride) at higher (>60°) energy even though a lower-energy process is consistent with the previous discussion.

General consistency with pseudorotation should not be taken for granted. As shown in Fig. 7, careful examination of the carbonyl resonances observed in (methyl acrylate)$Fe(CO)_4$ show that all the resonances do not broaden to the same degree at the onset of averaging (61, 78). Solvent variations do not alter this effect, and it is therefore unlikely to arise from averaging with an intermediate (see Section II). The Berry pseudorotation mechanism demands that each of the four carbonyl ligands move to a different environment; hence, all should broaden to the same extent. This anomalous result can be accommo-

FIG. 7. The metal carbonyl region of the 67.9 MHz ^{13}C spectrum of (methyl acrylate)$Fe(CO)_4$ at $-77°C$.

dated if one assumes that three of the ligands rotate relative to the other two with subsequent processes averaging all sites. This mechanism, as

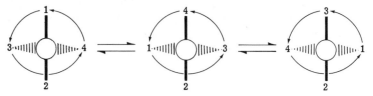

well as others that account for the differential broadening, are difficult to reconcile with the averaging reported for **1**. Furthermore, the carbonyls broaden equally in the $Ru(CO)_4$ analog (*78*). Details of this rearrangement still need further elucidation, but it appears that even though carbonyl rearrangement is indisputably coupled to olefin rotation in some manner, Berry pseudorotation may not be the only path of rearrangement for the carbonyls.

B. *Fluxional Polyene Metal Tricarbonyl Complexes*

Carbonyl scrambling is a well-recognized phenomenon in complexes containing metal–metal bonds (*1, 33, 34, 70, 85*). Although carbonyl ligands are often interchanged between metals, situations exist, particularly in clusters, in which a metal tricarbonyl fragment appears to rotate. The fluxional behavior of the $(diene)Fe(CO)_3$, $(triene)Mo(CO)_3$, and

TABLE IV

FLUXIONAL BEHAVIOR OF DIENE AND TRIENE METAL TRICARBONYLS

Compound	Activation parameter[a]	Limiting spectrum	References
$(Butadiene)Fe(CO)_3$	$E_a = 9.5$	$-78°$	(*76a, 77*)
$(1,3\text{-Hexadiene})Fe(CO)_3$	$E_a = 7.4$	$-93°$	(*77*)
$(Cycloheptatriene)Fe(CO)_3$	$E_a = 11.6$	$-73°$	(*77*)
$(1\text{-Methoxy-1,3-cyclohexadiene})Fe(CO)_3$	$\Delta G^* = 7.3$	$-61°$	(*81*)
$(3,5\text{-Heptadien-2-ol})Fe(CO)_3$		$-63°$	(*76a*)
$(3,5\text{-Heptadien-2-one})Fe(CO)_3$		$-51°$	(*76a*)
$[(CF_3)_4\text{-cyclopentadienone}]Fe(CO)_3$	$\Delta G^* = 11.1$	$-70°$	(*79*)
$(Butadiene)Fe(CO)_2(PF_3)$	$E_a = 6.7$	$-107°$	(*107*)
$(Cyclohexadiene)Fe(CO)_2(PF_3)$	$\Delta G^* < 7$	$< -148°$	(*108*)
$(Cyclooctatetraene)Fe(CO)_3$	$E_a = 8.3$	$-134°$	(*35*)
$(Cyclooctatetraene)Ru(CO)_3$	$E_a = 8.9$	$-126°$	(*35*)
$(Cycloheptatriene)Mo(CO)_3$	$E_a \sim 12$	$-51°$	(*76*)
$(Cycloheptatriene)Cr(CO)_3$	$E_a \sim 11$	$-59°$	(*76*)

[a] Most values are subject to large experimental errors.

(triene)Cr(CO)$_3$ complexes given in Table IV have been interpreted in terms of rotation of the polyene fragment relative to the tricarbonyl fragment. This motion was first observed in trifluorophosphine derivatives by Clark *et al.* (*107, 108*). The diene fragment is presumed to occupy two basal positions of a square pyramidal structure. Low temperature spectra of tricarbonyl derivatives exhibit resonances in a ratio of 1:2 for the single apical and two basal carbonyls. The rotation of the diene relative to the tricarbonyl fragment would presumably be accompanied by flexing of the tricarbonyl fragment to interchange apical and basal designations of the carbonyls, as indicated below by Takats *et al.* (*77*). Although these motions might be interpreted in terms of Berry pseudorotations (*10*), attempts to describe the carbonyl motions in too much detail are unwarranted. A similar rotation has also been suggested

in the triene analogs (*76*). Although all the data published thus far are consistent with rotation and bending of the tricarbonyl fragment, there is no evidence that the cyclic order of the carbonyl ligands is maintained. Therefore, a more complex pseudorotation process could occur. There is also a possibility that a low-concentration intermediate might be involved; e.g., one might wish to consider the following possibility.

The migration of the Fe(CO)$_3$ group about the ring in (cyclooctatetraene)Fe(CO)$_3$ is accompanied by averaging of the carbonyl resonances (*35, 61, 96*). The line-shape analysis, however, is sufficiently difficult that the intervention of a concerted path is still questionable. It appears that an intermediate may be present in sufficiently high concentrations to affect the spectrum (see Section III).

The quality of the ^{13}C spectra generally used in line-shape analysis for the tricarbonyl species often leaves something to be desired; hence the activation energies given in Table IV should be viewed with some suspicion. The temperatures at which low-temperature limiting spectra are reported indicate that accurate activation energies would all fall within a range of 2 kcal/mole. It is clear, however, that the bonding requirements of a 1,3-diene significantly increase the barrier to carbonyl rearrangement as compared to $Fe(CO)_5$, (norbornadiene)$Fe(CO)_3$, or (1,5-cyclooctadiene)$Fe(CO)_3$ (77). Thus, one may anticipate that appropriate constraints on a 1,3-diene would raise the barrier substantially.

ACKNOWLEDGMENT

The original research reported in this review was supported in part by a grant from the National Science Foundation. The high-field NMR spectra were obtained on a Bruker HX-270 supported in part by the National Institutes of Health Research Grant No. 1-P07-PR00798 from the Division of Research Resources.

REFERENCES

1. Aime, S., Milone, L., Sappa, E., *J. Chem. Soc., Dalton Trans.* 838 (1976).
2. Allerhand, A., Gutowsky, J. S., Jones, J., and Meinzer, R. A., *J. Am. Chem. Soc.* **88**, 3185 (1966).
3. Alt, H., Herberhold, M., Kreiter, C. G., and Strack, H., *J. Organomet. Chem.* **77**, 353 (1974).
4. Anastasiou, A. G., *Chem. Commun.* 15 (1968).
5. Ashley-Smith, S., Douek, I., Johnson, B. F. G., and Lewis, J., *J. Chem. Soc., Dalton Trans.* 1776 (1972).
6. Bailey, N. A., Kita, W. G., McCleverty, J. A., Murray, A. J., Mann, B. E., and Walker, N. W. J., *Chem. Commun.* 592 (1974).
7. Burton, R., Pratt, W., and Wilkinson, G., *J. Chem. Soc.* 594 (1961).
8. Bennett, M. A., Bramley, R., and Watt, R., *J. Am. Chem. Soc.* **91**, 3089 (1969).
9. Bennett, M. A., Cotton, F. A., Davison, A., Faller, J. W., Lippard, S. J., and Morehouse, S. M., *J. Am. Chem. Soc.* **88**, 4371 (1966).
10. Berry, R. S., *J. Chem. Phys.* **32**, 933 (1960).
11. Binsch, G., *Top. Stereochem.* **3**, 97 (1968).
12. Binsch, G., *J. Am. Chem. Soc.* **91**, 1304 (1969).
13. Binsch, G., and Kleier, D. A., "The Computation of Complex Exchange-Broadened-NMR-Spectra," Program 140 Quantum Chemistry Program Exchange. Indiana Univ., Bloomington, Indiana, 1969.
14. Binsch, G., *in* "Dynamic Nuclear Magnetic Resonance Spectroscopy" (L. M. Jackman and F. A. Cotton, eds.), p. 45. Academic Press, New York, 1975.
15. Bosnich, B., and Fryzuk, M. D., *Abstr., Am. Chem. Soc. Nat. Meet., Fall 1976*, Inor #108.
16. Bratton, W. K., Cotton, F. A., Davison, A., Musco, A., and Faller, J. W., *Proc. Nat. Acad. Sci. U.S.* **58**, 1324 (1967).
17. Byrne, J. W., Blaser, H. U., and Osborn, J. A., *J. Am. Chem. Soc.* **97**, 3871 (1975).
18. Calderon, J. L., Cotton, F. A., and Shaver, A., *J. Organomet. Chem.* **42**, 419 (1972).
19. Chisholm, M. H., and Clark, H. C., *Inorg. Chem.* **12**, 991 (1973).
20. Ciappenelli, D., and Rosenblum, M., *J. Am. Chem. Soc.* **91**, 6873 (1969).
21. Collin, J., Savegnac, M., and Lambert, P., *J. Organomet. Chem.* **82**, C19 (1974).
22. Cooke, M., Goodfellow, R. J., Green, M., Maher, J. P., and Yondle, J. R., *Chem. Commun.* 565 (1970).

23. Cooke, M., Goodfellow, R. J., and Green, M., *J. Chem. Soc. A* 16 (1971).
24. Cotton, F. A., *Acc. Chem. Res.* 1, 257 (1968).
25. Cotton, F. A., Faller, J. W., and Musco, A., *J. Am. Chem. Soc.* 90, 1438 (1968).
26. Cotton, F. A., and Marks, T. J., *J. Am. Chem. Soc.* 91, 7523 (1969).
27. Cotton, F. A., and Marks, T. J., *Inorg. Chem.* 9, 2802 (1970).
28. Cotton, F. A., and Takats, J., *J. Am. Chem. Soc.* 92, 2353 (1970).
29. Cotton, F. A., and Ciappenelli, D. J., *Syn. Inorg. Metal-Org. Chem.* 2, 197 (1972).
30. Cotton, F. A., Hunter, D. L., and Lahuerta, P., *J. Am. Chem. Soc.* 96, 4723 (1974).
31. Cotton, F. A., *in* "Dynamic Nuclear Magnetic Resonance," (L. M. Jackman and F. A. Cotton, eds.), p. 377. Academic Press, New York, 1975.
32. Cotton, F. A., *J. Organometal. Chem.* 100, 29 (1975).
33. Cotton, F. A., Hunter, D. L., and Lahuerta, P., *Inorg. Chem.* 14, 511 (1975).
34. Cotton, F. A., Hunter, D. L., Lahuerta, P., *J. Organometal. Chem.* 87, C42 (1975).
35. Cotton, F. A., and Hunter, D. L., *J. Am. Chem. Soc.* 98, 1413 (1976).
36. Cotton, F. A., and Kolb, J. R., *J. Organometal. Chem.* 107, 113 (1976).
37. Cramer, R., *J. Am. Chem. Soc.* 86, 217 (1964).
38. Cramer, R., Kline, J. B., and Roberts, J. D., *J. Am. Chem. Soc.* 91, 2519 (1969).
39. Curtiss, M., and Fink, R., *J. Organometal. Chem.* 38, 299 (1972).
40. Cutler, A., Ehntholt, D., Lemmon, N. K., Morten, D. F., Madhauarao, M., Raghu, S., Rosan, A., and Rosenblum, M., *J. Am. Chem. Soc.* 97, 3149 (1975).
41. Cutler, A., Ehntholt, D., Giering, W. P., Lennon, P., Raghu, S., Rosan, A., Rosenblum, M., Tancrede, J., and Wells, D., *J. Am. Chem. Soc.* 98, 3495 (1976).
42. Dalton, J., and McAuliffe, C. A., *J. Organomet. Chem.* 39, 351 (1972).
43. Davison, A., Rakita, P. E., *J. Organomet. Chem.* 23, 407 (1970).
44. Davison, A., and Rakita, P. E., *Inorg. Chem.* 9, 289 (1970).
45. Deeming, A. J., Kimber, R. E., and Underhill, M., *J. Chem. Soc., Dalton Trans.* 2589 (1973).
46. Faller, J. W., *Inorg. Chem.* 8, 767 (1969).
47. Faller, J. W., and Incorvia, M. J., *J. Organomet. Chem.* 19, P13 (1969).
48. Faller, J. W., and Anderson, A. S., *J. Am. Chem. Soc.* 92, 5852 (1970).
49. Faller, J. W., Thomsen, M. E., and Mattina, M. J., *J. Am. Chem. Soc.* 93, 2642 (1971).
50. Faller, J. W., *in* "Determination of Organic Structures by Physical Methods" (F. C. Nachod and J. J. Zuckerman, eds.), Vol. 5, p. 75. Academic Press, New York, 1973.
51. Faller, J. W., Chen, C. C., Mattina, M. J., and Jakubowski, A., *J. Organomet. Chem.* 52, 361 (1973).
52. Faller, J. W., Adams, M. A., and LaMar, G. N., *Tetrahedron Lett.* 699 (1974).
53. Faller, J. W., Johnson, B. V., and Dryja, T. P., *J. Organomet. Chem.* 65, 395 (1974).
54. Faller, J. W., and Johnson, B. V., *J. Organomet. Chem.* 88, 101 (1975).
55. Faller, J. W., and Johnson, B. V., *J. Organomet. Chem.* 90, 99 (1975).
56. Faller, J. W., Johnson, B. V., and Schaeffer, C. D., Jr., *J. Am. Chem. Soc.* 98, 1395 (1976).
57. Faller, J. W., and Rosan, A. M., *J. Am. Chem. Soc.* 98, 3388 (1976).
58. Faller, J. W., Haitko, D. A., Adams, R. D., and Chodosh, D. F., *J. Am. Chem. Soc.* 99, 1654 (1977).
59. Faller, J. W., and Adams, M. A., unpublished observations.
60. Faller, J. W., and Rosan, A. M., unpublished observations.
61. Faller, J. W., Osborn, J. A., and Schaeffer, C. D., Jr., unpublished work.
62. Foxman, B. M., *Chem. Commun.* 221 (1975).
63. Gordon, R. G., and McGinnis, R. P., *J. Chem. Phys.* 49, 2455 (1968).
64. Green, M. L. H., and Nagy, P. L. I., *J. Organomet. Chem.* 1, 58 (1963).

65. Holloway, C. E., Kelley, J. D., and Stiddard, M. H. B., *J. Chem. Soc. A* 931 (1969).
66. Herberhold, M., Alt, H., and Kreiter, C. G., *Liebigs Ann. Chem.* 300 (1976).
67. Hine, J., *J. Am. Chem. Soc.* **88**, 5525 (1966).
68. Hoffman, R. A., and Forsen, S., *Progr. Nucl. Magn. Reson. Spectrosc.* **1**, 15 (1966).
69. Iwao, T., Saika, A., and Segal, J. A., *Chem. Commun.* 1312 (1972).
70. Jackman, L. M., and Cotton, F. A., (eds.), "Dynamic Nuclear Magnetic Resonance Spectroscopy." Academic Press, New York, 1975.
71. Johnson, C. S., Jr., *Adv. Magn. Reson.* **1**, 33 (1965).
72. Johnson, C. S., Jr., and Moreland, C. G., *J. Chem. Educ.* **50**, 477 (1973).
73. Johnson, B. F. G., and Segal, J. A., *Chem. Commun.* 1312 (1972).
74. Kang, J. W., Childs, R. F., and Maitlis, P. M., *J. Am. Chem. Soc.* **92**, 721 (1970).
75. King, R. B., Zipperer, W. C., and Ishaq, M., *Inorg. Chem.* **11**, 1361 (1972).
76. Kreiter, C. G., and Lang, M. J., *Organometal. Chem.* **55**, C27 (1973).
76a. Kreiter, C. G., Stueber, S., and Wackerle, L., *J. Organometal. Chem.* **66**, C49 (1974).
77. Kruczynski, L., and Takats, J., *J. Am. Chem. Soc.* **96**, 932 (1974).
78. Kruczynski, L., Lishingman, L. K. K., and Takats, J., *J. Am. Chem. Soc.* **96**, 4006 (1974).
79. Kruczynski, L., Martin, J. L., and Takats, J., *J. Organomet. Chem.* **80**, C9 (1974).
80. LaMar, G. N., and Faller, J. W., *J. Am. Chem. Soc.* **95**, 3817 (1973).
81. Lallemand, J. Y., Laszlo, P., Muzette, C., and Stockis, A., *J. Organomet. Chem.* **91**, 71 (1975).
82. Larrabee, R. B., *J. Am. Chem. Soc.* **93**, 1510 (1970).
83. Larrabee, R. B., *J. Organomet. Chem.* **74**, 313 (1974).
84. Lazzaroni, R., Salvadori, P., Bertucci, C., and Veracini, C. A., *J. Organomet. Chem.* **99**, 475 (1975).
85. Lewis, J., and Johnson, B. F. G., *Pure Appl. Chem.* 43 (1975).
85a. Lishing Man, L. K. K., and Takats, J., *J. Organomet. Chem.* **117**, C104 (1976).
86. Mann, B. E., Shaw, B. L., and Shaw, G., *J. Chem. Soc. A* 3536 (1971).
87. Meakin, P., Trofimenko, S., and Jesson, J. P., *J. Am. Chem. Soc.* **94**, 5677 (1972).
88. Morrison, J. D., Masler, W. F., and Neuberg, M. K., *Adv. Catal.* **25**, 81 (1976).
89. Moseley, K., Kang, J. W., and Maitlis, P. M., *J. Chem. Soc. A* 2875 (1970).
90. Nesmeyanov, A. N., and Kritskaya, I. I., *J. Organometal. Chem.* **14**, 387 (1968).
91. Nesmeyanov, A. N., Ustynyuk, Y. A., Kritskaya, I. I., and Sheembelov, G. A., *J. Organomet. Chem.* **14**, 395 (1968).
92. Pfeiffer, E., Kuyper, J., and Vrieze, K., *J. Organomet. Chem.* **105**, 371 (1976).
93. Powell, J., and Shaw, B. L., *J. Chem. Soc. A* 583 (1968).
94. Randall, E. W., Rosenberg, E., and Milone, L., *J. Chem. Soc., Dalton Trans.* 1672 (1973).
95. Reeves, L. W., Shaddion, R. C., and Shaw, K. N., *Can. J. Chem.* **49**, 3683 (1971).
96. Rigatti, G., Boccalon, G., Ceccon, A., and Giacometti, G., *Chem. Commun.* 1165 (1972).
97. Robinson, B. H., and Spencer, J., *J. Organometal. Chem.* **33**, 97 (1971).
98. Roesch, N., and Hoffmann, R., *Inorg. Chem.* **13**, 2656 (1974).
99. Roesch, N., Messmer, R. P., and Johnson, K. H., *J. Am. Chem. Soc.* **96**, 3855 (1974).
100. Sack, R. A., *Mol. Phys.* **1**, 163 (1958).
101. Saunders, M., *Tetrahedron Lett.* 1699 (1963).
102. Saunders, M., "Magnetic Resonance in Biological Systems," p. 85. Pergamon, Oxford, 1967.

103. Sergeyev, N. M., Avramenko, G. I., and Ustynyak, Y. A., *J. Organomet. Chem.* **24,** C39 (1970).
104. Su, C. C., *J. Am. Chem. Soc.* **93,** 5653 (1971).
105. Teyssie, P., Julemont, M., Thomassin, J. M., Walckiers, E., and Waring, R., "Coordination Polymerization," p. 327. Academic Press, New York, 1975.
106. Vrieze, K., *in* "Dynamic Nuclear Magnetic Resonance Spectroscopy" (L. M. Jackman and F. A. Cotton, eds.), p. 441. Academic Press, New York, 1975.
107. Warren, J. D., and Clark, R. J., *Inorg. Chem.* **9,** 373 (1970).
108. Warren, J. D., Busch, M. A., and Clark, R. J., *Inorg. Chem.* **11,** 452 (1972).
109. Weidenborner, J. E., Larrabee, R. B., and Bednowitz, A. L., *J. Am. Chem. Soc.* **94,** 4140 (1972).
110. West, P., Woodville, M. C., and Rausch, M. D., *J. Am. Chem. Soc.* **91,** 5649 (1969).
111. Wheeloch, K. S., Nelson, J. H., Kelly, J. D., Jonassen, H. B., and Cusachs, L. C., *J. Chem. Soc., Dalton Trans.* 1457 (1973).
112. Whitesides, G. M., and Fleming, J. S., *J. Am. Chem. Soc.* **89,** 2855 (1967).
113. Wilson, S. T., Coville, N. J., Shapley, J. R., and Osborn, J. A., *J. Am. Chem. Soc.* **96,** 4038 (1974).
114. Woodward, R. B., and Hoffmann, R., "The Conservation of Orbital Symmetry." Academic Press, New York, 1970.

σ–π Rearrangements of Organotransition Metal Compounds

MINORU TSUTSUI and ARLENE COURTNEY

Department of Chemistry
Texas A&M University
College Station, Texas

I

SCOPE AND LIMITATION

Since the isolation of triphenyltris(tetrahydrofuran)chromium(III) (*1*) as an intermediate in the synthesis of Hein's π-arenechromium complex (*2*), many examples of an important class of organometallic reactions, σ–π rearrangements, have been reported. These rearrangements are of both theoretical and practical interest, since they appear to be very important in homogeneous catalytic processes; therefore, a knowledge of the physical and chemical factors that govern this phenomenon is essential.

A σ–π rearrangement is an intramolecular reaction in which an organic group σ-bonded (h^1) to a metal becomes π-bounded (h^n) to the metal. The reverse of this process, π–σ ($h^n \rightarrow h^1$), also occurs.

An attempt will be made to understand the σ–π interconversion by looking at the factors that influence the static modes, i.e., the σ- and the π-complex, and also, those factors that allow the static mode to become dynamic, causing a rearrangement to occur. Also, various catalytic processes will be studied in an effort to gain insight into the possible role of the $\sigma-\pi$ rearrangement in catalysis.

II
FACTORS INFLUENCING THE STATIC MODES

A σ-complex involves a metal–ligand bond that results from overlap of a metal d-orbital with a carbon sp^n hybrid. In these bonds there is little or no π-bonding character. The π-complex involves a metal–ligand bond having two components—a σ-bond between a metal $d-\sigma$ acceptor orbital and a ligand π molecular orbital and a π-bonding portion between a metal d-orbital or a $p\pi-d\pi$ hybrid and the ligand π^* molecular orbital in which electron density flows from the metal to the ligand. These two bonding modes are illustrated in Fig. 1.

The stabilities of the two bonding modes will be somewhat dependent on the electron density present on the metal. The electron density on the metal, in turn, is determined to a large extent by the other ligands present in the complex (3–7). Ligands that form strong dative π-bonds with the metal exert an electron-withdrawing effect, allowing the metal to better accept electrons. Less electron density will be present for back-donation, thus stabilizing the σ-bonding mode. Similarly, ligands possessing little dative π-bonding capacity will leave more electron density on the metal stabilizing the π-bonding mode. These effects are summarized in Fig. 2.

From the molecular orbital standpoint, π-bonding ligands (L_1) that are trans to the ligand of interest (L_2) stabilize filled and partially filled metal d-orbitals that might otherwise interact with π-orbitals of the appropriate

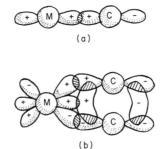

FIG. 1.   Metal-ligand bonding in (a) σ-complex; (b) π-complex.

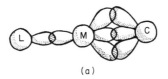

(a)

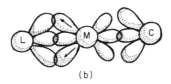

(b)

FIG. 2.   Effect of ligand L on metal orbitals when L is (a) non π-bonding; (b) π-bonding.

energy and symmetry present on L_2. Thus, the net effect is stabilization of the σ-bonding mode of L_2. Conversely, if L_1 is not capable of providing such stabilization (i.e., it is non π-bonding), π interaction with L_2 is enhanced. This is depicted in Fig. 3. Generally, this trans effect dictates the following stabilities: $\sigma,\pi \simeq \sigma,\sigma > \pi,\pi$.

The instability of "π,π" complexes is evidenced by the rapid isomerization of trans-$(h^2$-$C_2H_4)_2PtCl_2$ to the cis isomer (8) (Fig. 4).

Since the relative stability of the two species is dependent on the polarization of the metal center, it may also be correlated to the concept of "hard" and "soft" acids and bases (9). If σ-organic groups are "harder," owing to the small polarizability of the localized σ-orbitals, than π-groups having the delocalized π-electron cloud, "hardening" the metal by placing electron-withdrawing groups on it will favor the σ-complex. Conversely, "softening" the metal with nonelectron-withdrawing substituents will favor the π-complex.

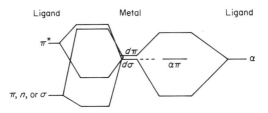

FIG. 3.   Schematic representation of molecular orbitals involving trans σ,π ligands.

$$trans\text{-}(h^2-C_2H_4)_2PtCl_2 \xrightarrow[-10°C]{ether} cis\text{-}(h^2-C_2H_4)_2PtCl_2$$

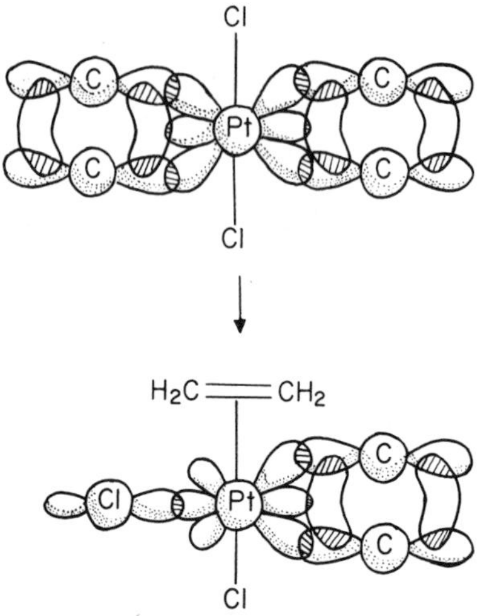

FIG. 4.   Isomerization of $trans$-$(h^2$-$C_2H_4)_2PtCl_2$ to the cis isomer.

III

INITIATION OF THE σ–π REARRANGEMENT

Most σ–π rearrangement reactions are initiated in one of two ways— reaction of the metal center or reaction on the ligand itself. However, some molecules are "fluxional," requiring no chemical promotion for rearrangement.

A. Reaction at the Metal Center

It is the change in electron density in the coordination sphere caused by addition or abstraction of a ligand that acts as the initiator. Loss of a ligand capable of dative π-bonding generally results in a σ–π rearrangement, and addition of such a ligand causes the reverse rearrangement. If loss of a ligand leaves a metal coordinatively unsaturated, and a rearrangement will satisfy the coordination requirement by donation of more than one pair, rearrangement will occur.

When σ-allyl complexes of h^5-cyclopentadienyltricarbonylmolybdenum (10) and tungsten (11) are irradiated with ultraviolet light, a σ–π

rearrangement occurs, yielding the π-allyl complex. Upon addition of carbon monoxide a $\sigma-\pi$ rearrangement is observed.

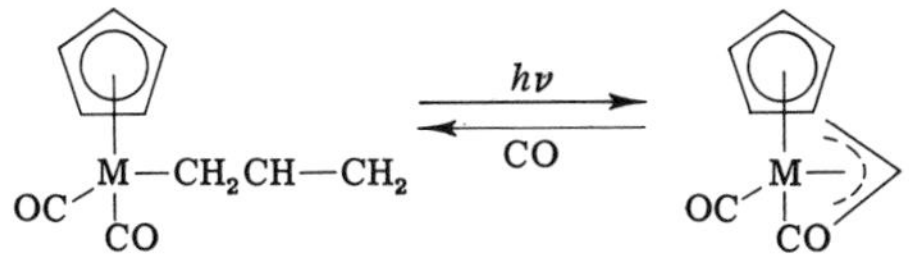

A similar rearrangement is observed with allyl complexes of cobalt cyanides (12).

Some interesting results were obtained when the system $L_2Rh(h^3\text{-}C_3H_4R)Cl_2$ [$L = C_6H_5P$, C_6H_5As, or C_6H_5Sb; $R = H$ or CH_3] was allowed to react with sulfur dioxide, ethylene, and carbon monoxide (13).

Reaction with sulfur dioxide yields a σ-allylic rhodium–sulfur dioxide complex (1).

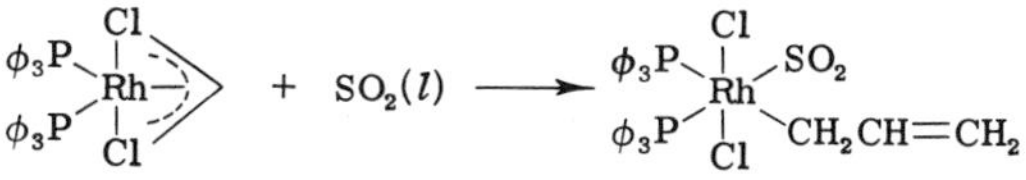

(1)

Reaction of the methallyl rhodium complex with ethylene leads to formation of a mixture of butenes, with isobutene formed from the methallyl group. The starting material is converted initially to an asymmetric π-methyallyl ethylene complex (2) which then reacts with another mole of ethylene to give a σ-methallylbis(ethylene) complex (3).

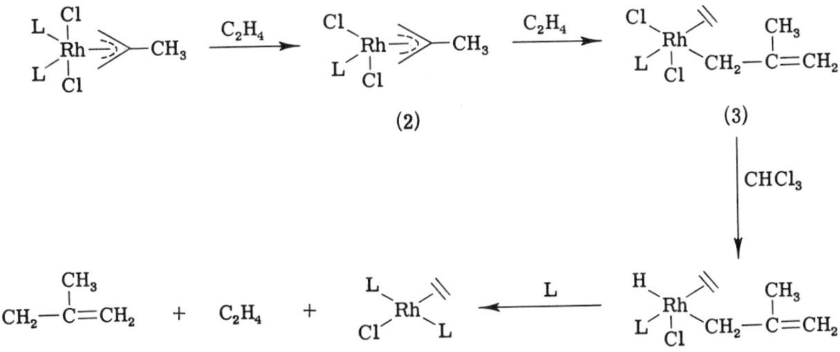

$L = \phi_3As$

FIG. 5.   Reaction of $(\phi_3As)_2Rh(h^3\text{-}C_4H_7)Cl_2$ with ethylene.

Solvent reaction may form a hydride (**4**), which, upon readdition of the ligand, L, yields isobutene, ethylene, and a rhodium–ethylene complex (Fig. 5).

Reaction with carbon monoxide yields methallyl chloride and a rhodium–carbonyl complex (Fig. 6).

σ-Cyclopentadienyl iron derivatives have been reported as intermediates in the formation of ferricenium chloride and ferrocene (*14*). The reaction of sodium cyclopentadienide with iron(III) chloride in a 2:1 molar ratio yields a σ-dicyclopentadienyl complex (**5**). This complex can be rearranged either thermally at $-50°C$ or chemically by addition of organic solvents, such as ether or pentane, to ferricenium chloride (**6**).

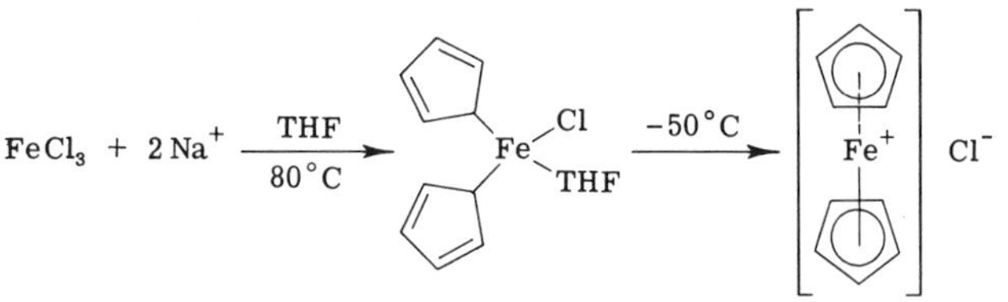

(6)

This rearrangement is initiated by loss of the labile tetrahydrofuran (THF) ligand, which leaves an open coordination site on the metal.

When the reactants are mixed in a 3:1 ratio, an intermediate (**7**) appearing to contain three σ-cyclopentadienyl groups is formed at $-80°C$. Upon allowing the temperature to rise to $-60°C$, rearrangement occurs that yields ferrocene and a polymeric cyclopentadiene. The proposed mechanism for this reaction is similar to that proposed in the formulation of Hein's π-arene chromium complex (*2*) (Fig. 7).

An irreversible π–σ rearrangement of cyclopentadienyl ligands has been observed in the reaction of titanocene dichloride and dimethyl sulfoxide (DMSO) (*15*).

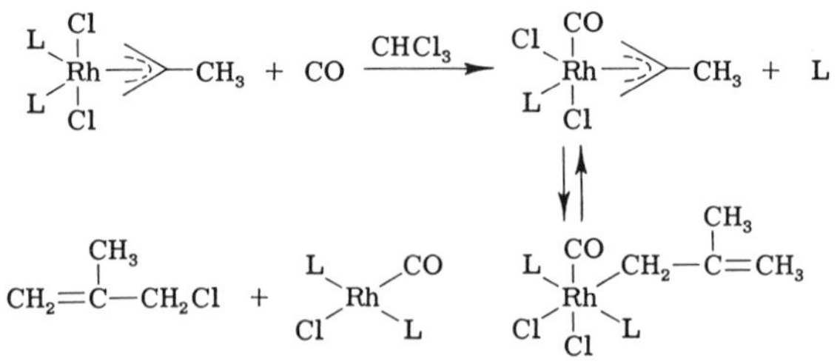

Fɪɢ. 6.   Reaction of $(\phi_3As)_2Rh(h^3-C_4H_7)Cl_2$ with carbon monoxide.

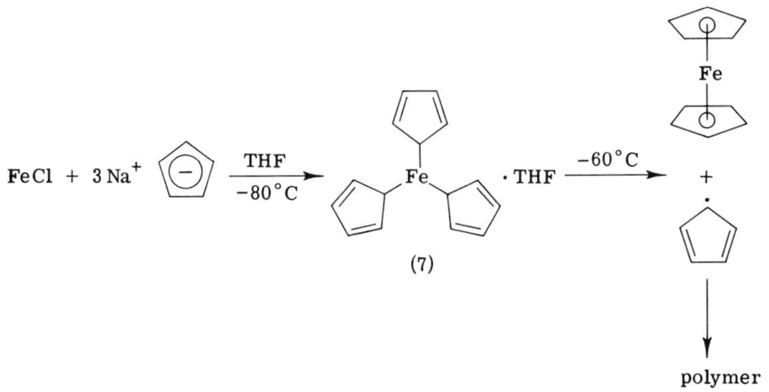

FIG. 7.   Formation of ferrocene.

Evidence for such a rearrangement product as (8) is found in its nuclear magnetic resonance (NMR) spectrum and in isolation of a maleic anhydride addition product (9).

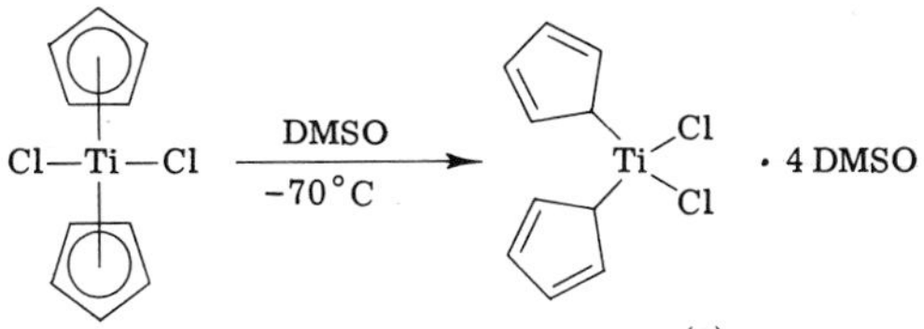

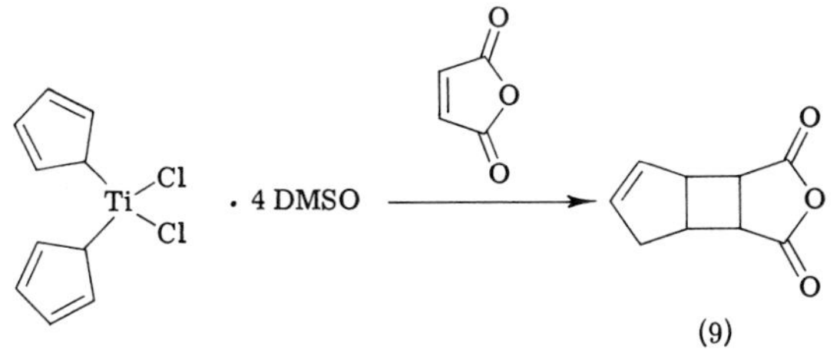

B. *Reaction on the Ligand*

Creating or removing unsaturation on the ligand can initiate a $\sigma-\pi$ rearrangement. Hydride abstraction and protonation reactions are perhaps the best-elucidated examples.

Some transition metal σ-alkyl complexes, with the alkyl group ethyl, *n*-propyl, or isopropyl, react with the triphenylmethyl cation forming an olefinic complex (*16*) (**10**).

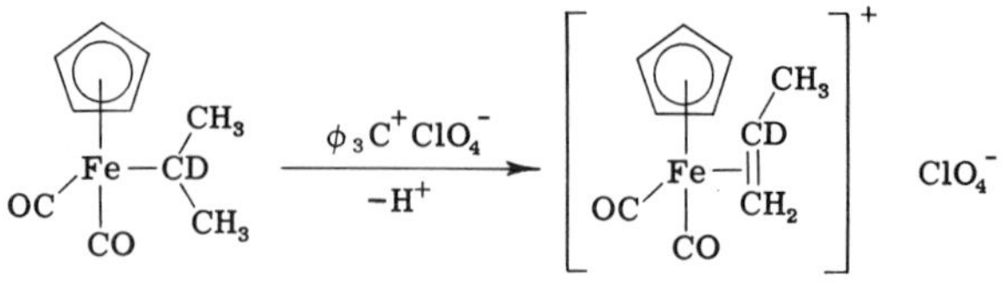

$\phi_3C^+ClO_4^-$
$-H^+$
ClO_4^-

(10)

An interesting complex having a vinyl alcohol moiety, chloro-(acetylacetonato)(h^2-ethenol)platinum(II) (**12**), was originally synthesized via a trimethylsilyl ether complex (**11**) (*17*).

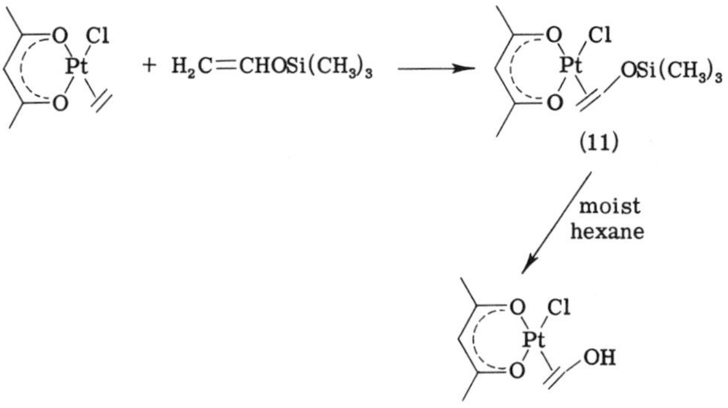

$+ H_2C{=}CHOSi(CH_3)_3$
$OSi(CH_3)_3$
(11)
moist hexane

(12)

This complex and the analogous π-propen-2-ol complex have since been more easily prepared through the formation of the σ-bonded carbonyl complex and subsequent protonation, which causes a σ–π rearrangement (*18, 19*).

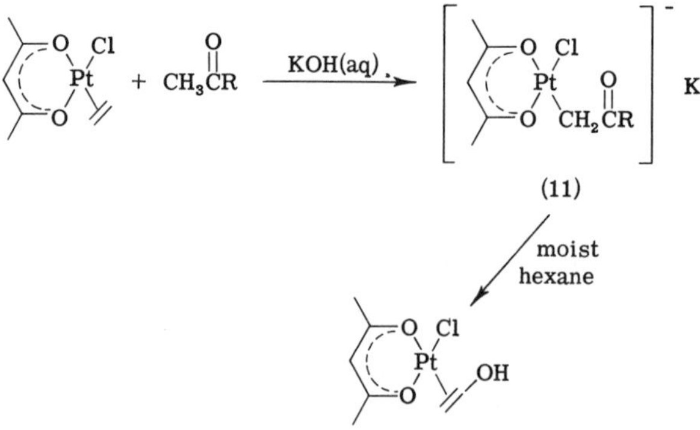

KOH(aq)
K^+
(11)
moist hexane

(12)

C. *Nonchemical Initiation*

Not all σ–π rearrangements require chemical initiation. Some metal complexes exist in a dynamic equilibrium between the σ and π forms.

NMR spectroscopy has shown that the allyl ligand in both platinum and rhodium allyl complexes at room temperature in deuterochloroform solution has all terminal hydrogens magnetically equivalent (20). This phenomenon may result from an interchange of the four allyl protons via a short-lived σ-allyl intermediate or transition state. As seen in Fig. 8, for such a rearrangement to take place a rotation around the C(1)–C(2) bond occurs, interchanging protons 1 and 2 concurrent with a rotation around the C(2)–C(3) bond interchanging protons 3 and 4.

Recently, the first organolanthanide complexes containing an allyl ligand $(h^5\text{-}C_5H_5)_2LnC_3H_5$, (where Ln = Sm Er, Ho) were reported (21).

$$(h^5\text{-}C_5H_5)_2LnCl + C_3H_5MgBr \xrightarrow[\,-78°C\,]{\text{THF—ether}} (h^5\text{-}C_5H_5)_3LnC_3H_5 + MgBrCl$$

The bonding scheme in these complexes is very interesting. Spectral data indicate h^3-bonding in these lanthanide complexes in contrast to the h^1-bonding which is observed in the analogous actinide complex $(h^5\text{-}C_5H_5)_3UC_3H_5$ (22).

In transition metal complexes, where the 18-electron rule is essentially obeyed, the bonding mode of the allyl ligand is mainly determined by the electronic requirements of the metal. However, this does not appear to be the case in the rare earth complexes. Instead, the bonding mode may be largely influenced by steric requirements. In the case of uranium, the coordination site necessary for π-bonding is highly constrained relative to that in the lanthanide complexes.

Temperature-dependent NMR studies have shown that the uranium allyl complex is fluxional. This fluxionality has been proposed as a $\sigma \rightleftarrows$

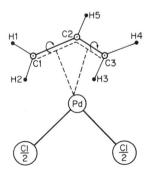

FIG. 8. Rotation necessary for interchange of allyl protons in the idealized structure of one half of the dimer $[(\pi\text{-allyl})PdCl]_2$.

$\pi \rightleftharpoons \sigma$ interconversion in which the h^3-configuration lies approximately 8–9 kcal/mole higher in energy than the h^1-configuration (22). Conversely, the lanthanide complexes do not appear to easily undergo π–σ rearrangements, as evidenced by the fact that addition of tetrahydrofuran, which coordinates to the complex, does not cause rearrangement. Apparently, the steric requirements are not such that coordinative saturation is reached.

Thus, preliminary results indicate that in lanthanides and actinides σ–π rearrangements are dominated by a combination of electrostatic and steric requirements whereas in the transition metals electrostatic considerations predominate.

IV

MECHANISTIC ASPECTS OF THE REARRANGEMENT

Although no one general mechanism has been established for the σ–π rearrangement, it appears that intermediates resulting from transition metal-β-interactions are often involved.

Even when a transition metal is coordinatively saturated, either filled or unfilled d-orbitals of suitable energy and orientation are available, which may interact with orbitals at the β-position on the ligand. The existence of such interactions has been established by spectral studies (23). In the systems where the β-position on the ligand is unsaturated, metal d-orbitals of suitable symmetry may be found that can weakly overlap with the π-orbitals of the β-substituent. If the β-position is saturated, metal d-orbitals may overlap with either an empty sp^3 lobe of the β-carbon or one of the sp^3 carbon-substituent orbitals. These interactions are summarized in Fig. 9.

The NMR spectrum (17, 19) of chloro(acetylacetonato)(h^2-ethenol) platinum (II) in polar solvents shows an A_2X pattern for the vinyl protons rather than the expected ABX pattern, such as that found for the π-vinylsilyl ether complex.

The π-vinyl alcohol complex is a moderately strong acid in rapid equilibrium ($k > 50$ sec^{-1}) (19) with the β-oxoethyl complex.

All the structural features obtained from X-ray crystallography (24)

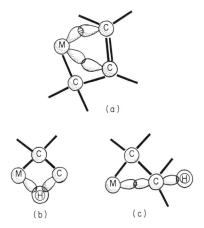

FIG. 9. Possible transition metal β-interactions: (a) ligand with unsaturated β-carbon; (b) and (c) ligand with saturated β-carbon.

can be interpreted in terms of a bonding model that is intermediate between a conventional π-olefin complex and a σ-bonded aldehyde (Fig. 10); i.e., the π-vinyl alcohol complex is probably a direct result of a β-interaction. This scheme is supported by the following observations: (1) The principal coordination plane of the platinum atom does not bisect the C—C bond or include the methylene carbon atom. (2) Both carbon atoms are within bonding distance of the platinum, but the two Pt—C distances are significantly different. (3) The C—C bond length implies that the bond order is intermediate between a single and a double bond, as is the C—O bond.

The overall result of the β-interaction is the formation of a somewhat electron-deficient β-atom or a cyclic system of orbital overlap having fewer electrons than orbitals. Thus, in going from the σ-species to the

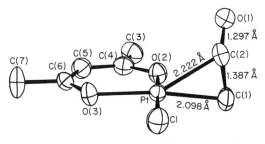

FIG. 10. Structure of (acac)Pt(h²-C_2H_3OH)Cl. acac = acetylacetonate group.

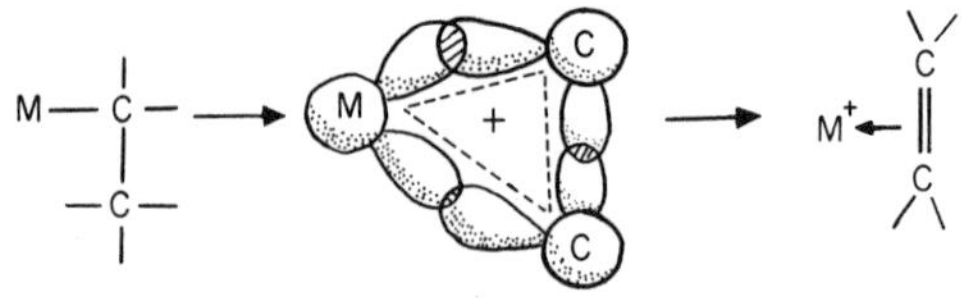

FIG. 11.   Possible route from σ-bonded ligand to π-bonded ligand.

π-species, the electron-deficient transition state will be stabilized by electron donation from the metal forming the π-bond and partial oxidation of the metal as shown in Fig. 11.

The reversible insertion of ethylene into the platinum–hydrogen bond (25, 26) may proceed by this type of mechanism (Fig. 12).

V

THE ROLE OF THE $\sigma-\pi$ REARRANGEMENT IN CATALYSIS

A. Oxidation of Olefins

Olefins can be oxidized by a variety of metal ions, such as platinum(II), palladium(II), rhodium(II), and the isoelectronic series mer-

$$trans\text{-}[PtHCl(PEt_3)_2] \xrightarrow[180°C]{\overset{CH_2=CH_2}{\overset{90°C, 40 \text{ atm}}{}}} trans\text{-}[PtClEt(PEt_3)_2]$$

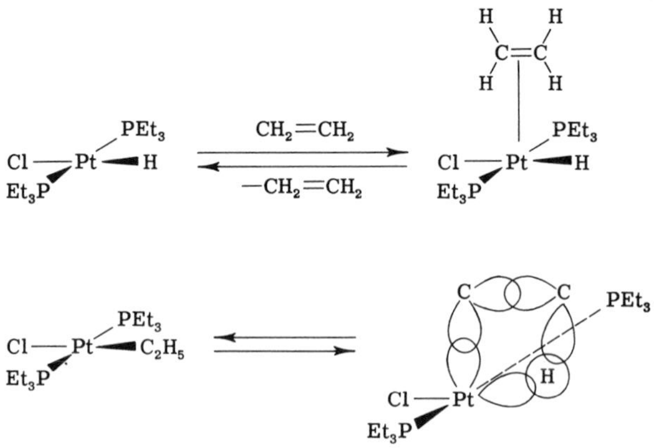

FIG. 12.   Insertion of ethylene into platinum–hydrogen bonds.

cury(II), thallium(III), and lead(IV). This process has been generally described as a cis-ligand insertion between a metal and a hydroxide ion bond followed by a rearrangement. The metal catalyst with appropriate oxidizing agents participates in a two-electron oxidation–reduction cycle.

Oxidation of olefins was first indirectly observed in the hydrolysis of the trichloro(h²-ethylene)platinum(II) anion (27).

$$[C_2H_4PtCl_3]^- + H_2O \rightarrow CH_3CHO + 2HCl + Pt° + Cl^-$$

Palladium(II) catalysts have been found to be more effective than platinum(II), presumably owing to the lesser stability of the olefin complexes of the former resulting from minimized back-bonding in palladium. By reducing the electron density around the double bond, nucleophilic attack on the double bond is facilitated.

The palladium chloride–copper(II) chloride couple (28, 29) used industrially in the Wacker process oxidizes olefins to carbonyl compounds. Experimental kinetic and isotope effect data (30) seem to indicate that a π-olefin complex is initially formed in a series of preequilibrium steps. The rate-determining step is postulated to be a rearrangement of the π-olefin complex to a σ-complex followed by the final breakdown of the σ-complex to products. Figure 13 depicts the widely accepted Henry mechanism (31).

This mechanism involves cis "insertion" of the olefin into the Pd—OH bond. Since the reaction rate has only a slight dependence on the olefin structure, it has been suggested that the transition state has little carbonium ion character, and the π-σ rearrangement may proceed via a concerted, nonpolar, four-center transition state (32) (13).

(13)

Recently, complete neglect of differential overlap (CNDO) calculations have been performed on this system to help elucidate the mechanism (33) (Fig. 14). These calculations indicate that a palladium–ethylene complex is formed initially. As the reaction progresses, a C—O bond is formed at the expense of Pd—C and C—C bonding. The Pd—C(1) interaction becomes nonbonding, and the double bond character diminishes. This step—the π-σ rearrangement—has a relatively small activation energy and is the rate-determining step. It is interesting to note that metal d-orbitals act as the transfer agent so that, as the hydroxide moves

$$C_2H_4 + [PdCl_4]^{2-} \rightleftharpoons [C_2H_4PdCl_3]^- + Cl^-$$

$$[C_2H_4PdCl_3]^- + H_2O \rightleftharpoons [C_2H_4PdCl_2(OH)]^- + Cl^-$$

$$[C_2H_4PdCl_2(H_2O)] \rightleftharpoons [C_2H_4PdCl_2(OH)]^- + H_3O^+$$

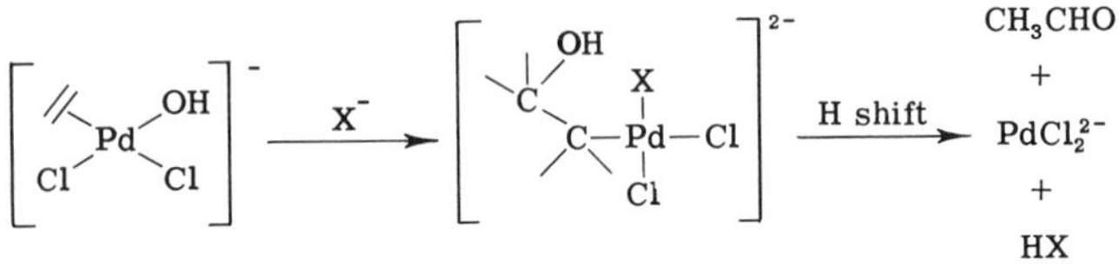

$$\left[\begin{array}{c} \overset{//}{\underset{Cl}{\diagdown}} Pd \overset{OH}{\diagdown} \\ Cl \end{array} \right]^- \xrightarrow{X^-} \left[\begin{array}{c} OH \\ \underset{\diagdown}{\overset{/}{C}} \overset{X}{\diagdown} \\ C - Pd - Cl \\ | \\ Cl \end{array} \right]^{2-} \xrightarrow{\text{H shift}} \begin{array}{c} CH_3CHO \\ + \\ PdCl_2^{2-} \\ + \\ HX \end{array}$$

FIG. 13. The Henry mechanism for the Wacker process.

toward the ethylene, it remains in the coordination sphere of the metal. Acetaldehyde is then formed by a hydrogen shift in which metal orbitals aid in the transfer.

The major difference between this mechanism and that of Henry is the fact that a cis arrangement of hydroxide and olefin, which would be unlikely owing to the trans directing influence of ethylene on the entering water molecules, is not necessary.

The ions mercury(II) (34), thallium(III) (35), and lead(IV) (36) oxidize olefins to carbonyls and glycols with a metal reduction, as in the case of palladium.

Thallium(III) in acetic acid oxidizes 2-hexene to 2-hexanone, 2,3-hexanediol, and a monoester with reduction to thallium(I) (35). The

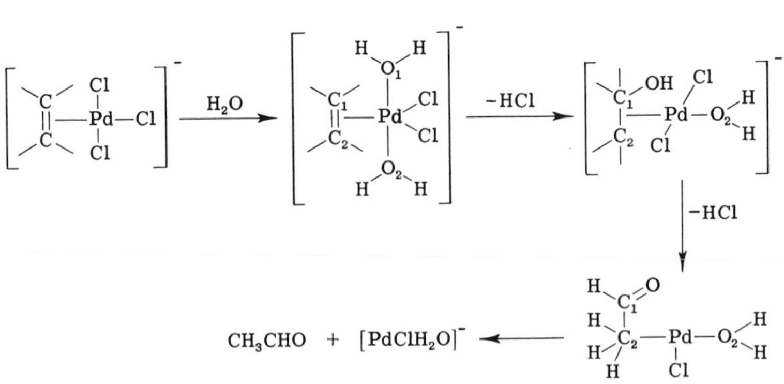

FIG. 14. Mechanism for oxidation of olefins based on complete neglect of differential overlap (CNDO) calculations.

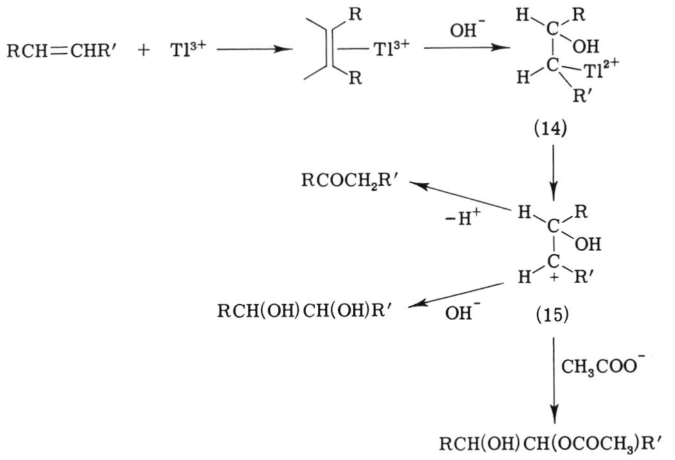

$$RCH{=}CHR' + Tl^{3+} \longrightarrow \text{(π-complex)}{-}Tl^{3+} \xrightarrow{OH^-} \underset{\substack{R'}}{\overset{\substack{H\diagdown C\diagup R \\ \quad\quad OH \\ H\diagup C{-}Tl^{2+}}}{}}$$

$$(14)$$

$$RCOCH_2R' \xleftarrow{-H^+} \underset{\substack{H\diagup C^+\diagdown R'}}{\overset{\substack{H\diagdown C\diagup R \\ \quad\quad OH}}{}}$$

$$RCH(OH)CH(OH)R' \xleftarrow{OH^-} \quad (15)$$

$$\downarrow CH_3COO^-$$

$$RCH(OH)CH(OCOCH_3)R'$$

FIG. 15.   Oxidation of 2-hexenes by thallium(III) in acetic acid.

mechanism (Fig. 15) for this reaction is essentially the same as that of palladium for the first two steps—formation of a π-complex and reaction with hydroxide inducing a $\pi{-}\sigma$ rearrangement. Intermediates such as the β-hydroxy complex (14) have been isolated (37–39). This intermediate may dissociate into a carbonium ion (15), which then reacts to form the mixture of products. A similar mechanism has been proposed by Henry (Fig. 16) (40, 41).

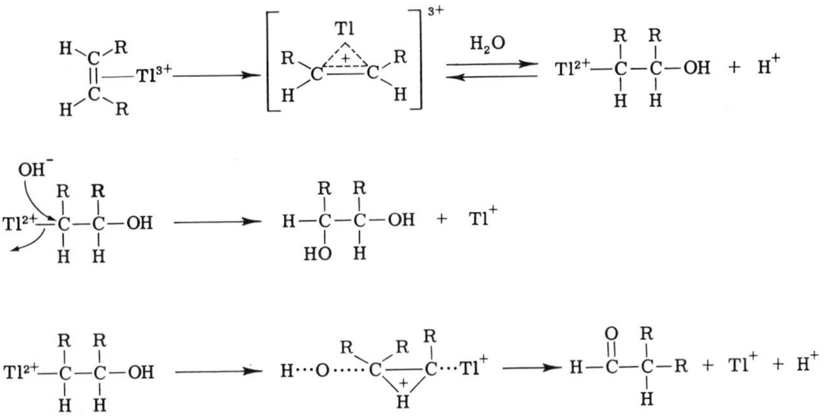

$$\underset{\substack{H\diagup C\diagdown R}}{\overset{\substack{H\diagdown C\diagup R \\ \parallel}}{}}{-}Tl^{3+} \longrightarrow \left[\overset{Tl}{\underset{\substack{H}}{\overset{R}{C}}{\cdots}\underset{\substack{H}}{\overset{R}{C}}} \right]^{3+} \xrightarrow{H_2O} Tl^{2+}{-}\underset{\substack{H\ H}}{\overset{R\ R}{C{-}C}}{-}OH + H^+$$

$$\underset{\substack{OH^-}}{} Tl^{2+}{-}\underset{\substack{H\ H}}{\overset{R\ R}{C{-}C}}{-}OH \longrightarrow H{-}\underset{\substack{HO\ H}}{\overset{R\ R}{C{-}C}}{-}OH + Tl^+$$

$$Tl^{2+}{-}\underset{\substack{H\ H}}{\overset{R\ R}{C{-}C}}{-}OH \longrightarrow H{\cdots}O{\cdots}\underset{\substack{H}}{\overset{R\quad R}{C{-}C}}{\cdots}Tl^+ \longrightarrow H{-}\underset{\substack{H}}{\overset{O\ R}{C{-}C}}{-}R + Tl^+ + H^+$$

FIG. 16.   Mechanism proposed by Henry for oxidation of 2-hexene by thallium(III).

B. *Olefin Polymerization: The Ziegler–Natta Process*

The discovery of a catalyst system (*42*) consisting of a transition metal halide, such as $TiCl_4$, and a nontransition metal alkyl, such as $AlEt_3$, for the low-temperature polymerization of olefins by Ziegler and Natta has led to extensive experimental work in this area.

Although many mechanisms have been proposed, perhaps the most widely favored is that of Cossee (*43, 44*). Generally, as depicted in Fig. 17, the process is seen as proceeding by initial alkylation of the titanium halide by the alkyl aluminum yielding an octahedral titanium complex having a vacant coordination site. An incoming ethylene molecule coordinates to the vacant site, which is followed by alkyl migration to the ethylene and a $\pi-\sigma$ rearrangement.

Recently, elucidation of the mechanism of this catalytic process has been attempted by using self-consistent field, all-valence electron calculations in an effort to obtain a quantitative theoretical rationalization of the electronic structures and geometries of the intermediate complexes. Perkins *et al.* (*45*) provided the initial work in this area on a "soluble" catalyst system formed by combination of $TiCl_4$ and $MeAlCl_2$. In general, the results obtained were in agreement with the Cossee mechanism. Some of the main features presented were as follows: (1) at all the intermediate stages of the process both the alkyl group and the ethylene ligand can remain bonded to the titanium atom, and therefore, no large bond-breaking energy is required; and (2) a metal *d*-orbital is utilized as a transfer agent, allowing the alkyl group to migrate and link to the ethylene ligand.

More recently, calculations have been done to show the characteris-

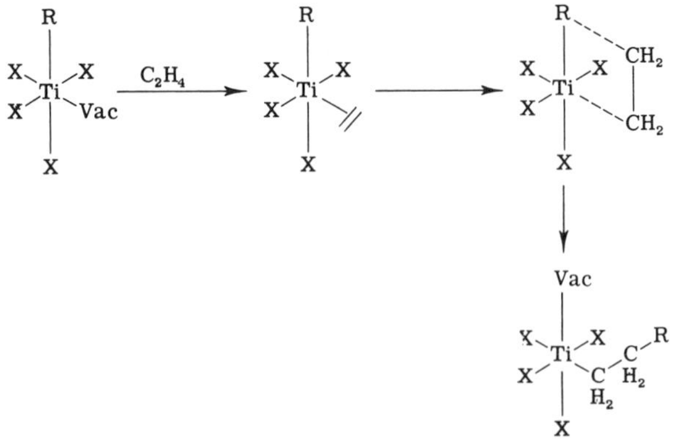

FIG. 17.   Cossee mechanism for Ziegler–Natta process.

tics of the addition of ethylene to the catalyst, the nature of the driving force of the chain propagation, and the manner in which the end products are liberated (46, 47). The system used in this study was the catalyst formed from combination of $Ti(OCH_3)_4$ and $AlEt_3$, a model based on the actual catalytic system formed by the reaction of $AlEt_3$ and $Ti(O\phi$—$pCH_3)_3O$—nBu, which oligomerizes ethylene to 1-butene and 1-hexene.

The formation of the initial catalyst complex has been postulated to follow the scheme in Fig. 18, which has a 2:1 Al/Ti and a reduced titanium(III) atom. This catalytic structure has been experimentally confirmed by its EPR spectrum (46, 48).

Calculation has shown that the most stable configuration of the initial catalyst is not the octahedral complex having a vacant site, as the Cossee mechanism proposes, but rather, a trigonal–bipyramidal complex which is 1.43 eV lower in energy. This complex has an unpaired electron strongly localized in the highest stable orbital (d_{xz}), which interacts with the titanium–alkyl bond.

Upon addition of ethylene, the complex assumes an octahedral configuration in which the d_{xz}-orbital interacts with both the ethylene and alkyl groups. Thus, the titanium d_{xz}-orbital acts as a transfer agent, as it favors movement of the labile alkyl ligand toward the electron-rich α-carbon of the ethylene. Therefore, an appreciable energy barrier is not present. The final oligomer liberation has been attributed to a β-hydrogen abstraction from the alkyl chain forming a carbon–carbon double bond while breaking the titanium–carbon bond. This is a direct consequence of the fact that, as the chain length increases, the alkyl

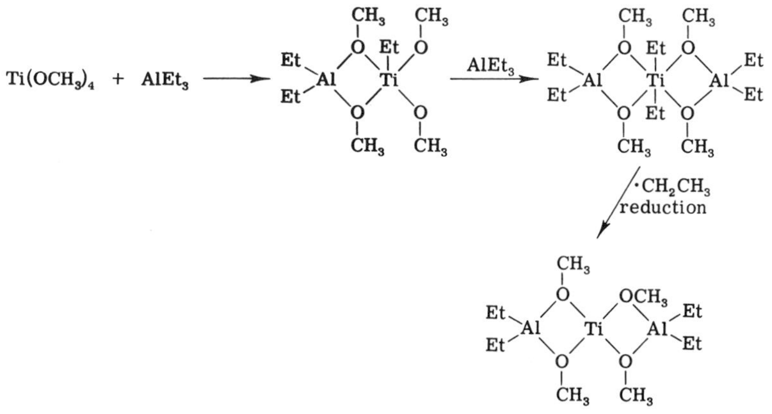

FIG. 18.   Postulated formation of initial catalytic complexes.

β-hydrogens approach close enough to the C—2 ethylene carbon to form a pseudohydrogen bond. Then, the negative character of the ethylene carbon abstracts the hydrogen, forming a carbon–carbon single bond and weakening the titanium–alkyl bond. Thus, the oligomer is eliminated, and the catalyst is regenerated. The important stages of the oligomerization are depicted in Fig. 19.

The driving force of this catalytic process appears to be the low-energy pathway provided by the titanium d_{xz}-orbital. Also, both the propagation and liberation steps are strongly related; i.e., the propagation sequence has an alternative at each point of liberating an α-olefin or lengthening the chain. This relationship fits the observed kinetic data (49), since rate constants for both chain propagation and chain liberation depend on the concentration of active sites, Ti*, and monomer.

$$r_p = k_p[\text{Ti}^*][\text{C}_2\text{H}_4]$$

$$r_l = k_l[\text{Ti}^*][\text{C}_2\text{H}_4]$$

The type of mechanism utilized by the Ziegler–Natta process also

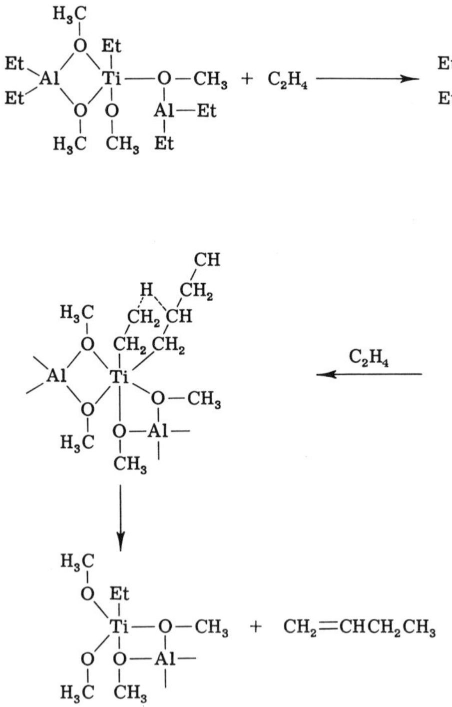

Fig. 19. Important stages of oligomerization.

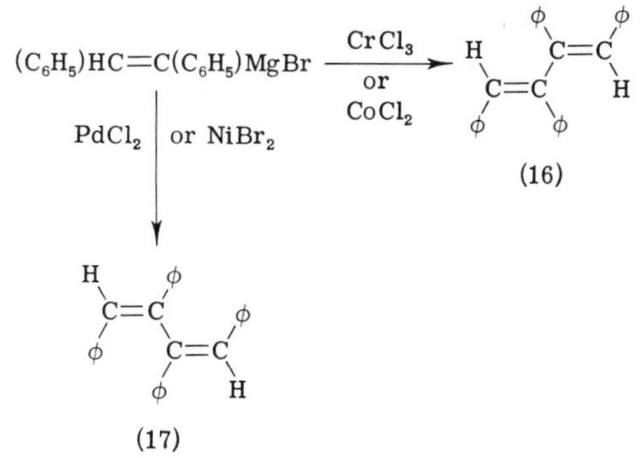

$$(C_6H_5)HC{=}C(C_6H_5)MgBr \xrightarrow[\substack{or\\CoCl_2}]{CrCl_3}$$

(16)

(17)

Fig. 20. Stereospecific coupling of 1,2-disubstituted vinyl groups.

seems to explain coupling and polymerization reactions of a wide variety of alkenes by various transition metal compounds.

Stereospecific coupling of 1,2-disubstituted vinyl groups is attained by reaction of a vinyl Grignard with $CrCl_3$, $CoCl_2$, $PdCl_2$, and $NiBr_2$ catalysts (50, 51). The stereospecificity of the products appears to result from the selective configuration of the transition metal halide (Fig. 20).

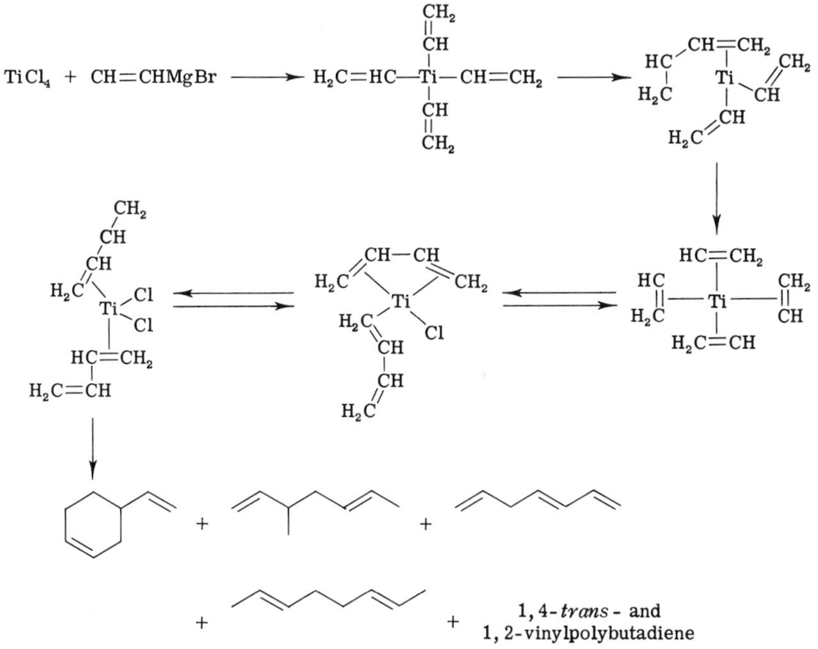

$$TiCl_4 + CH{=}CHMgBr \longrightarrow H_2C{=}HC{-}Ti{-}CH{=}CH_2 \longrightarrow$$

1,4-*trans*- and
1,2-vinylpolybutadiene

Fig. 21. The "double coupling" reaction.

A cis-coupling product (16) is obtained from the $CrCl_3$ or $CoCl_2$ catalyst, which can take on an octahedral geometry (d^2sp^3 hybridization), whereas a trans-coupling product (17) is obtained from $PdCl_2$ or $NiBr_2$, which can take on a square-planar geometry (dsp^2 hybridization). The reaction probably proceeds from a σ-vinyl metal complex, which rearranges to a coupled π-butadiene complex.

A "double coupling" reaction occurs when vinyl Grignard and titanium tetrachloride are combined (52). One possible mechanism is shown in Fig. 21 (53).

C. Olefin Isomerization

Isomerization of olefins—redistribution of carbon–carbon double bonds—often occurs in the presence of transition metal complexes such as metal carbonyls, platinum group metal ions, and silver(I) compounds. It appears that such catalytic activity has two requirements. (1) The metal complex must contain labile ligands that can easily be displaced by an olefin to form a π-complex or must be able to easily expand its coordination. (2) The metal complex must be kinetically labile so that the isomerized olefin can be displaced by another olefin to continue the cycle.

Presumably, the isomerization occurs via olefin interaction with the transition metal to give an intermediate in which the metal atom labilizes an hydrogen attached to the β-carbon.

The most attractive mechanism for this isomerization is a metal hydride addition–elimination mechanism (Fig. 22). Initial complexing occurs between an olefin and the metal complex, which is followed by the addition of a hydrogen causing a π–σ rearrangement. Next, elimination of an hydrogen occurs in the opposite direction by a β-interaction with eventual release of the isomerized olefin.

Nickel-catalyzed isomerizations are thought to proceed by this mechanism. In general, in these reactions, isomerization occurs in a stepwise manner and the final isomer concentrations approach the thermodynamic equilibrium values. The ease of isomerization seems to follow the order of the complexing ability of olefins: terminal > disubstituted >

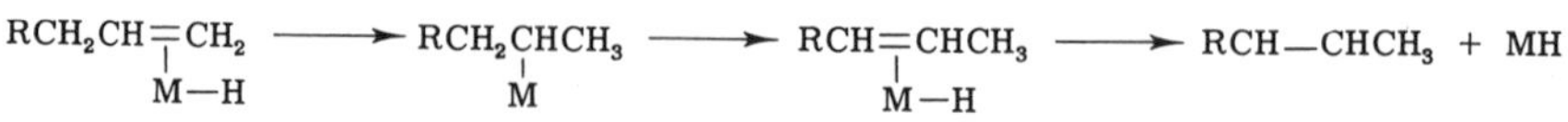

$$RCH_2CH{=}CH_2 \longrightarrow RCH_2CHCH_3 \longrightarrow RCH{=}CHCH_3 \longrightarrow RCH{-}CHCH_3 + MH$$
$$\underset{M-H}{\vert} \qquad\qquad \underset{M}{\vert} \qquad\qquad \underset{M-H}{\vert}$$

FIG. 22.   Elimination of a hydrogen by a β-interaction.

trisubstituted. These points are illustrated by the isomerization of 4-methyl-1-pentene by π-C$_3$H$_5$NiCl[P(CH$_3$)$_3$]-C$_2$H$_5$AlCl$_2$ (54).

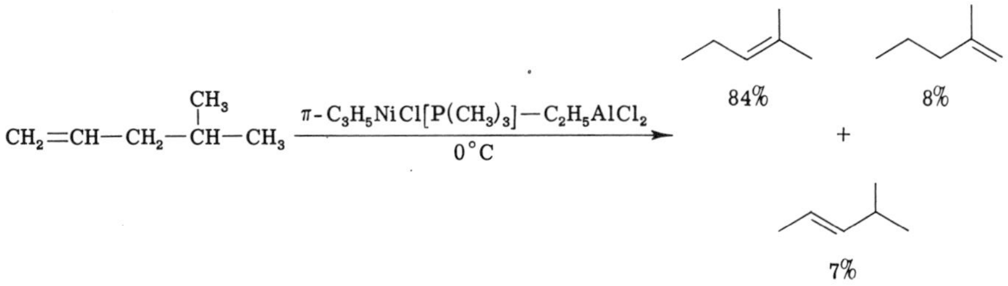

$$CH_2=CH-CH_2-\underset{\underset{CH_3}{|}}{CH}-CH_3 \xrightarrow[0°C]{\pi-C_3H_5NiCl[P(CH_3)_3]-C_2H_5AlCl_2}$$

84%   8%

+

7%

If the reaction is carried out at low temperatures, the individual isomerization steps can be observed (Fig. 23).

1-Pentene is isomerized to *cis*- and *trans*-2-pentene by tetrakis-(triethylphosphite)nickel (55).

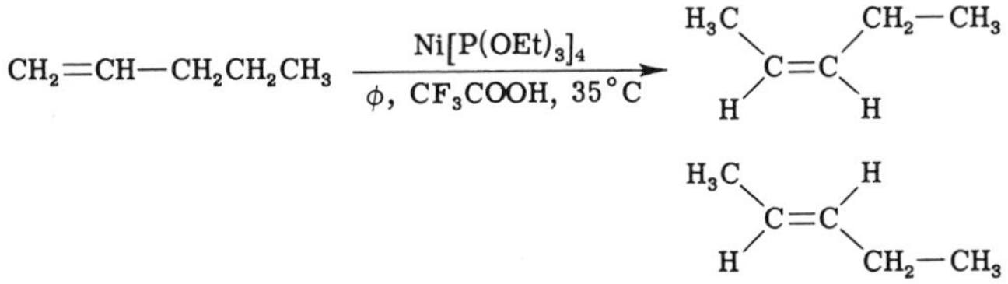

$$CH_2=CH-CH_2CH_2CH_3 \xrightarrow[\phi,\ CF_3COOH,\ 35°C]{Ni[P(OEt)_3]_4}$$

$$\underset{H}{\overset{H_3C}{>}}C=C\underset{H}{\overset{CH_2-CH_3}{<}}$$

$$\underset{H}{\overset{H_3C}{>}}C=C\underset{CH_2-CH_3}{\overset{H}{<}}$$

This reaction has provided strong evidence that the hydride addition-elimination pathway is utilized.

The active catalyst in the reaction is believed to be the cationic nickel hydride NiH[P(OEt$_3$)]$_4$$^+$, which has been observed to form in the solvent system (56).

$$Ni[P(OEt)_3]_4 + CF_3COOH \xrightarrow{\hspace{1cm}} NiH[P(OEt)_3]_4 + CF_3CO_2^-$$

$$CH_2=CHCH_2\underset{\underset{CH_3}{|}}{CH}CH_3 \xrightarrow[-58°C]{\pi-C_3H_5NiCl[P(CH_3)_3]-C_2H_5AlCl_2}$$

$$\underset{H}{\overset{H_3C}{>}}C=C\underset{H}{\overset{\underset{CH_3}{|}}{\overset{CHCH_3}{}}}$$

$\downarrow$

$$\xleftarrow{-40°C} \quad \underset{H}{\overset{H_3C}{>}}C=C\underset{\underset{\underset{CH_3}{|}}{CHCH_3}}{\overset{H}{}}$$

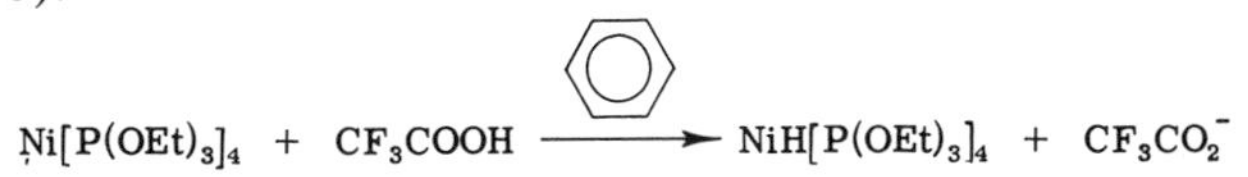

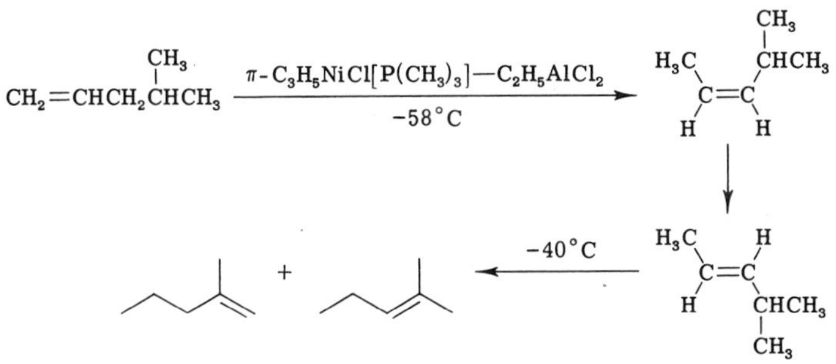

+

Fig. 23. Isomerization steps observed if the reaction is carried out at low temperatures.

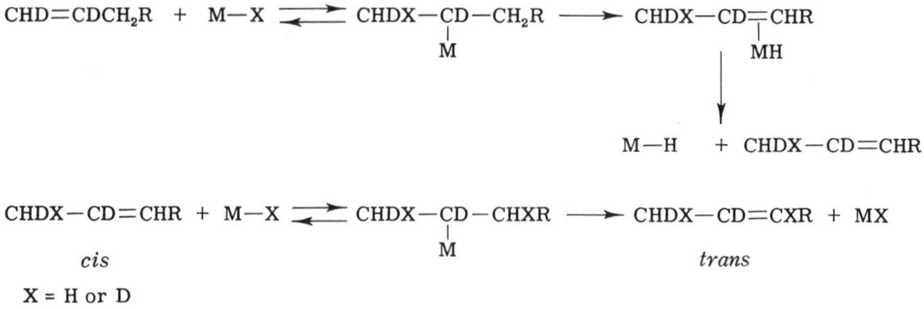

$$CHD{=}CDCH_2R \; + \; M{-}X \; \rightleftharpoons \; \underset{M}{CHDX{-}CD{-}CH_2R} \; \longrightarrow \; CHDX{-}\underset{MH}{CD{=}CHR}$$

$$M{-}H \; + \; CHDX{-}CD{=}CHR$$

$$\underset{cis}{CHDX{-}CD{=}CHR} \; + \; M{-}X \; \rightleftharpoons \; \underset{M}{CHDX{-}CD{-}CHXR} \; \longrightarrow \; \underset{trans}{CHDX{-}CD{=}CXR} \; + \; MX$$

X = H or D

FIG. 24.   Mechanism formulated on the basis of studies using $1,2\text{-}d_2\text{-}1\text{-}pentene$.

Studies using $1,2\text{-}d_2\text{-}1\text{-}pentene$ have led to the formulation of the mechanism shown in Fig. 24.

Further support for this mechanism comes from the fact that butadiene, which reacts with the nickel-hydride forming a stable π-allyl complex, hinders the isomerization (57).

It has been observed that the isomerization of 1-olefins by palladium complexes shows simultaneous appearance of all the possible internal isomers at comparable rates (58, 59) rather than the stepwise formation shown in the nickel systems. It has been proposed that perhaps the reaction proceeds by some type of π-allyl mechanism (60). Two mechanistic schemes have been proposed for the isomerization of olefins by iron pentacarbonyl (61, 62). One is analogous to that of the nickel isomerizations, and the other involves a π-allyl intermediate (Fig. 25).

The metal center abstracts a hydrogen from the coordinated olefin to form a π-allyl complex. Readdition of hydrogen and concurrent double-bond migration cause the isomerization. In support of this mechanism, it has been shown that allyl alcohol is isomerized by iron pentacarbonyl to propionaldehyde (63) (Fig. 26).

Thus, it appears that different catalyst systems may utilize different reaction pathways for isomerization. It is also possible that different substrates or different reaction media may cause mechanistic variations for the same catalyst.

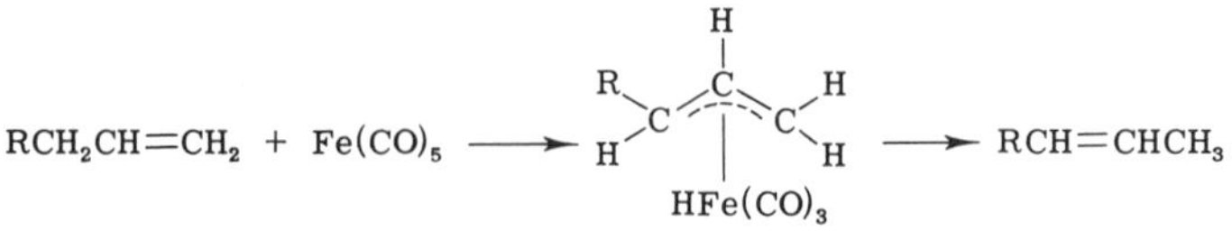

$$RCH_2CH{=}CH_2 \; + \; Fe(CO)_5 \; \longrightarrow \; \underset{HFe(CO)_3}{\overset{\displaystyle R{-}C{\cdots}\overset{H}{\overset{|}{C}}{\cdots}C{-}H}{}} \; \longrightarrow \; RCH{=}CHCH_3$$

FIG. 25.   Proposed scheme for the isomerization of olefins by iron pentacarbonyl involving a π-allyl intermediate.

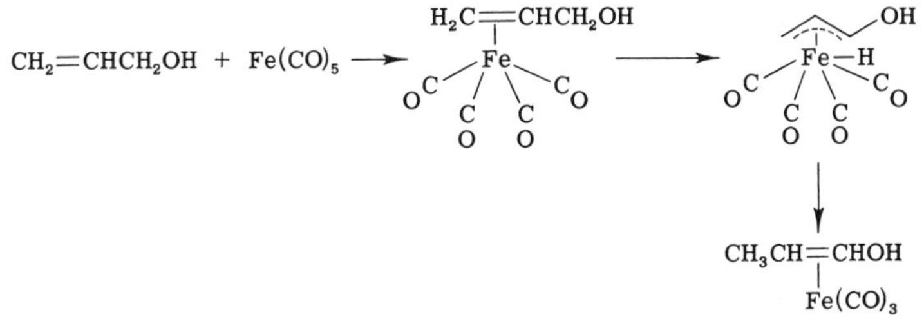

$$CH_2\!=\!CHCH_2OH \ + \ Fe(CO)_5 \ \longrightarrow$$

$$CH_3CH\!=\!CHOH$$
$$\underset{Fe(CO)_3}{|}$$

Fɪɢ. 26.   Isomerization of allyl alcohol to propionaldehyde by iron pentacarbonyl.

D. *Hydrogenation*

It has been found that molecular hydrogen can be activated for the hydrogenation of unsaturated compounds by transition metal complexes that have an electronic configuration in the d^5–d^{10} range, although Mn^{2+}, Co^{2+}, Ni^{2+}, and Fe^{3+} are inactive. It is apparent that catalytic activity is related to the ability of the ion to form labile hydrido complexes in solution.

Maleic and fumaric acids can be catalytically hydrogenated to succinic acid with ruthenium(II) chloride in aqueous solution *(64)*.

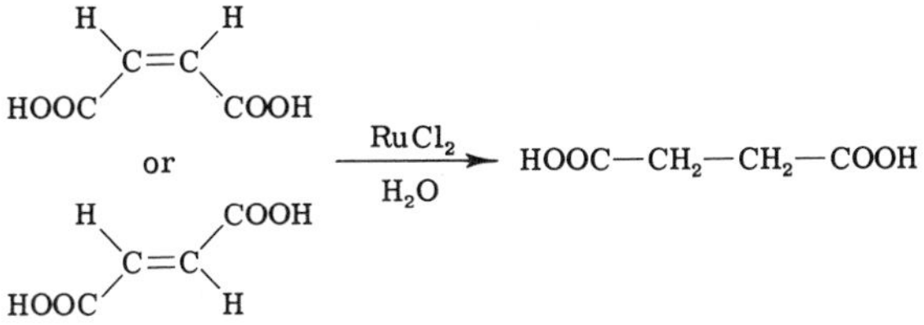

The addition of olefinic compounds to the catalyst in solution corresponds to an uptake of 1 mole of olefin per mole of ruthenium(II), which apparently results in the formation of a 1:1 ruthenium–olefin π-complex.

$$Ru^{2+} + olefin \rightleftarrows Ru^{2+} (h^2\text{-olefin})$$

Kinetic measurements of the uptake of H_2 by the olefin shows that the rate of reaction (r) is first order in both hydrogen and olefin complex.

$$r = k[H_2][Ru^{2+}(h^2\text{-olefin})]$$

This is the rate-determining step involving heterolytic splitting of the hydrogen molecule and formation of an hydridoruthenium(II) complex. The next step involves rearrangement of the hydrido–π-olefin complex to a σ-alkyl complex via "insertion" of the olefin into the metal hydride bond. Finally, electrophilic attack occurs on the metal-bonded carbon atom by a proton, completing the hydrogenation reaction and regenerating the original catalyst. This mechanism is depicted in Fig. 27.

It has been observed that only olefins possessing a carboxyl group adjacent to the double bond can be hydrogenated by this system. For example, 5-norbornene-2,3-dicarboxylic anhydride is not susceptible to hydrogenation. It has been suggested that the carboxyl substituent is associated with the activation of the π–σ rearrangement because the electron-withdrawing substituent will increase the rate of hydrometallation by favoring nucleophilic attack of the hydrido ligand on the double bond. Thus, it appears that the σ–π rearrangement is an integral part of the catalytic process.

Pentacyanocobaltate(II) has been found to catalyze the hydrogenation of conjugated unsaturated hydrocarbons (65–67). When butadiene is hydrogenated in the presence of $[Co(CN)_5]^-$ and excess cyanide ion, a mixture of isomers is formed that is dependent on the molar ratio of [CN]/[Co]. If a [CN]/[Co] ratio of less than 5.5 is used, trans-2-butene is formed in about 86% yield. At a ratio of 6.0 or more, 1-butene predominates. This change in product ratios indicates that perhaps two different butadiene complexes are formed—a π-allyl complex at lower [CN]/[Co] values, which is intermediate in the formation of trans-2-butene, and a σ-complex at high [CN]/[Co] values, which is intermediate in the formation of 1-butene. These two complexes probably exist in an equilibrium that shifts with changes in cyanide concentration. Thus, the dependency of the product on [CN] is apparently related to differences in the steric requirements of π-allyl and σ-butenyl groups. At low [CN]/[Co] ratios, there is a lesser hindrance for the replacement of a cyano group by a π-allyl group that will occupy two coordination sites than at high [CN]/[Co] ratios. A mechanism for this process is given in Fig. 28.

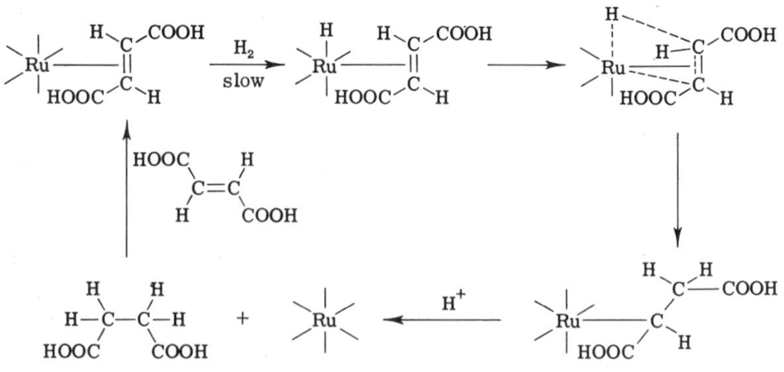

FIG. 27.   Mechanism for the hydrogenation of fumaric acid.

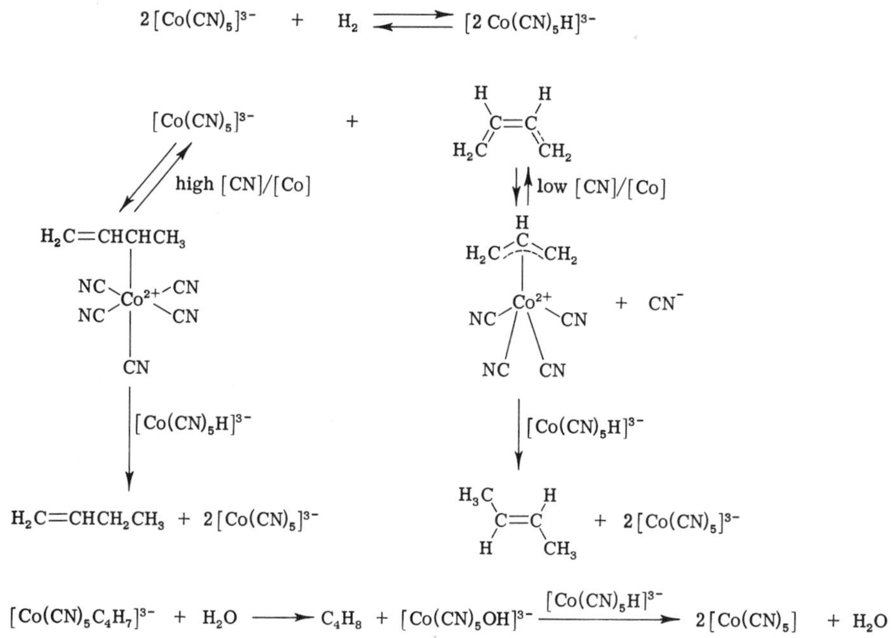

$$2\,[\mathrm{Co(CN)_5}]^{3-} \quad + \quad \mathrm{H_2} \;\rightleftharpoons\; [2\,\mathrm{Co(CN)_5H}]^{3-}$$

$$[\mathrm{Co(CN)_5}]^{3-} \quad + \quad \overset{\text{H}\quad\text{H}}{\underset{\text{H}_2\text{C}\quad\text{CH}_2}{\text{C}=\text{C}}}$$

high [CN]/[Co]   low [CN]/[Co]

$$\mathrm{H_2C=CHCHCH_3}$$

$$\underset{\text{CN}}{\overset{\text{NC}\quad\text{CN}}{\text{Co}^{2+}}}\text{-CN}$$

$$\underset{\text{NC}\quad\text{CN}}{\overset{\text{H}_2\text{C}\quad\text{CH}_2}{\text{Co}^{2+}}} + \text{CN}^-$$

$$[\mathrm{Co(CN)_5H}]^{3-}$$

$$\mathrm{H_2C=CHCH_2CH_3} + 2\,[\mathrm{Co(CN)_5}]^{3-}$$

$$\underset{\text{H}\quad\text{CH}_3}{\overset{\text{H}_3\text{C}\quad\text{H}}{\text{C}=\text{C}}} + 2\,[\mathrm{Co(CN)_5}]^{3-}$$

$$[\mathrm{Co(CN)_5C_4H_7}]^{3-} + \mathrm{H_2O} \longrightarrow \mathrm{C_4H_8} + [\mathrm{Co(CN)_5OH}]^{3-} \xrightarrow{[\mathrm{Co(CN)_5H}]^{3-}} 2\,[\mathrm{Co(CN)_5}] + \mathrm{H_2O}$$

FIG. 28. Mechanism of hydrogenation of butadiene with pentacyanocobaltate(II).

Complexes of the type $(R_3Q)_2MX_2$ (R=alkyl or aryl; Q=P, As, Sb; M=Ni, Pd, Pt; and X=halogen or pseudohalogen) have been found to catalyze hydrogenation of all but one double bond (68–70) in polyolefins. In the palladium and platinum complexes, catalytic behavior is enhanced by the addition of a compound of the type $M'X_2$ or $M'X_4$ (M' = Si, Ge, Sn, or Pb). The hydrogenation process occurs with very rapid isomerization to the conjugated isomer followed by a rapid reduction to the monoene product. The mechanism shown in Fig. 29 has been proposed for this system.

First, the transition metal catalyst is activated by the group IV halide, which reacts with hydrogen, forming the metal hydride catalyst. Hydrido–metal–olefin complex formation occurs followed by stepwise double-bond migration to the conjugated species. One of the conjugated double bonds adds to the metal hydride, forming a π-complex, which then rearranges to the σ-complex bonded in a position to the remaining double bond. This double bond attacks the metal center, forming a complex that has one σ- and one π-bond to the metal. The π-bond effectively weakens the metal–carbon σ-bond, making it susceptible to attack by hydrogen. In this manner a π-metal–monoene–hydrido complex is formed, which can be displaced by another molecule of polyene.

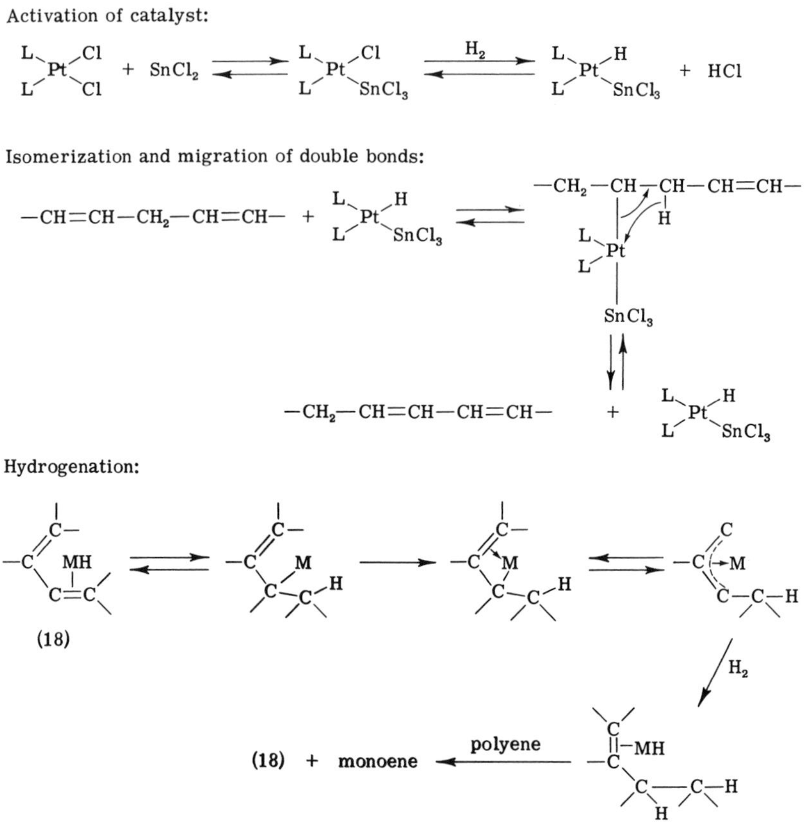

Activation of catalyst:

$$\underset{L}{\overset{L}{\diagdown}}Pt\underset{Cl}{\overset{Cl}{\diagup}} + SnCl_2 \;\underset{\longleftarrow}{\longrightarrow}\; \underset{L}{\overset{L}{\diagdown}}Pt\underset{SnCl_3}{\overset{Cl}{\diagup}} \;\overset{H_2}{\underset{\longleftarrow}{}}\; \underset{L}{\overset{L}{\diagdown}}Pt\underset{SnCl_3}{\overset{H}{\diagup}} + HCl$$

Isomerization and migration of double bonds:

$$-CH=CH-CH_2-CH=CH- \;+\; \underset{L}{\overset{L}{\diagdown}}Pt\underset{SnCl_3}{\overset{H}{\diagup}} \;\underset{\longleftarrow}{\longrightarrow}$$

$$-CH_2-CH\underset{\displaystyle \underset{L}{\overset{L}{\diagdown}}Pt}{\diagup}CH-CH=CH-$$

$$\Big\updownarrow$$

$$SnCl_3$$

$$-CH_2-CH=CH-CH=CH- \;+\; \underset{L}{\overset{L}{\diagdown}}Pt\underset{SnCl_3}{\overset{H}{\diagup}}$$

Hydrogenation:

(18)

(18) + monoene $\xleftarrow{\text{polyene}}$

FIG. 29. Mechanism for hydrogenation of polyenes.

E. *Hydroformylation*

The oxo- or hydroformulation reaction involves the addition of hydrogen and formyl groups across the double bond of alkenes.

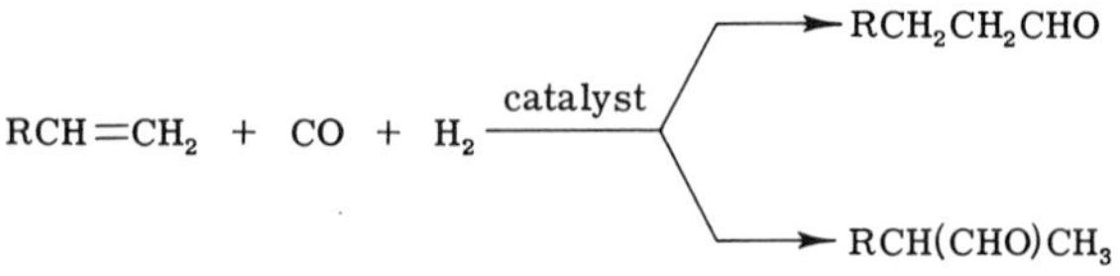

$$RCH=CH_2 + CO + H_2 \;\xrightarrow{\text{catalyst}}\; \begin{cases} RCH_2CH_2CHO \\ \\ RCH(CHO)CH_3 \end{cases}$$

The reaction has been found to be catalyzed by a number of transition metal complexes such as cobalt *(71)*, rhodium *(72)*, iron *(73)*, and

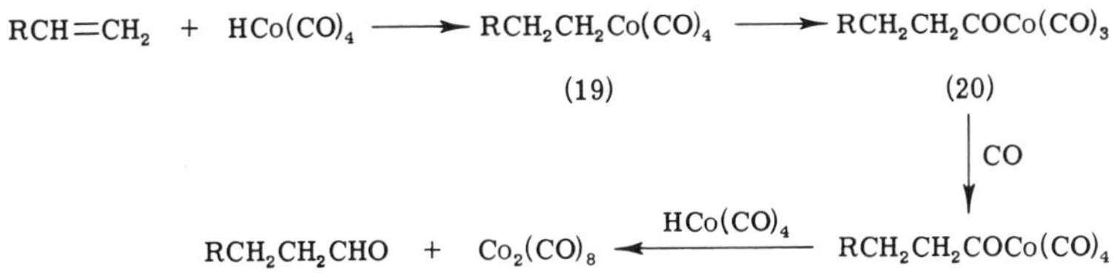

$$RCH{=}CH_2 \;+\; HCo(CO)_4 \longrightarrow RCH_2CH_2Co(CO)_4 \longrightarrow RCH_2CH_2COCo(CO)_3$$

$$(19) \hspace{4cm} (20)$$

$$\Big\downarrow CO$$

$$RCH_2CH_2CHO \;+\; Co_2(CO)_8 \xleftarrow{\;\; HCo(CO)_4 \;\;} RCH_2CH_2COCo(CO)_4$$

FIG. 30.   Hydroformylation mechanism suggested by Kirch and Orchin (78).

manganese (73) carbonyls, bis(triphenylphosphine)rhodium(I) (74, 75), and hydridocarbonyltris(triphenylphosphine)rhodium(I) (76, 77).

As in the hydrogenation reactions, metal hydrides are believed to play an important role. When terminal olefins are added to solutions of cobalt hydrocarbonyl in the presence of carbon monoxide, aldehydes and dicobalt octacarbonyl can be isolated from the reaction mixture (78).

$$2HCo(CO_4) + CO + olefin \rightarrow Co_2(CO)_8 + aldehyde$$

Kirch and Orchin (78) have suggested that the reaction follows the mechanism shown in Fig. 30.

Evidence for the formation of (19) and its rearrangement to (20) has been presented based on spectral studies (79) of the equilibrium:

$$RCo(CO)_4 \rightleftarrows RCOCo(CO)_3$$

In the presence of hydrogen, the reaction is believed to follow the mechanism in Fig. 31.

Hydridocarbonyltris(triphenylphosphine)rhodium is an efficient catalyst for the hydroformylation of alkenes at 25°C and 1 atm. The mechanism proposed for this reaction is shown in Fig. 32 (80).

The catalyst $RhH(CO)_2(P\phi_3)_2$ reacts in either an associative or dissociative fashion to form an hydrido–alkene complex which then undergoes a π–σ rearrangement to a five-coordinate alkyl complex (21). Alkyl transfer to the coordinated CO gives a square acyl complex (22) which oxidatively adds H_2 yielding a dihydroacylrhodium(III) complex (23). This then eliminates the aldehyde and $RhH(CO)(P\phi_3)_2$.

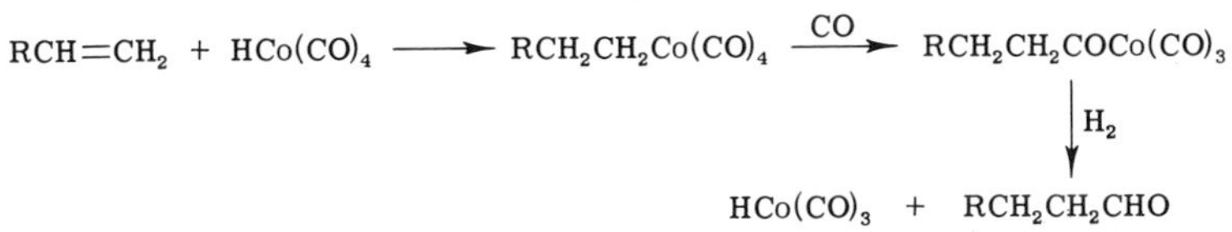

$$RCH{=}CH_2 \;+\; HCo(CO)_4 \longrightarrow RCH_2CH_2Co(CO)_4 \xrightarrow{\;CO\;} RCH_2CH_2COCo(CO)_3$$

$$\Big\downarrow H_2$$

$$HCo(CO)_3 \;+\; RCH_2CH_2CHO$$

FIG. 31.   Proposed hydroformylation mechanism in the presence of hydrogen.

$$RhH(CO)(P\phi_3)_3 \rightleftarrows RhH(CO)(P\phi_3)_2 \overset{CO}{\rightleftarrows} RhH(CO)_2P\phi_3)_2$$

Associative pathway:

$$RhH(CO)_2(P\phi_3)_2 \overset{olefin}{\rightleftarrows} RhH(olefin)(CO)(P\phi_3)_2 \rightleftarrows RhR(CO)_2(P\phi_3)_2$$

(21)

Dissociative pathway:

$$RhH(CO)_2(P\phi_3)_2 \overset{-P\phi_3}{\rightleftarrows} RhH(CO)_2(P\phi_3) \overset{olefin}{\rightleftarrows} RhH(olefin)(CO)_2(P\phi_3)$$

$$\updownarrow P\phi_3$$

$$RhR(CO)_2(P\phi_3)_2$$

(21)

$$\underset{(21)}{RhR(CO)_2(P\phi_3)_2} \rightleftarrows \underset{(22)}{Rh(COR)(CO)P\phi_3)_2} \overset{CO}{\rightleftarrows} Rh(COR)CO)_2(P\phi_3)_2$$

$$\updownarrow H_2$$

$$\underset{(23)}{Rh(COR)H_2(CO)(P\phi_3)_2} \longrightarrow RhH(CO)(P\phi_3)_2 + RCHO$$

FIG. 32.   Proposed mechanism for hydroformylation of alkenes with a rhodium catalyst.

To verify this mechanism, analogs of the suspected intermediates having less reactive metal–carbon bonds have been prepared and studied. Square alkyls and square acyls similar to (21) and (22), respectively, having the formula $MR(CO)(P\phi_3)_2$ ($M{=}Rh$, $R{=}C_2H_4H$ + $M{=}Ir$, $R{=}EtCO$ or ϕ) have been characterized.

NMR studies of the reaction of $RhH(CO)(P\phi_3)_3$ with an equilibrium mixture of ethylene and carbon monoxide have shown the formation of the alkyl complex and the subsequent rearrangement to the acyl complex (Fig. 33) (51).

$$RhH(CO)(P\phi_3)_3 \overset{-P\phi_3}{\rightleftarrows} RhH(CO)(P\phi_3)_2 \overset{CO/C_2H_4}{\rightleftarrows} RhEt(CO)_2(P\phi_3)_2$$

$$\updownarrow$$

$$RhH(CO)(P\phi_3)_2 \overset{H_2}{\underset{-EtCHO}{\longleftarrow}} Rh(COEt)(CO)(P\phi_3)_2$$

FIG. 33.   Formation of the alkyl complex and rearrangement to the acyl complex.

F. *Olefin Metathesis*

Olefin metathesis—the catalytically induced redistribution of alkylidene groups between olefins—is a rather recently discovered process. The first example was the conversion of propylene into ethylene and 2-butene in the presence of molybdenum hexacarbonyl on alumina (82).

$$2CH_2{=}CH{-}CH_3 \; \underset{150°C, \ 30 \ atm}{\overset{Mo(CO)_6/Al_2O_3}{\rightleftharpoons}} \; CH_2{=}CH_2 + CH_3{-}CH{=}CH{-}CH_3$$

The overall metathesis reaction is

$$R'CH{=}CHR' + RCH{=}CHR \; \overset{catalyst}{\rightleftharpoons} \; R'CH{=}CHR + R'CH{=}CHR$$

One early mechanism, which appeared to satisfy labeling studies, is the "quasi-cyclobutane" mechanism involving a metal–cyclobutane transition state (83, 84) (Fig. 34).

Although the simultaneous making and breaking of double bonds is symmetry forbidden (85), it has been suggested that metal–olefin orbital interactions provide a low-energy pathway (86, 87). However, experimental studies have failed to find much support for this mechanism (88).

A second mechanism involving a π–σ rearrangement from the olefin complex to a metallocycle intermediate in which a "pairwise" exchange between the two olefins occurs has been proposed (Fig. 35) (89, 90).

Support for this mechanism comes from isolation of metallocycles and subsequent reactions (91).

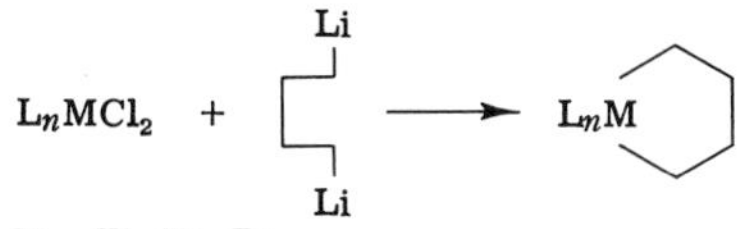

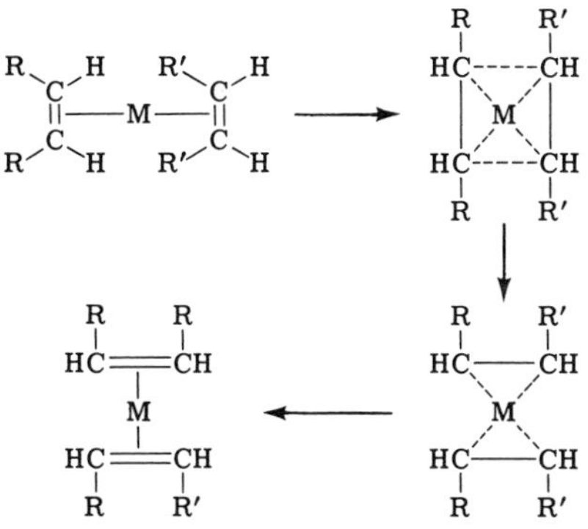

FIG. 34.   "Quasi-cyclobutane" mechanism.

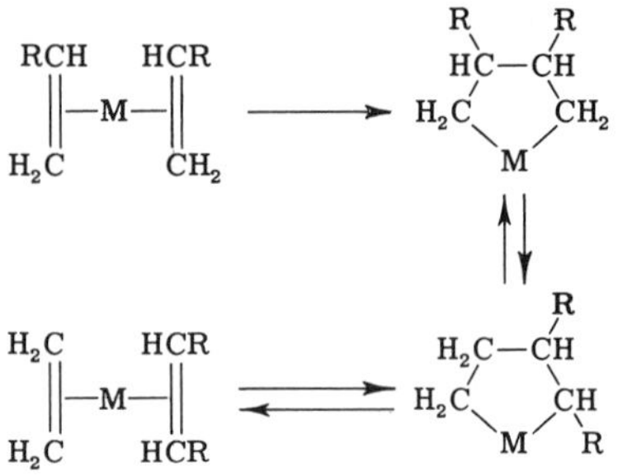

$$
\begin{array}{cc}
\text{RCH} & \text{HCR} \\
\| \;\;-\text{M}-\;\; \| \\
\text{H}_2\text{C} & \text{CH}_2
\end{array}
\quad\longrightarrow\quad
\begin{array}{c}
\text{R} \qquad \text{R} \\
\text{HC}-\text{CH} \\
\text{H}_2\text{C} \qquad \text{CH}_2 \\
\text{M}
\end{array}
$$

$$
\begin{array}{cc}
\text{H}_2\text{C} & \text{HCR} \\
\| \;\;-\text{M}-\;\; \| \\
\text{H}_2\text{C} & \text{HCR}
\end{array}
\quad\rightleftharpoons\quad
\begin{array}{c}
\text{R} \\
\text{H}_2\text{C}-\text{CH} \\
\text{H}_2\text{C} \qquad \text{CH} \\
\text{M} \qquad \text{R}
\end{array}
$$

FIG. 35. Metallocycle mechanism.

A third mechanism involving a chain reaction having a carbene–metal complex as the active catalyst (Fig. 36) has recently been investigated using 1,7-octadiene and 1,7-octadiene-1,1,8,8-d_4 as the olefin in three different catalyst systems (*92*).

By comparing the ratios of ethylene -d_4, -d_2, and -d_6 expected from each of the various mechanisms with the experimentally obtained ratios, results were obtained that appear to be most consistent with the carbene mechanism. It has also been observed that interchange of the carbene carbon occurs with olefin carbons (*93*).

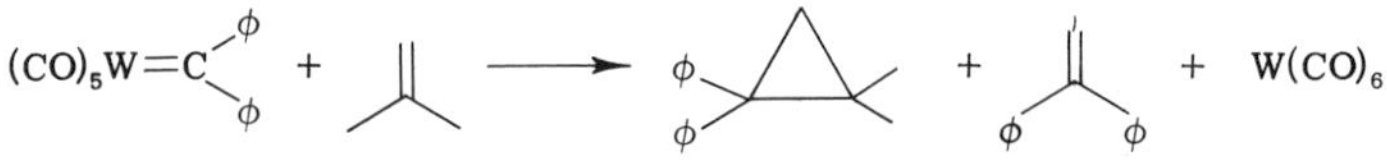

$$
(\text{CO})_5\text{W}{=}\text{C}\overset{\phi}{\underset{\phi}{\diagup}} \;+\; \parallel \longrightarrow \phi\text{-}\triangle\text{-}\phi \;+\; \parallel \;+\; \text{W(CO)}_6
$$

Since there are experimental data supporting both the metallocycle and carbene mechanism, it is possible that either different catalysts and reaction conditions may proceed by different mechanisms or that competing mechanisms may occur.

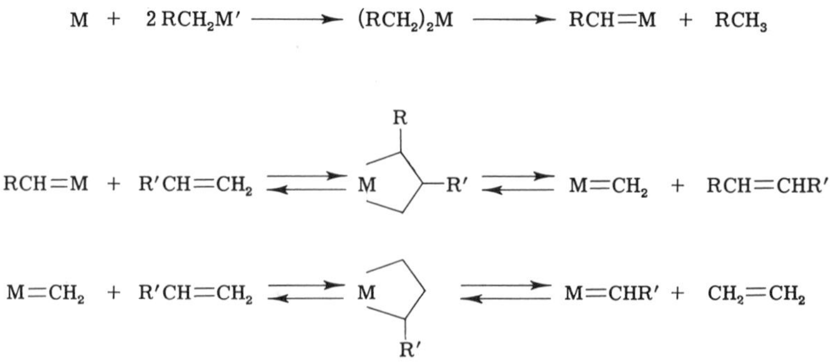

$$
\text{M} \;+\; 2\,\text{RCH}_2\text{M}' \longrightarrow (\text{RCH}_2)_2\text{M} \longrightarrow \text{RCH}{=}\text{M} \;+\; \text{RCH}_3
$$

$$
\text{RCH}{=}\text{M} \;+\; \text{R}'\text{CH}{=}\text{CH}_2 \;\rightleftharpoons\; \text{M}\!\!\underset{}{\diagdown}\!\!\text{-R}' \;\rightleftharpoons\; \text{M}{=}\text{CH}_2 \;+\; \text{RCH}{=}\text{CHR}'
$$

$$
\text{M}{=}\text{CH}_2 \;+\; \text{R}'\text{CH}{=}\text{CH}_2 \;\rightleftharpoons\; \text{M}\!\!\underset{\text{R}'}{\diagdown}\!\! \;\rightleftharpoons\; \text{M}{=}\text{CHR}' \;+\; \text{CH}_2{=}\text{CH}_2
$$

FIG. 36. Carbene mechanism.

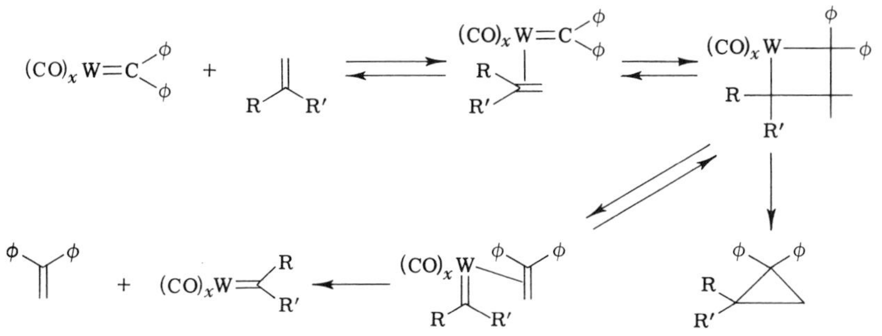

FIG. 37.   Interchange of carbene carbon with olefin carbons.

VI

END-ON AND SIDE-ON COORDINATION

One area that has recently become of interest is the activation of small, readily available molecules, such as N_2 and O_2, by metal complexes.

The activation of dinitrogen for NH_3 synthesis, oxidative fixation to nitric acid and fertilizers, organic amine and heterocyclic synthesis, and amino acid synthesis is a topic that has generated much interest. The sequence of filled molecular orbitals of the nitrogen is ($^1\Sigma_g{}^+$): $1\sigma_g{}^2$, $1\sigma_u{}^2$, $2\sigma_g{}^2$, $2\sigma_u{}^2$, $1\pi_u{}^4$, and $3\sigma_g{}^2$. The highest occupied molecular orbital (HOMO), $3\sigma_g$, has an ionization energy of 15.58 eV (similar to the 15.75 eV of argon). The lowest unoccupied molecular orbital (LUMO), $1\pi_g$, lies 8.6 eV higher than the $3\sigma_g$. The low reactivity of molecular nitrogen has been attributed to the absence of orbitals in the energy range between the HOMO and LUMO (94). Although many dinitrogen transition metal complexes have been isolated, no efficient homogeneous catalyst for N_2 fixation has been developed. The bonding of dinitrogen to transition metals can be separated into "end-on" and "side-on" categories, as shown in Fig. 38.

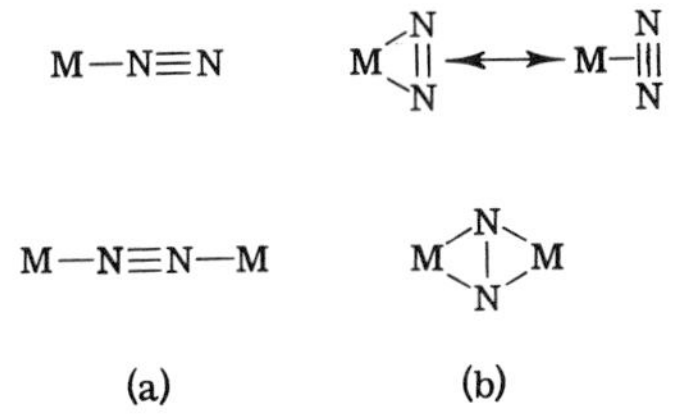

FIG. 38.   (a) "End-on" bonding mode; (b) "side-on" bonding mode.

Most stable complexes of dinitrogen have end-on structures, only a few compounds showing reductive chemical reactivity. However, the compound $\{C_6H_5[Na \cdot O(C_2H_5)_2][(C_6H_5)_2Ni]_2N_2NaLi_6(OC_2H_5)_4 \cdot P(C_2H_5)_2\}$ having a side-on nitrogen has been isolated and characterized (95–97).

Theoretical calculations show that the lower stability of the side-on mode with respect to the end-on mode can be attributed to the more antibonding character of the nitrogen–metal π-bond and the decrease in the total electronic population of the bond in the former (98).

In the complex $[Ru(NH_3)_5(N_2)]^{2+}$, the dinitrogen ligand undergoes an end-over-end rotation

$$[(NH_3)_5Ru-{}^{15}N\equiv N^{14}]^{2+} \rightarrow [(NH_3)_5Ru-{}^{14}N\equiv N^{15}]^{2+}$$

This process has a half-life of 2 hours in aqueous solution at 25°C and occurs 45 times faster than the competing substitution reaction

$$[Ru(NH_3)_5(N_2)]^{2+} + H_2O \rightarrow [Ru(NH_3)_5(H_2O)]^{2+} + N_2$$

Therefore, it appears that this isomerization occurs via a side-on coordinated intermediate without ligand dissociation. The energetics of the isomerization and dissociation reactions indicate that the energy of activation of the dissociation of the nitrogen ligand in the side-on mode is only 7 kcal/mole as compared with 29 kcal/mole in the end-on mode (99).

Recently, the unusual behavior of reaction systems that involve titanocene toward the normally quite inert molecule N_2 has been the subject of considerable interest. The fixation and reduction of molecular nitrogen has been observed in mixtures of titanocene dichloride and ethyl magnesium bromide and in other systems containing titanocene derivatives (100–108).

It has been observed that, when solutions of permethyl titanocene are exposed to molecular nitrogen, three different dinitrogen complexes may be formed (**24, 25, 26**) (109). The sequence of possible reactions leading to these complexes is shown in Fig. 39.

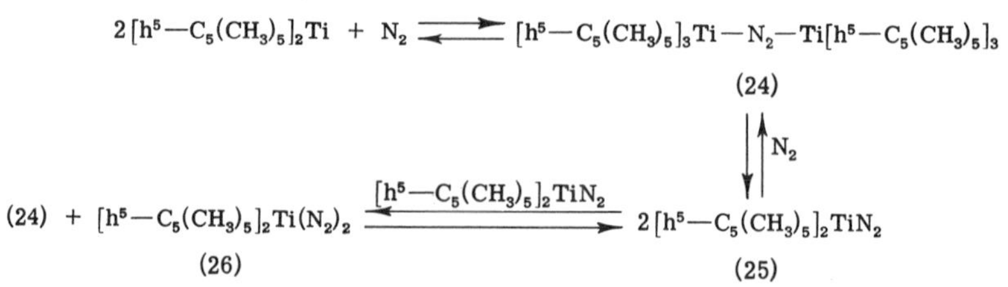

$$2\,[h^5-C_5(CH_3)_5]_2Ti \;+\; N_2 \;\rightleftharpoons\; [h^5-C_5(CH_3)_5]_3Ti-N_2-Ti[h^5-C_5(CH_3)_5]_3$$

(24)

$$\Big\updownarrow N_2$$

$$(24) \;+\; [h^5-C_5(CH_3)_5]_2Ti(N_2)_2 \xrightarrow{\;[h^5-C_5(CH_3)_5]_2TiN_2\;} 2\,[h^5-C_5(CH_3)_5]_2TiN_2$$

(26) (25)

FIG. 39. Reactions of permethyl titanocene with N_2.

Spectral studies are indicative of a centrosymmetric Ti—N$_2$—Ti substructure (**27, 28, 29**) for compound (**24**).

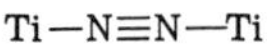

$$Ti-N\equiv N-Ti \qquad Ti-\overset{N}{\underset{N}{|||}}-Ti \qquad \overset{Ti}{\underset{Ti}{N=N}}$$

$$(27) \qquad\qquad (28) \qquad\qquad (29)$$

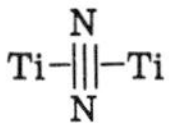

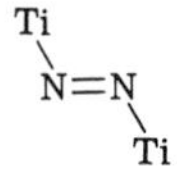

^{13}C NMR data indicate the presence of an end-on (h^1—N$_2$)–side-on (h^2—N$_2$) equilibrium for (**25**). There are three possible forms in which this equilibrium may occur.

In the first, a dinitrogen complex, [h^5—C$_5$(CH$_3$)$_5$]$_2$TiN$_2$, which contains an end-on N$_2$ ligand, is in rapid equilibrium with a side-on isomer (Fig. 40).

An alternative explanation involves equilibrium between the h^1-species and a dimeric species.

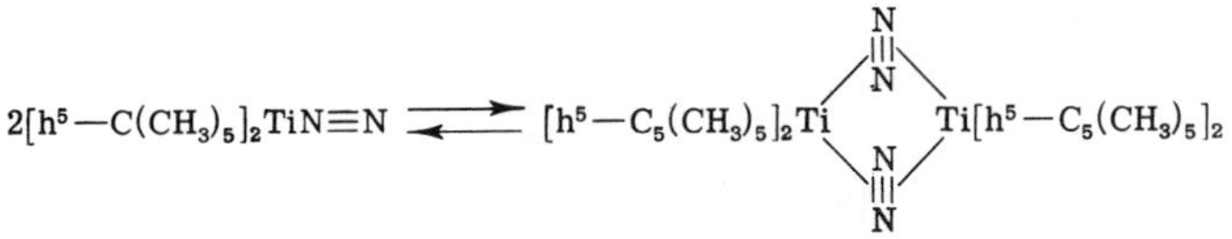

$$2[h^5-C(CH_3)_5]_2TiN\equiv N \rightleftharpoons [h^5-C_5(CH_3)_5]_2Ti\overset{\overset{N}{\underset{N}{|||}}}{\underset{\underset{N}{\underset{N}{|||}}}{}}Ti[h^5-C_5(CH_3)_5]_2$$

Another alternative is an intramolecular fluxional equilibrium:

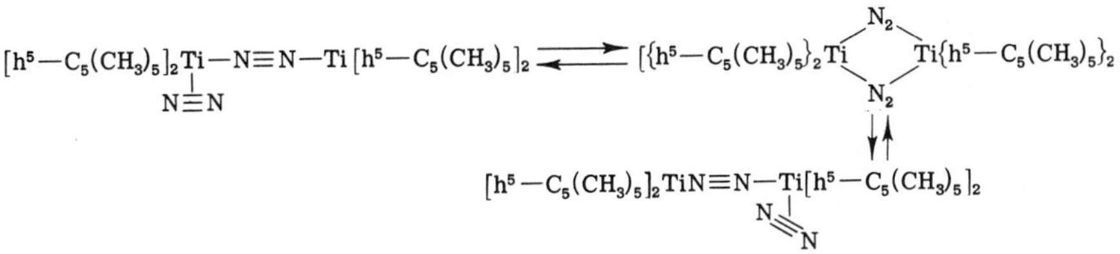

$$\underset{N\equiv N}{[h^5-C_5(CH_3)_5]_2Ti-N\equiv N-Ti}[h^5-C_5(CH_3)_5]_2 \rightleftharpoons [\{h^5-C_5(CH_3)_5\}_2Ti\overset{N_2}{\underset{N_2}{}}Ti\{h^5-C_5(CH_3)_5\}_2$$

$$[h^5-C_5(CH_3)_5]_2TiN\equiv N-Ti[h^5-C_5(CH_3)_5]_2$$

or

$$[h^5-C_5(CH_3)_5]_2Ti\overset{Ti[h^5-C_5(CH_3)_5]_2}{\underset{N}{}} \rightleftharpoons [\{h^5-C_5(CH_3)_5\}_2Ti\overset{N_2}{\underset{N_2}{}}Ti\{h^2-C_5(CH_3)_5\}_2]^{\ddagger}$$

$$[h^5-C_5(CH_3)_5]_2Ti\overset{N}{\underset{}{}}Ti[h^5-C_5(CH_3)_5]_2$$

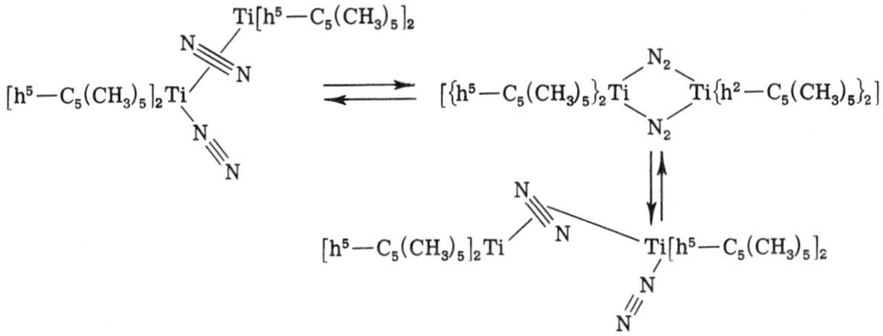

Presently, no clear distinction between these possibilities has been made.

After a solution of the permethyltitanocene dinitrogen complex (**25**) is

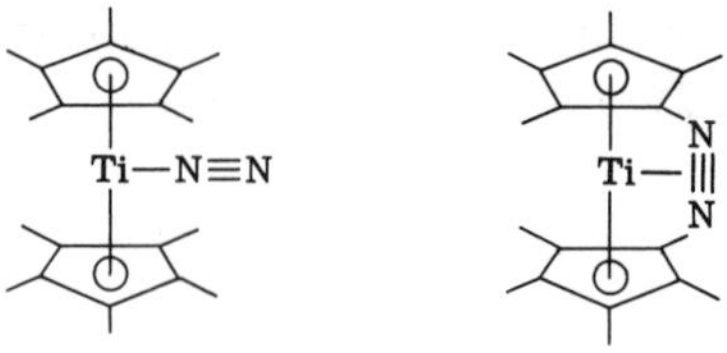

(a)   (b)

Fig. 40.   (a) N_2 is "end-on"; (b)N_2 is "side-on."

treated with HCl at $-80°C$, extraction of the reaction yields hydrazine.

$$(25) + 2HCl \rightarrow [h-{}^5C_5(CH_3)_5]_2TiCl_2 + \tfrac{1}{2}N_2 + \tfrac{1}{2}N_2H_4$$

Reaction of the dimeric dinitrogen complex (24) with HCl gives a very small yield ($<5\%$) of the ammonium chloride.

Although the mechanisms of these reactions are not thoroughly understood, it has been suggested on the basis of NMR and IR results that the reactive configuration that ultimately leads to hydrazine is that of the h^2-dinitrogen complex of (25).

Therefore, it remains to be clarified in which way such N_2 complexes enter into reduction reactions—i.e., in which bonding mode the dinitrogen is activated, and in what way the two bonding modes are interrelated.

Presently, catalysts are being sought that can cause selective, direct oxygenation of organic substrates. Direct incorporation of one or both atoms of the oxygen molecule into organic substrates with great selectivity occurs in nature via mono- and dioxygenases. Generally, these oxygenases are metalloenzymes, and activation of the O_2 involves coordination of dioxygen to the transition metal center. Knowledge of how metal species bind O_2 is essential in understanding such processes.

Currently, two geometrical structures for coordinated dioxygen are known—the side-on, which corresponds to a π-bonding mode, and the end-on, corresponding to a σ-bonding mode, as seen in Fig. 41.

For example, the dioxygen complexes from iron(II)porphyrins (110) have the end-on configuration whereas the dioxygen complex of titanium

(a)   (b)

Fig. 41.   (a) "Side-on" O_2 coordination; (b) "end-on" coordination.

octaethylporphyrin (*111*) has the side-on configuration. However, it is not known whether one bonding mode is more catalytically favorable or whether the two modes are interconvertible.

VII

VITAMIN B$_{12}$ COENZYME-CATALYZED REARRANGEMENTS

Although σ-π rearrangements of organometallic complexes have been known and studied for only the last two decades, perhaps nature has known and utilized them for a long time.

The vitamin B$_{12}$ coenzyme (Fig. 42) has been implicated as a cofactor for ten different enzyme reactions (*112*). All these reactions involve a substrate rearrangement in which a hydrogen and an adjacent alkyl, acyl, or electronegative group are interchanged.

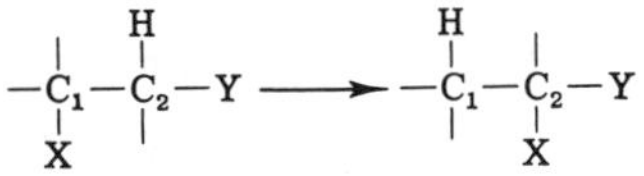

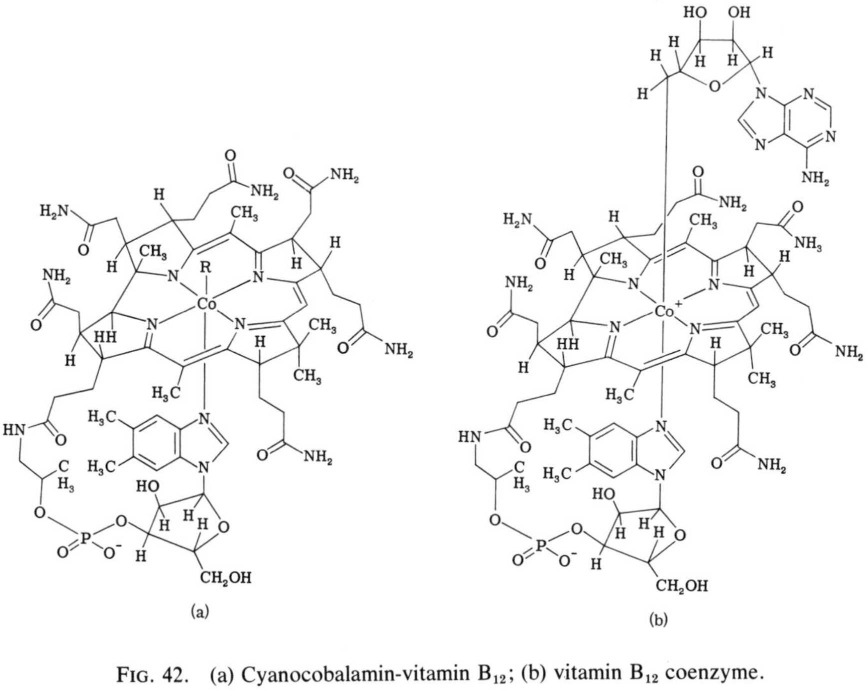

$$-\underset{X}{\overset{}{C_1}}-\underset{}{\overset{H}{C_2}}-Y \longrightarrow -\underset{}{\overset{H}{C_1}}-\underset{X}{\overset{}{C_2}}-Y$$

FIG. 42. (a) Cyanocobalamin-vitamin B$_{12}$; (b) vitamin B$_{12}$ coenzyme.

To study the mechanism of this type of rearrangement, cobaloximes (Fig. 43) have served as models for the cobalamins.

The ethanolysis of 2-acetoxyethyl and 2-acetoxypropyl(pyridine)-cobaloxime yields 2-ethoxyethyl and 2-ethoxypropyl(pyridine)-cobaloxime, respectively (113). Kinetic data suggest that ionization to one of three possible intermediates (30, 31, 32) may occur in the rate-determining step followed by capture of solvent.

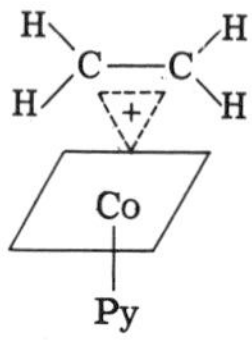

(30) 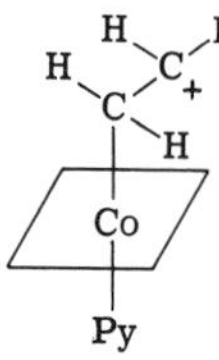   (31) 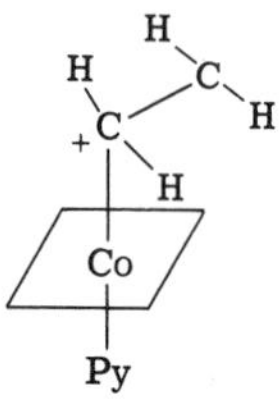   (32)

π-Complexes such as (30) might be generated by two independent routes: the loss of a nucleofugal group from a carbon β to the metal, or the addition of an olefin to an unalkylated Co(III) complex (Fig. 44).

The methanolysis of 2-acetoxyethyl-[2-^{13}C]-(pyridine)cobaloxime yielded an equal mixture of 2-methoxyethyl-[1-^{13}C]- and 2-methoxyethyl-[2-^{13}C]-(pyridine)cobaloxime (114). This indicates that at some point along the reaction coordinate the metal must interact equally with the

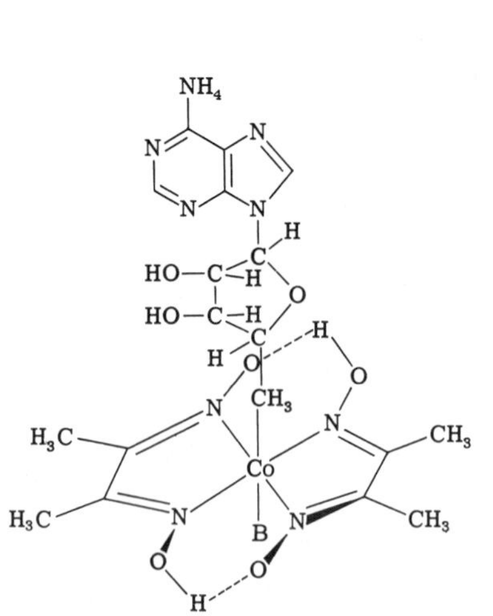

FIG. 43. The cobaloxime model of vitamin B_{12} coenzyme.

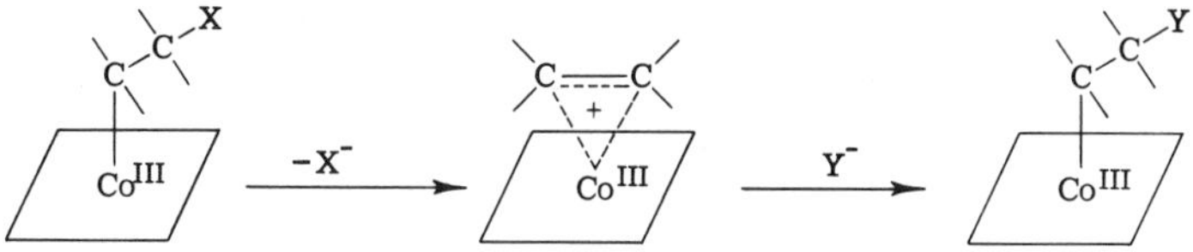

(a)

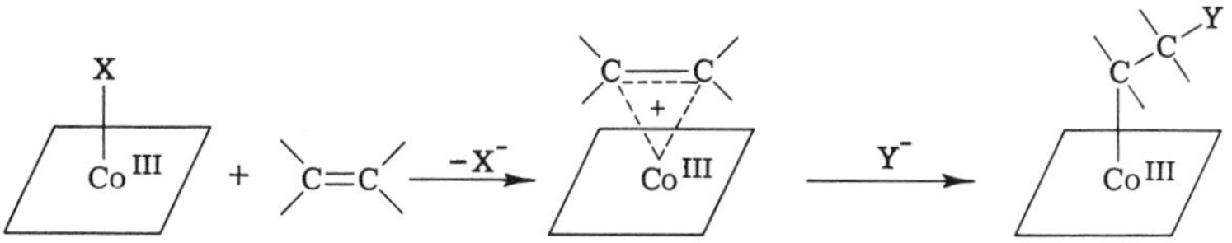

(b)

FIG. 44. Possible mechanisms for formation of π-complex. (a) σ-π rearrangement of alkyl complex; (b) addition of olefin to an unalkylated complex.

two carbon atoms, i.e., form a π-complex. A pathway involving (31) or (32) would result in a product containing the label in only one position.

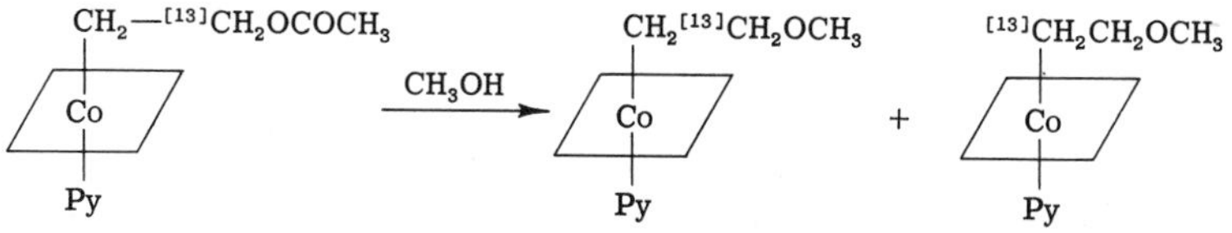

It has been shown that olefins that are electron rich, such as enol ethers, give products formed via an olefin complex. An olefin complex between an enol ether and a trivalent cobalt complex should in the presence of a nucleophile (such as an alcohol) undergo addition of the nuclephile to the oxygen-bearing carbon, forming a σ-bonded acetal.

Bromo(pyridine)cobaloxime reacts with ethyl vinyl ether, ethanol, and an amine to yield 2,2-diethoxyethyl (33) and formylmethyl(pyridine)-cobaloxime (34) (115).

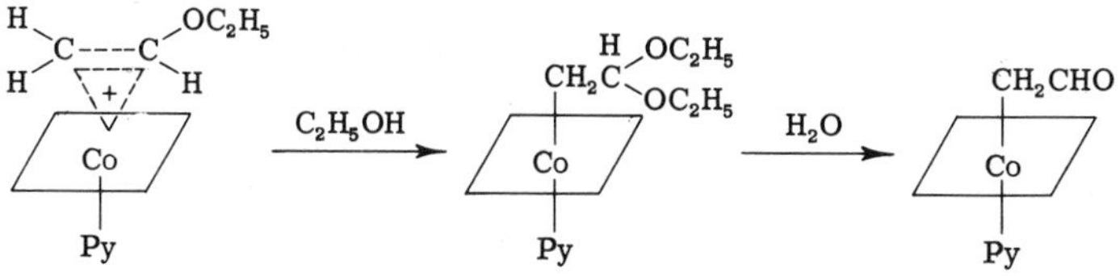

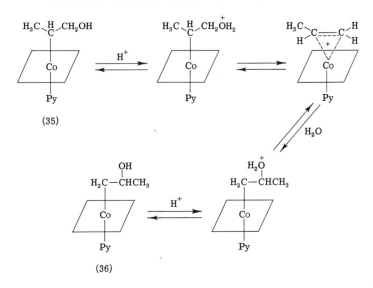

FIG. 45. Rearrangement in the presence of acid.

Treatment of hydroxocobalamin (B_{12b}) with ethyl vinyl ether in ethanol containing triethylamine yields 2,2-diethoxyethylcobalamin, and treatment with excess 2-hydroxyethyl vinyl ether in water gives the cyclic acetal 1,3-dioxa-2-cyclopentylmethylcobalamin (*115*).

Additional support for the π-complex intermediate comes from the observation that β-hydroxyisopropyl(pyridine)cobaloxime (**35**) rearranges to β-hydroxy-*n*-propyl(pyridine)cobaloxime (**36**) in the presence of acid (Fig. 45) (*116*).

Thus, the experimental data presented are consistent with the intermediacy of π-olefin–cobalt(III) complexes. By invoking a σ–π rearrangement mechanism, rearrangements catalyzed by the vitamin B_{12} coenzyme can be generalized.

VIII

CONCLUSION

It appears that σ–π rearrangements are basic to the chemistry of organometallic compounds. In the preceding discussion, we have presented some of the factors believed to influence these rearrangements.

Although many parallels have been drawn between the σ–π rearrangement and catalytic activity, the exact nature of this relationship is still unclear.

The catalytic systems which have been surveyed have the possibility of two common mechanistic steps—catalyst interaction with the π-system of the substrate, and reaction on the coordinated substrate, which removes the unsaturation-causing rearrangement.

Thus, academic studies on this chemical phenomenon may well lead to discoveries that will have great industrial impact.

ACKNOWLEDGMENT

The preparation of this manuscript was made possible by the support of the Robert A. Welch Foundation, Grant No. A-420.

REFERENCES

1. Herwig, W., and Zeiss, H. H., *J. Am. Chem. Soc.* **79**, 6561 (1957).
2. Zeiss, H. H., and Tsutsui, M., *J. Am. Chem. Soc.* **79**, 3062 (1957).
3. Chatt, J., Duncanson, L. A., and Venanzi, L. M., *J. Chem. Soc.* 4456 (1955).
4. Chatt, J., Duncanson, L. A., and Venanzi, L. M., *J. Chem. Soc.* 4461 (1955).
5. Chatt, J., and Shaw, B. L., *J. Chem. Soc.* 705 (1959).
6. Orgel, L. E., *J. Inorg. Nucl. Chem.* **2**, 137 (1956).
7. Armstrong, D. R., Fortune, R., and Perkins, P. G., *Inorg. Chem. Acta* **9**, 9 (1974).
8. Coates, G. E., Green, M. L. H., and Wade, K., "Organometallic Compounds: The Transition Elements," Vol. II, p. 32. Methuen, London, 1968.
9. Pearson, R. G., *J. Am. Chem. Soc.* **85**, 3533 (1963).
10. Green, M. L. H., and Nagy, P. L. I., *J. Chem. Soc.* 189 (1963).
11. Green, M. L. H., and Stear, A. N., *J. Organomet. Chem.* **1**, 230 (1964).
12. Hancock, M., Levy, M. N., and Tsutsui, M., "Organometallic Reactions" (E. I. Becker, and M. Tsutsui, eds.), Vol. 4, Chapter 1. Wiley, New York, 1972.
13. Volger, H. C., and Vrieze, K., *J. Organomet. Chem.* **13**, 479 (1968).
14. Tsutsui, M., Hancock, M., Ariyoshi, J., and Levy, M. N., *J. Am. Chem. Soc.* **91**, 5233 (1969).
15. Tsutsui, M., and Hudman, C. E., *Chem. Lett.* 777 (1972).
16. Green, M. L. H., and Nagy, P. L. I., *J. Organomet. Chem.* **1**, 58 (1963).
17. Tsutsui, M., Ori, M., and Francis, J., *J. Am. Chem. Soc.* **94**, 1414 (1972).
18. Hillis, J., and Tsutsui, M., *J. Am. Chem. Soc.* **95**, 7907 (1973).
19. Hillis, J., Francis, J., Ori, M., and Tsutsui, M., *J. Am. Chem. Soc.* **96**, 4800 (1974).
20. Vrieze, K., MacLean, C., Cossee, P., and Hilbers, C. W., *Rec. Trav. Chem.* **85**, 1077 (1966).
21. Tsutsui, M., and Ely, N., *J. Am. Chem. Soc.* **97**, 3551 (1975).
22. Marks, T. J., Seyam, A. M., and Kolb, J. R., *J. Am. Chem. Soc.* **95**, 5529 (1973).
23. Winkhaus, G., Pratt, L., and Wilkinson, G., *J. Chem. Soc.* 3807 (1961).
23. Cotton, F. A., Francis, J. N., Frenz, B. A., and Tsutsui, M., *J. Am. Chem. Soc.* **95**, 2483 (1973).
25. Chatt, J., and Shaw, B. L., *J. Chem. Soc.* 5075 (1962).
26. Armstrong, D. R., Fortune, R., and Perkins, P. G., *J. Catal.*, **41**, 51 (1976).

27. Anderson, J. S., *J. Chem. Soc.* 971 (1934).
28. Smidt, J., Hafner, W., Jira, R., Sieber, R., Sedlmeier, J., and Sabel, A., *Angew. Chem.* **74,** 93 (1962).
29. Hafner, W., Jira, R., Sedlmeier, J., and Smidt, J., *Ber.* **95,** 1575 (1962).
30. Henry, P. M., *J. Am. Chem. Soc.* **86,** 3246 (1964).
31. Henry, P. M., *J. Am. Chem. Soc.* **88,** 1595 (1966).
32. Henry, P. M., *J. Am. Chem. Soc.* **94,** 4437 (1972).
33. Perkins, D. R., Fortune, R., and Perkins, P. G., personal communication.
34. Wright, G. F., *Ann. N.Y. Acad. Sci.* **65,** 436 (1957).
35. Grinstead, R. R., *J. Org. Chem.* **26,** 238 (1961).
36. Criegee, R., *Angew Chem.* **70,** 173 (1958).
37. Anderson, C. B., and Winstein, S., *J. Chem. Soc.* **28,** 605 (1963).
38. Kabbe, H. J., *Ann.* 656, 204 (1962).
39. Criegee, R., *Angew Chem.* **70,** 173 (1958).
40. Henry, P. M., *J. Am. Chem. Soc.* **87,** 990 (1965).
41. Henry, P. M., *J. Am. Chem. Soc.* 4423 (1965).
42. Ziegler, K., Holzkamp, E., Breil, H., and Martin, H., *Angew. Chem.* **67,** 541 (1955).
43. Cossee, P., *J. Catal.,* **3,** 80 (1964).
44. Cossee, P., "The Stereochemistry of Macromolecules" (A. D. Ketley, ed.), vol. 1, Chapter 3. Dekker, New York, 1967.
45. Armstrong, D. R., Perkins, P. G., and Stewart, J. J. P., *J. Chem. Soc., Dalton Trans.* 1972 (1972).
46. Novaro, O., Chow, S., and Magnovat, P., *J. Catal.* **41,** 91 (1976).
47. Novaro, O., Chow, S., and Magnovat, P., *J. Catal.* **42,** 131 (1976).
48. Djabiev, T. S., Sabirova, R. D., and Shilov, A. D., *Kinet. Katal.* **5,** 441 (1964).
49. Henrici-Olivé, G., and Olivé, S., *Polymer Lett.* **12,** 39 (1974).
50. Tsutsui, M., *Trans. N. Y. Acad. Series II,* **26,** 423 (1964).
51. Tsutsui, M., Ariyoshi, J., Koyano, T., and Levy, M. N., *Adv. Chem. Ser.* **70,** (1968).
52. Tsutsui, M., and Ariyoshi, J., *Trans. N. Y. Acad. Sci.* **26,** 431 (1964).
53. Gorewit, B., and Tsutsui, M., *Adv. Catal.*, in press.
54. Jolly, P. W., and Wilke, G., "The Organic Chemistry of Nickel," Vol. 2, p. 56. Academic Press, New York, 1975.
55. Bingham, D., Webster, D. W., and Wells, P. B., *J. Chem. Soc., Dalton Trans.* 1928 (1972).
56. Drinkard, W. C., Eaton, D. R., Jesson, J. P., and Lindsay, Jr., R. V., *Inorg. Chem.* **9,** 392 (1970).
57. Tolman, C. A., *J. Am. Chem. Soc.* **92,** 6785 (1970).
58. Davies, N. R., *Aust. J. Chem.* **17,** 212 (1964).
59. Bond, G. C., and Hellier, M., *J. Catal.* **4,** 1 (1965).
60. Harrod, J. F., and Chalk, A. J., *J. Am. Chem. Soc.* **88,** 3491 (1966).
61. Manuel, T. A., *J. Org. Chem.* **27,** 3941 (1962).
62. Manuel, T. A., *Trans. N. Y. Acad. Sci.* **26,** 442 (1964).
63. Emerson, G. F., and Pettit, R., *J. Am. Chem. Soc.* **84,** 4591 (1962).
64. Halpern, J., Harrod, J. F., and James, B. R., *J. Am. Chem. Soc.* **88,** 5150 (1966).
65. Kwiatek, J., Mador, I. L., and Seylor, J. K., *"Adv. in Chem. Series, No. 37"*, American Chemical Society, 1963, p. 201.
66. Kwiatek, J., Mador, I. L., and Seylor, J. K., *J. Am. Chem. Soc.* **84,** 304 (1962).
67. Taqui Khan, M. M., and Martell, A. E., "Homogeneous Catalysis by Metal Complexes," pp. 33–42. Academic Press, New York, 1974.
68. Bailar, J. C., and Itatani, H., *J. Am. Chem. Soc.* **89,** 1592 (1967).
69. Bailar, J. C., and Itatani, H., *J. Am. Chem. Soc.* **89,** 1600 (1967).
70. Tayim, H. A., and Bailar, J. C., *J. Am. Chem. Soc.* **89,** 4330 (1967).

71. Winder, I., Sternberg, H. W., Freidel, R. A., Metlin, J., and Markby, R., *U. S. Bur. Mines Bull.* **600** (1962).
72. Wakamatsu, H., *J. Chem. Soc., Jpn., Pure Chem. Sect.* **85**, 227 (1964).
73. Zachry, J. B., *Ann. N. Y. Acad. Sci.* **125**, 154 (1965).
74. Gardine, F. H., Osborn, J. A., Wilkinson, G., and Young, J. F., *Chem. Ind. (London)* 560 (1965).
75. Osborn, J. A., Gardine, F. H., Young, J. F., and Wilkinson, G., *J. Chem. Soc., A* 1711 (1965).
76. Yagupsky, G., Brown, C. K., and Wilkinson, G., *J. Chem. Soc., A* 1392 (1970).
77. Brown, C. K., and Wilkinson, G., *J. Chem. Soc., A* 2753 (1970).
78. Kirch, L., and Orchin, M., *J. Am. Chem. Soc.* **81**, 3597 (1959).
79. Bredlow, D. S., and Heck, R. F., *Chem. Ind. (London)* **83**, 467 (1960).
80. Brown, C. K., and Wilkinson, G., *J. Chem. Soc., A* 2753 (1970).
81. Yagupsky, G., Brown, C. K., and Wilkinson, G., *J. Chem. Soc., A* 1392 (1970).
82. Banks, R. L., and Bailey, G. C., *Ind. Eng. Chem.* **3**, 170 (1964).
83. Bailey, G. C., *Catal. Rev.* **3**, 37 (1969).
84. Bradshaw, C. P., Howman, E. J., and Turner, L., *J. Catal.* **7**, 269 (1967).
85. Hoffmann, R., and Woodward, R. B., *Acc. Chem. Res.* **1**, 17 (1968).
86. Mango, F. O., and Schachtschneider, J. H., *J. Am. Chem. Soc.* **89**, 2484 (1967).
87. Lewandos, G. S., and Pettit, R., *Tetrahedron Lett.* 780 (1967).
88. Lewandos, G. S., and Pettit, R., *J. Am. Chem. Soc.* **93**, 7087 (1971).
89. Grubbs, R., and Brunck, T. K., *J. Am. Chem. Soc.* **94**, 2538 (1972).
90. Biefield, C. G., Eick, H. A., and Grubbs, R. H., *Inorg. Chem.* **12**, 2166 (1973).
91. Grubbs, R. H., Carr, D. D., and Burk, P. L., in "Organotransition-Metal Chemistry" (Y. Ishii and M. Tsutsui, eds.), p. 135. Plenum, New York, 1975.
92. Grubbs, R. H., Carr, D. D., Hoppin, C., and Burk, P. L., *J. Am. Chem. Soc.* **98**, 3478 (1976).
93. Casey, C. P., and Burkhardt, T. J., *J. Am. Chem. Soc.* **96**, 7808 (1974).
94. Chatt, J., and Richards, R. L., in "The Chemistry and Biochemistry of Nitrogen Fixation" (J. R. Postgate, ed.), pp. 57–103. Plenum, New York, 1971.
95. Jonas, K., *Angew. Chem.* **85**, 1050 (1973).
96. Jonas, K., *Angew. Chem., Int. Ed. Engl.* **12**, 997 (1973).
97. Jonas, K., Brauer, D. J., Kruğer, C., Roberts, P. J., and Tsay, Y. H., *J. Am. Chem. Soc.* **98**, 74 (1976).
98. Yatsimirskiii, K. B., and Kruglyak, Yu. A., *Dokl. Akad. Nauk SSSR* **186**, 885 (1969).
99. Armor, J. N., and Taube, H., *J. Am. Chem. Soc.* **92**, 2560 (1970).
100. Volpin, M. E., and Shur, V. B., *Nature (London)* **209**, 1236 (1966).
101. Volpin, M. E., and Shur, V. B., *Dokl. Akad. Nauk SSSR* **156**, 1102 (1964).
102. Henrici-Olivé, G., and Olivé, S., *Angew. Chem., Int. Ed. Engl.* **8**, 650 (1969).
103. Henrici-Olivé, G., and Olivé, S., *Angew. Chem.* **80**, 398 (1968).
104. Shilov, A. E., Shilova, A. K., and Kvashina, E. F., *Kinet. Katal.* **10**, 1402 (1969).
105. Shilov, A. E., Shilova, A. K., Kvashina, E. F., and Vorontsova, T. A., *Chem. Commun.* 1590 (1971).
106. van Tamelen, E. E., *Acc. Chem. Res.* **3**, 361 (1970).
107. Brintzinger, H., *J. Am. Chem. Soc.* **89**, 6871 (1967).
108. Brintzinger, H., *Biochemistry* **5**, 3947 (1966).
109. Bercaw, J. E., *J. Am. Chem. Soc.* **96**, 5087 (1974).
110. Collman, J. P., Gagne, R. R., Reed, C. A., Halber, T. R., Yang, G., and Robinson, W. T., *J. Am. Chem. Soc.* **97**, 1427 (1975).
111. Guilard, R., Fontessee, M., and Fournari, P., *J. Chem. Soc., Chem. Commun.* 161 (1976).
112. Abeles, R. H., and Dolphin, D., *Acc. Chem. Res.* **9**, 114 (1976).

113. Golding, B. T., Holland, H. L., Horn, V., and Sakrikar, S., *Angew. Chem., Int. Ed. Engl.* **9**, 959 (1970).
114. Silverman, R. B., and Dolphin, D., *J. Am. Chem. Soc.* **94**, 4028 (1972).
115. Silverman, R. B., and Dolphin, D., **98**, 4626 (1976).
116. Brown, K. L., and Ingraham, L. L., *J. Am. Chem. Soc.* **96**, 7681 (1974).

The Olefin Metathesis Reaction

THOMAS J. KATZ

Department of Chemistry
Columbia University
New York, New York

I

INTRODUCTION

When olefins are combined with a variety of catalysts, most commonly containing tungsten, molybdenum, or rhenium, they undergo the transformation generalized as Eq. (1) and known as the olefin metathesis

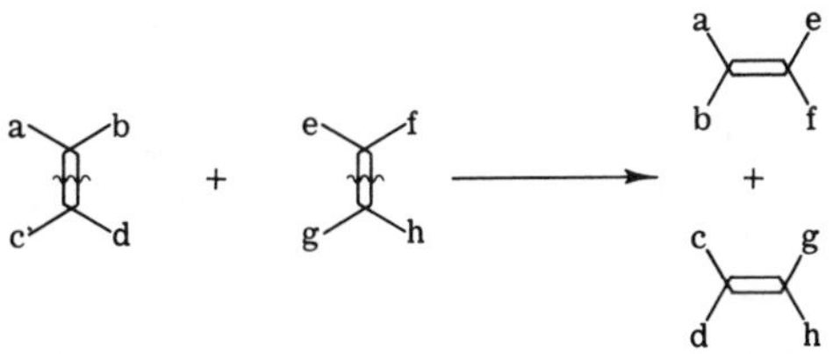

$$\tag{1}$$

reaction (*5, 6, 12, 16a, 51, 55, 58, 59, 70, 88, 112, 113*). When the olefinic bonds are included in rings, metatheses result in polymers called polyalkenamers, illustrated in Eq. (2) (*15, 27*). The catalysts most often

$$\qquad\qquad\qquad\qquad\qquad\qquad\qquad\qquad\qquad (2)$$

reported to bring about such reactions are prepared by combining tungsten, molybdenum, or rhenium halides with organoaluminum compounds and sometimes an activator, usually containing oxygen. Sometimes the role of the organoaluminum compounds is taken by organic derivatives of other metals, notably tin, and sometimes just by Lewis acids like aluminum chloride. Sometimes very different catalyst preparations containing noble metals, most often ruthenium chloride, are effective.

The first olefin metathesis reactions described, in 1957, were the transformations of propene into ethene and butene by molybdenum oxide on alumina combined with triisobutyl aluminum (*106*), and of norbornene and cyclopentene into their polyalkenamers by a similar catalyst, molybdenum oxide on alumina combined with lithium aluminum hydride and hydrogen (*39*). There was a polymerization of norbornene by a mixture of titanium tetrachloride and either ethylmagnesium bromide or lithium tetrabutylaluminum recorded in 1954 (*3*), but the product was not recognized as a polyalkenamer until 1960 (*114*). Tungsten or molybdenum halides plus organoaluminum compounds were introduced as metathesis catalysts in 1963, when they were found to polymerize cyclobutene (*99*) and norbornene (*109*), and both the generality and significance of their application became apparent the following year when they were found to polymerize the unstrained cyclopentene to *trans*-polypentenamer (*98*), a polymer whose properties resemble those of natural rubber (*27, 48, 49, 50, 86, 98*). Ruthenium trichloride in ethanol was discovered to be a catalyst the following year (*84, 100*). Although for acyclic olefins numerous catalysts were at first developed that were heterogeneous, soluble catalysts, notably the combinations WCl_6—C_2H_5OH—$C_2H_5AlCl_2$ (*14, 17*) and $Mo[(C_6H_5)_3P]_2Cl_2$-$(NO)_2$—$(CH_3)_3Al_2Cl_3$ (*124, 125*), were later also-found to be effective. Catalysts containing metals from throughout the periodic table have now been reported and are summarized in a recent review (*51*).

That the transformation these catalysts bring about is the one summarized in Eq. (1), not one in which alkyl groups, for example a and e, interchange, was demonstrated by experiments in which the fates of the olefinic carbon atoms of propylene (*25, 89, 90*), 2-butene (*17*), cyclopentene (*31*), and cyclobutene (*32*) were traced with isotopic labels. The first explanation put forward for the transformation (*1, 10, 17*), [indicated in

Eq. (3) and called below the conventional mechanism] was that two

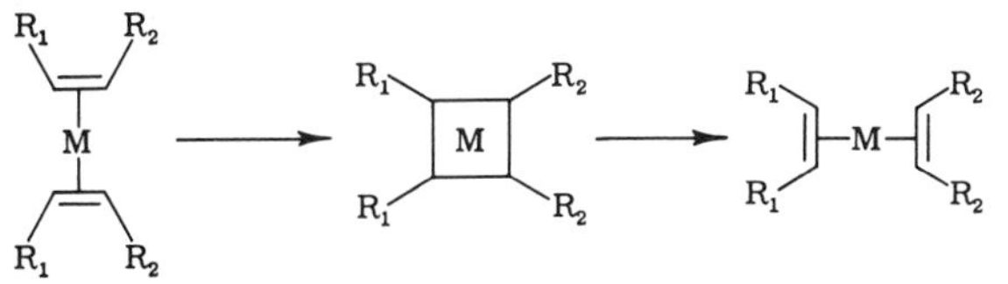

$$
\begin{array}{c}
\text{(structure of Eq. 3)}
\end{array}
\qquad (3)
$$

olefins united to form a cyclobutane whose cleavage completed the metathesis, an idea that was appealing because it accounted for the gross structural change and because it suggested that the role of the metal was to circumvent the orbital symmetry prohibitions that in the absence of metals prevent 2 + 2 cycloadditions from occurring (*18, 79, 80, 80a, 102, 103, 105*). This explanation suggests that cyclobutanes should either form as side-products of olefin metatheses, or when added to olefin metathesis reaction mixtures should cleave, but experiments to demonstrate either seemed to show that neither occurred (*77, 90, 107, 119*).

It now seems likely that the mechanism of the metathesis does not involve the steps summarized in Eq. (3), but instead the chain reaction propagated by metal-carbenes summarized in Eq. (4) (*20, 22, 36, 47, 52, 65, 68, 111*). This mechanism also accounts for the gross structural

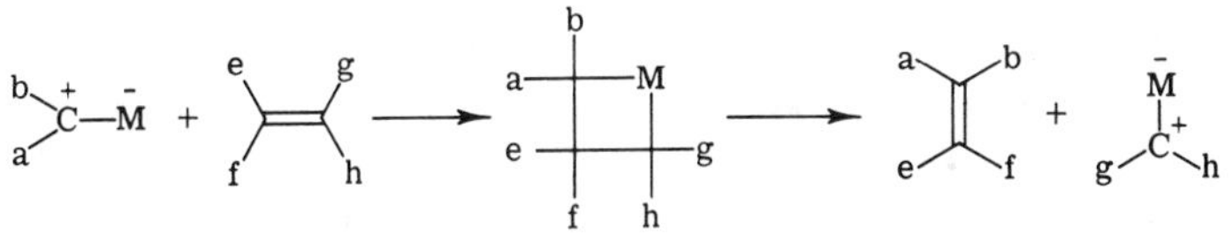

$$
\begin{array}{c}
\text{(structure of Eq. 4)}
\end{array}
\qquad (4)
$$

change, and according to it cyclobutanes should neither form nor react. In the discussion below are summarized (1) the reasons for believing this mechanism to be correct and (2) these three corollaries: (a) an account of why alkylidiene groups do not interchange randomly, (b) an account of the reaction's stereochemistry, and (c) the implication that rational initiators for olefin metathesis be found among metal-carbene species.

II

THE MECHANISM OF THE OLEFIN METATHESIS REACTION

A. *Double Cross Experiments*

Consider an experiment called the double cross in which cyclooctene, 2-butene, and 4-octene are allowed to react simultaneously and the amounts of the hydrocarbons in Eq. (5) are analyzed to determine the ratios of the concentrations $[C_{14}]/[C_{12}]$ and $[C_{14}]/[C_{16}]$ at the very beginning of reaction (65, 66). The idea is that if the mechanism of the

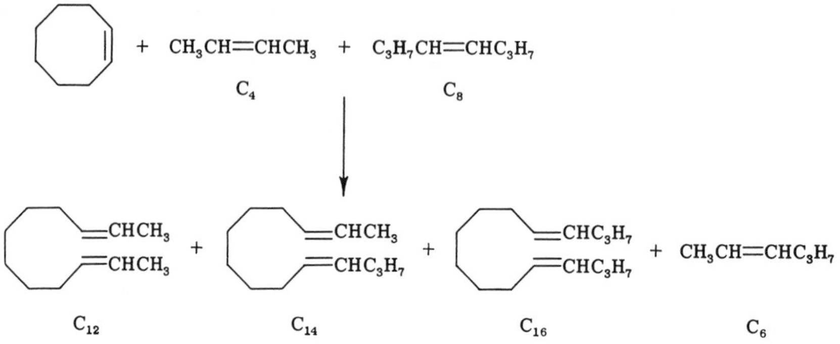

$$+ \quad CH_3CH{=}CHCH_3 \quad + \quad C_3H_7CH{=}CHC_3H_7$$
$$\qquad\qquad C_4 \qquad\qquad\qquad\qquad C_8$$

$$C_{12} \quad + \quad C_{14} \quad + \quad C_{16} \quad + \quad CH_3CH{=}CHC_3H_7$$
$$\qquad\qquad\qquad\qquad\qquad\qquad\qquad\qquad\qquad\qquad C_6$$

$$(5)$$

olefin metathesis were that summarized in Eq. (3), then at first both ratios would have to be zero, for C_{14} should not form from C_6 as quickly as C_{12} and C_{16} form from C_4 and C_8, since early in the reaction much less C_6 would be present than C_4 and C_8. Similarly, C_{14} could not form from C_{12} reacting with C_{16}, since neither would yet be present in appreciable quantity. If on the other hand the mechanism were that in Eq. (4), then C_{14} would have to predominate over C_{12} or C_{16} by at least a factor of 2. If the methyl and propyl groups that differentiate the butene and octene served only as labels and if the amounts of butene and octene were equal, the ratios of C_{12}, C_{14}, and C_{16} products should be $1:2:1$ because, as indicated in Eq. (6), reaction by path a would be just as likely as by path b.

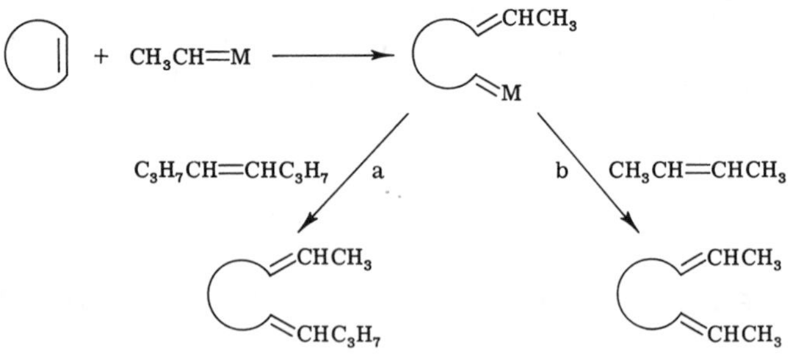

$$+ \quad CH_3CH{=}M \quad \longrightarrow$$

$$C_3H_7CH{=}CHC_3H_7 \Big/ a \qquad b \Big\backslash CH_3CH{=}CHCH_3 \qquad (6)$$

The fact is that in a number of such experiments in which cyclooctene was combined in chlorobenzene containing $Mo[(C_6H_5)_3P]_2Cl_2(NO)_2$ and $(CH_3)_3Al_2Cl_3$ with either *trans*-2-butene plus *trans*-4-octene or *cis*-2-butene plus *cis*-4-octene, and the composition of the products was

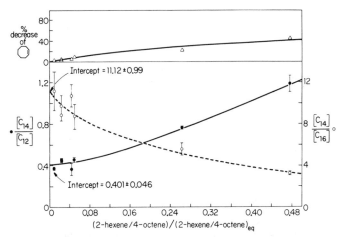

FIG. 1. Plot of product composition as a function of extent of reaction for the reaction at 0°C of cyclooctene (5.44 mmoles), *cis*-2-butene (2.64 mmoles), and *cis*-4-octene (3.51 mmoles) with $Mo[(C_6H_5)_3P]_2Cl_2(NO)_2$ (0.038 mmole) and $(CH_3)_3Al_2Cl_3$ (0.23 mmole) in chlorobenzene (2 ml). *n*-Heptane (352 mg) was the internal standard for the gas chromatographic analysis. Reprinted with permission from Katz and McGinnis (66); copyright by the American Chemical Society. The ordinate is labeled at the left for the solid curve and at the right for the dashed curve.

analyzed at short reaction times and extrapolated to zero time, the C_{14} hydrocarbon, which according to the conventional mechanism should initially not form at all, at the very start of reaction was always found to be formed in greater amount than either C_{12} or C_{16}. An example is given in Fig. 1.

These experiments would exclude the conventional mechanism were it not for one qualification, that the conventional mechanism would seemingly not be excluded if the rate-determining step were the olefin displacement, step 2 in Eq. (7), because, as shown in Scheme 1, the alkylidene groups would then be

$$
\begin{array}{c}
R_1CH{=}CHR_1 \\
| \\
M \\
| \\
R_2CH{=}CHR_2
\end{array}
\quad\xrightarrow{\text{step 1}}\quad
\begin{array}{c}
R_1CH \quad CHR_1 \\
\| \quad\quad \| \\
{=}M{-} \\
\| \quad\quad \| \\
R_2CH \quad CHR_2
\end{array}
\tag{7}
$$

$$
\xrightarrow[\text{step 2}]{R_1CH{=}CHR_1}\quad
\begin{array}{c}
R_1CH \quad CHR_1 \\
\| \quad\quad \| \\
{=}M{-} \\
\| \quad\quad \| \\
R_1CH \quad CHR_2
\end{array}
\;+\;
\begin{array}{c}
CHR_1 \\
\| \\
CHR_2
\end{array}
$$

scrambled while the initially formed diene was still attached to the metal

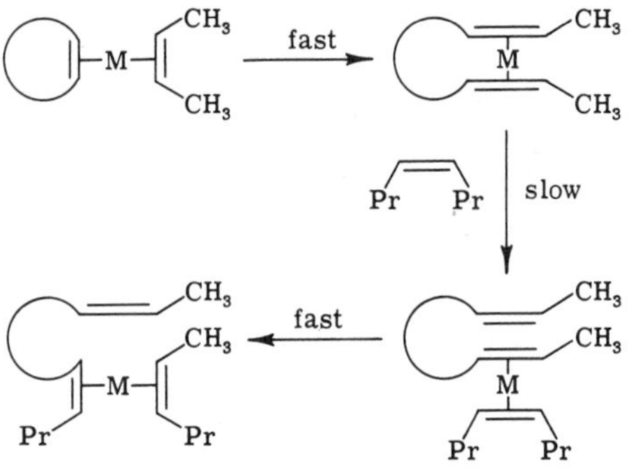

Scheme 1

and not yet released into solution. This possibility, which we call the sticky olefin hypothesis, was first suggested by Calderon (12) to account for the high molecular weights of the polymers formed initially in metatheses of cycloalkenes. It is considered and rejected below in Section II,C.

B. Single Cross Experiments

An experiment called the single cross, similar to that above, is one in which cyclooctene and 2-hexene are allowed to react and the products are analyzed to determine the ratios of concentrations $[C_{12}]/[C_{14}]$ and $[C_{16}]/[C_{14}]$ at zero time (65, 66). Now, according to the conventional mechanism and at zero time, C_{14}, which in the previous experiments was anomalous, would be the only product. C_{12} and C_{16} would be the anomalies. According to the metal-carbene chain mechanism, if the methyl and propyl groups served only as labels, the distribution of C_{12}, C_{14}, and C_{16} products should again be $1:2:1$. C_{14} should be the major product just as it is according to the conventional mechanism, so the change in product distribution with mechanism would be less striking.

As Fig. 2 shows, $[C_{12}]/[C_{14}]$ and $[C_{16}]/[C_{14}]$ at zero time are not zero, excluding the conventional mechanism if the metathesis step were rate determining, but not if the olefin-displacement reaction [step 2 in Eq. (7)] were. Notice that the values of $[C_{12}]/[C_{14}]$ and $[C_{16}]/[C_{14}]$ are not the 0.5 expected according to the assumptions made above. This is considered later below.

A similar experiment was performed earlier by Hérisson and Chauvin (52), who found that cyclopentene and 2-pentene with $WOCl_4$ and

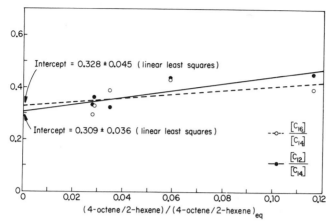

FIG. 2. Plot of product composition as a function of extent of reaction for the reaction at about 5°C of cyclooctene (1.7 mmoles) and 2-hexene (1.8 mmoles) with Mo[(C$_6$H$_5$)$_3$P]$_2$Cl$_2$(NO)$_2$ (0.012 mmole) and (CH$_3$)$_3$Al$_2$Cl$_3$ (0.08 mmole) in chlorobenzene (0.25 ml). Reprinted with permission from Katz and McGinnis (66); copyright by the American Chemical Society.

(n-C$_4$H$_9$)$_4$Sn in chlorobenzene after only partial reaction gave an analogous "triad" of C$_9$, C$_{10}$, and C$_{11}$ products in 1:2:1 ratio. They recognized the cross products as anomalous and proposed the carbene chain mechanism to account for them. However, they also found that cyclooctene with 1-pentene yields essentially no cross products, and similar experiments in other laboratories with cyclooctene and 1-hexene (75) or cyclopentene and 1-pentene (69) gave analogous results. Since the evidence for the carbene-chain mechanism cited above was the observation of cross products in reactions of cyclic with acyclic olefins, the failure to observe them in these experiments was taken as evidence against the mechanism (88).

That this absence of cross products is nevertheless a natural consequence can be seen (65) by considering the reaction of cyclooctene with 1-pentene in Scheme 2. Here the factors that stabilize more highly substituted carbenium ions over those less substituted should cause the more highly substituted carbene (26, 33, 72), and therefore only the conventional product, to form preferentially. The propylcarbene moiety adds to another molecule of cyclooctene and the cycle is repeated.

When the two ends of the acyclic olefin differ only a little, as in 2-hexene, the product ratios should be similar to 1:2:1, but any preference for one or the other of the paths analogous to those in Scheme 2 will increase the amount of conventional product.

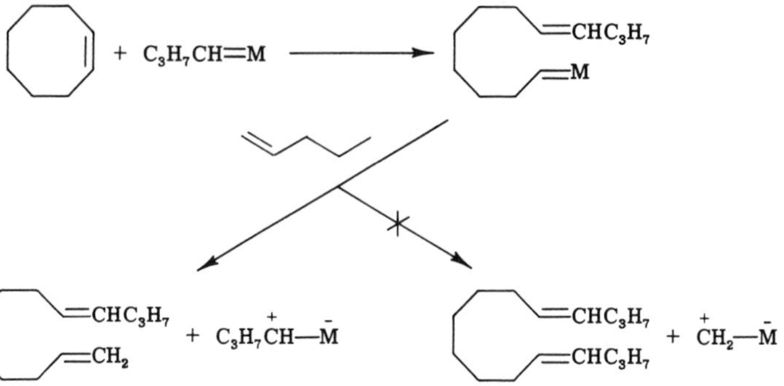

$$+ \; C_3H_7CH{=}M \longrightarrow$$

(Scheme 2)

Scheme 2

An analysis of the kinetics of the metal-carbene chain mechanism (66) shows that if rate constants are defined as in Eqs. (8) and (9) (here M is

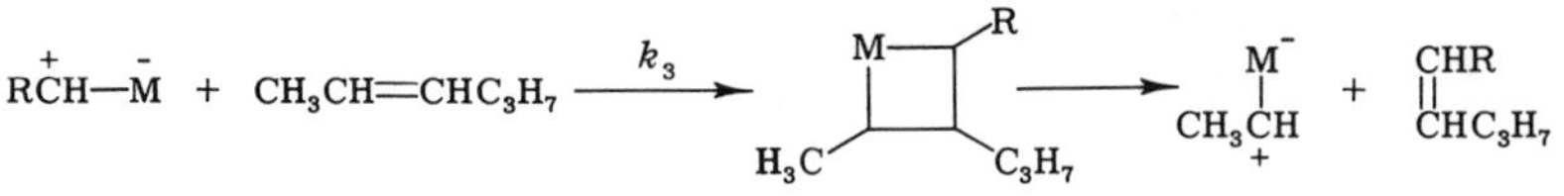

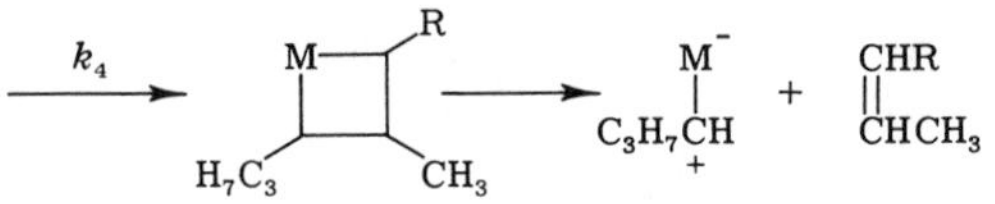

$$RCH{-}M + CH_3CH{=}CHC_3H_7 \xrightarrow{k_3} \qquad \longrightarrow \qquad + \qquad \tag{8a}$$

$$\xrightarrow{k_4} \qquad \longrightarrow \qquad + \qquad \tag{8b}$$

$$\xrightarrow{k_7} \qquad \longrightarrow \qquad \tag{9}$$

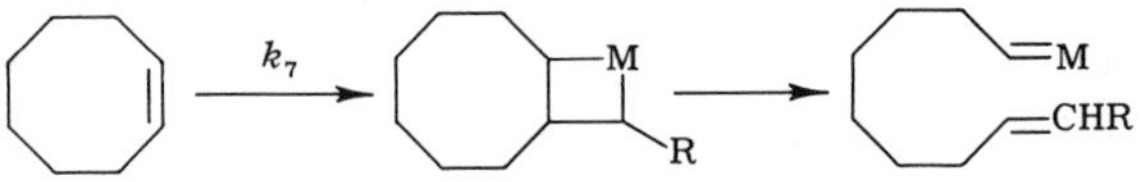

the metal) and C_{12+8n}, C_{14+8n}, and C_{16+8n} are defined as the homologs of C_{12}, C_{14}, and C_{16} in Eq. (5), the reaction of cyclooctene (Cy) with 2-hexene (Hex) will give the products C_{12+8n}, C_{14+8n}, and C_{16+8n} in the ratios $1:r:1$, where

$$r = k_3/k_4 + k_4/k_3 \tag{10}$$

Moreover,

$$\frac{d[C_{m+8n}]}{d[C_{m+8(n+1)}]} = 1 + \frac{(k_3 + k_4)[\text{Hex}]}{k_7[\text{Cy}]} \tag{11}$$

where $m = 12, 14, 16$. Since Eq. (10) requires that $r = 2$ if $k_3 = k_4$ and that $r > 2$ if $k_3 \neq k_4$, in the single cross experiments the ratios of C_{12}, C_{14}, and C_{16} products will be $1:r:1$ where $r \geq 2$.

The ratio r measures whether metal carbenes react with unsymmetrically substituted alkenes preferentially along one or the other of two paths [Eqs. (8)]. Since the stability of carbenium ions is sensitive to the substituents attached to the ion center, in any experiment like this there should always be some preference for one path over the other, and the three products, like C_{12}, C_{14}, C_{16}, should therefore essentially never form in the $1:2:1$ ratio that is likely at thermodynamic equilibrium. The ratio initially should always be $1:r:1$ with $r > 2$.

When the substituents at the two ends of the double bond are very different, as when terminal olefins like 1-pentene or 1-hexene are used in place of 2-hexene, the preference for one of the paths (presumably the one giving the more highly substituted carbenium ion center) should be large, and in place of triads of products only the conventional products should be, and, as discussed above, are observed.

This selectivity means that although for experiments like the single cross to work, the substituents at the ends of the acyclic olefin's double bond must be different, for otherwise there would be no cross products like C_{12} or C_{16}, they must not be very different, or no cross products would be observed either, since one path or the other in Eq. (6) would be preferred.

Quantitative aspects of the theory are these. Since for 2-hexene the single cross experiments show $r = 3.25 \pm 0.17$, Eq. (10) indicates that k_3/k_4 in Eqs. (8) or $k_4/k_3 = 2.91 \pm 0.19$. For terminal olefins this figure should be much larger, and in agreement with this expectation other measurements (described in Section III,A) using various tungsten catalysts have shown the ratio to be about 100 (83).

Equation (10) then requires the ratio r for the triads that form in metatheses of unsubstituted cyclic olefins with terminal olefins to be also about 100. Reported ratios, r, in experiments using various tungsten catalysts are: for cyclooctene plus propene, 10 (52); for cyclooctene plus 1-pentene, 20 (52); for cyclopentene plus 1-pentene, <38 (69); and for cyclooctene plus 1-hexene, about 4 (75). Since in none of these experiments were extrapolations made to zero time, and since with time the measured ratios r will decrease as thermodynamic equilibrium is approached, these figures are in reasonable agreement with expectation.

There are no data to determine $(k_3 + k_4)/k_7$ using Eq. (11), but published data that have not been extrapolated to zero time indicate that the analogous ratio for cyclopentene plus 2-pentene is approximately 1 (52) and for cyclopentene plus 1-pentene, as seems plausible, larger, about 16 (69). For terminal olefins plus cycloolefins, the ratio seems to be smaller for cyclooctene (about 2) (52, 75) than for cyclopentene (about 16) (69).

Two consequences (considered further in Section III) of the selectivity that makes $k_3 \neq k_4$ are (a) the implication that, as indicated in Eq. (12),

$$\text{etc.} \quad (12)$$

unsymmetrically substituted cyclic alkenes should yield polymers that are translationally invariant and (b) an explanation of why some catalysts, for example $WCl_6 + C_2H_5OH + C_2H_5AlCl_2$, that are effective in bringing about metatheses of 1,2-disubstituted ethylenes give low yields of metathesis products with monosubstituted ethylenes. This explanation would center about the idea that if the reaction of a metal-carbene with a terminal olefin were to proceed selectively accordingly to Eq. (13a) rather than Eq. (13b), the major effect of metathesis would be the

$$\underset{H}{\overset{R}{\diagdown}}C{=}M \ + \ \underset{H}{\overset{R'}{\diagdown}}C{=}CH_2 \ \xrightarrow{\ k\ } \ \underset{H}{\overset{R}{\diagdown}}C{=}CH_2 \ + \ \underset{H}{\overset{R'}{\diagdown}}C{=}M \quad (13a)$$

$$\xrightarrow{\ k'\ } \ \underset{H\quad H}{\overset{R\quad R'}{C{=}C}} \ + \ \underset{H}{\overset{H}{\diagdown}}C{=}M \quad (13b)$$

conversion of starting olefin into itself. If the chain terminated after a large number of metatheses, there still might be no observable metathesis product.

C. The Sticky Olefin Hypothesis

It is interesting to note that if the sticky olefin hypothesis were correct, the conventional mechanism would also account for both these consequences as well as for cyclic olefins and terminal olefins yielding essentially no cross products. This can be seen, for example, in Scheme 3, where the reaction of cyclic and acyclic olefins is considered. Here, if the end of the olefin to which R' is attached is more easily displaced from the metal than the end to which R is attached, reaction 2 will be faster than reaction 1. This will preferentially form conventional product—diene capped by the same groups, R' and R, as the starting olefin. The other product, **B**, upon rapid metathesis yields **A**, and the cycle is then repeated. The effect would be that if one of the paths in Scheme 3 were preferred, only the conventional product would form.

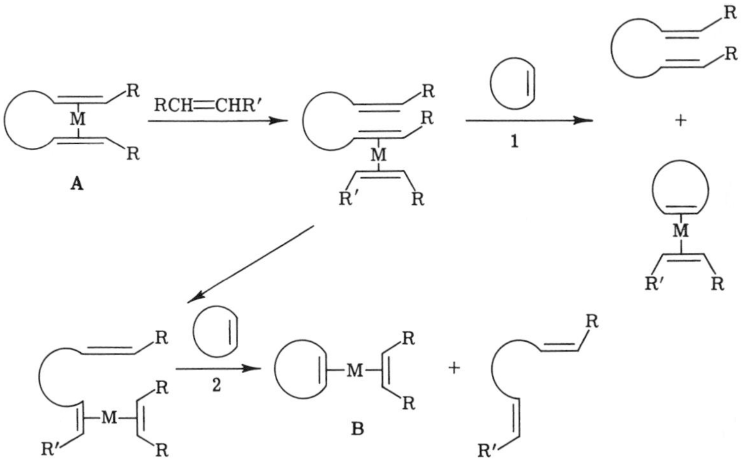

Scheme 3

Thus, although the sticky olefin hypothesis requires, as considered in Section II,D, an unacceptable account for the bonding in the four-membered ring intermediate, it is experimentally not easily distinguishable from the carbene-chain mechanism. However, the distribution of products formed in the double cross experiments should distinguish the two, as shown below, and so should the distribution of products in another version of the experiment that is essentially the reverse of the double cross.

In this version [Eq. (14)] a diene deuterated at both ends and the same diene undeuterated are converted into ethylenes-d_0, -d_2, and -d_4, whose

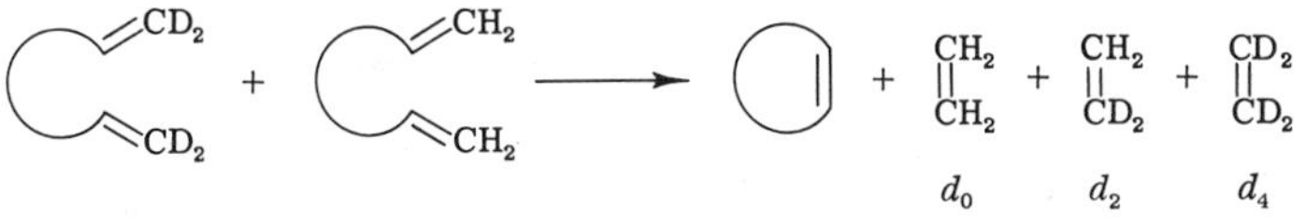

$$\begin{array}{ccccc} & & & CH_2 & CH_2 & CD_2 \\ & & & \| & \| & \| \\ & & & CH_2 & CD_2 & CD_2 \\ & & & d_0 & d_2 & d_4 \end{array} \qquad (14)$$

distributions are determined at short reaction times. Such experiments have been reported with 1,7-octadiene (46, 47) and with 2,2'-divinylbiphenyl (68).

In the latter, when equal amounts of deuterated and undeuterated 2,2'-divinylbiphenyls were combined with either molybdenum or tungsten catalysts, the ratios of undeuterated, dideuterated, and tetradeuterated molecules after about 1% reaction were found to be approximately 1:2:1 (24.3% d_0, 48.7% d_2, and 25.9% d_4) in the ethylenes formed, while in the divinylbiphenyl precursors they were still 1:0:1. (The amount of isotope scrambling in the remaining divinylbiphenyls was measured as

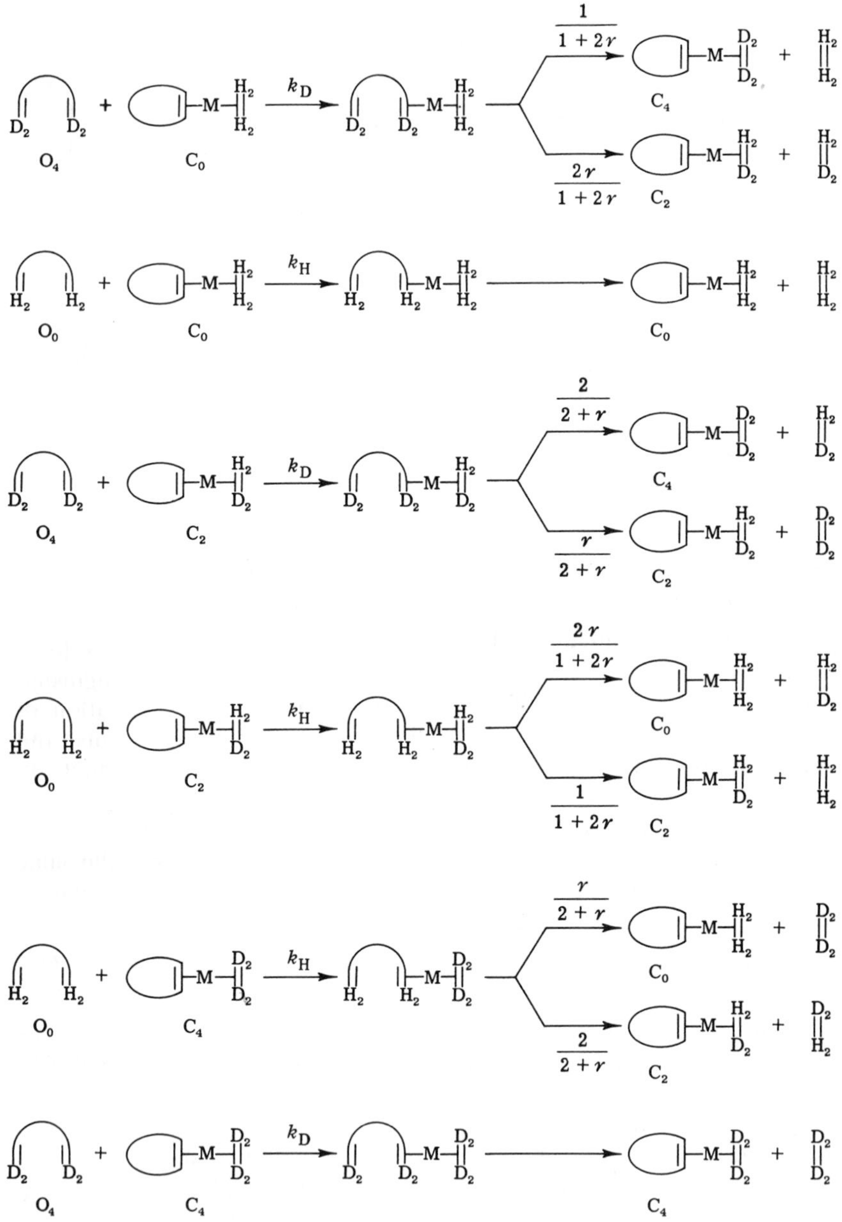

Scheme 4

about 1%.) The disparity could not be ascribed to rapid scrambling of the ethylenes after their initial formation but before analysis because similar metatheses of undeuterated divinylbiphenyl in the presence of *trans*-1,2-dideuterioethylene did not give appreciable amounts of mono-deuterated ethylene.

The observation that a $1:0:1$ mixture of d_0-, d_2-, and d_4-divinylbiphenyls gives a $1:2:1$ mixture of d_0-, d_2-, and d_4-ethylenes is again not in accord with the conventional mechanism if the metathesis reaction, step 1 of Eq. (7), is rate determining, but that it is also not in accord with the conventional mechanism if the olefin displacement reaction, step 2, is rate determining was shown in two analyses of the reaction sequences in Scheme 4 (*46, 47, 68*). Here the ethylene complex with n deuteriums is labeled C_n, the diene with n deuteriums O_n, and r is an isotope effect. The kinetics show what the ratios of the ethylenes formed should be as functions of the ratio (here called y) of $[O_0]$ and $[O_4]$ in the reaction mixture and of r, but the essential difference between these ratios and those according to the kinetics of the metal-carbene chain mechanism can be seen (*68*) by considering what the product of $d[C_2D_2H_2]/d[C_2H_4]$ and $d[C_2D_2H_2]/d[C_2D_4]$ should be. According to the metal-carbene chain mechanism the rates of formation of the d_0-, d_2-, and d_4-ethylenes should be $r^2y^2:2ry:1$, and the product $d[C_2D_2H_2]/d[C_2H_4] \times d[C_2D_2H_2]/d[C_2D_4]$ should therefore be $(2ry/r^2y^2)(2ry/1) = 4$, independent of r. But according to the mechanism of Scheme 4, if $y \approx 1$ the product should be 2.56 [essentially $(8/5)^2$] for all reasonable values of r, and since the measured product $(\%d_2)^2/(\%d_0)(\%d_4) = 3.80 \pm 0.31$, the mechanism in Scheme 4 is excluded. So too are related mechanisms in which the metals are always bonded to three olefins, to two ethylenes, and to one other (*68*).

This analysis suggests that in the double cross experiment the product of $[C_{14}]/[C_{12}]$—called r_1—and $[C_{14}]/[C_{16}]$—called r_2—should be 4, and Fig. 1 shows that experimentally this is approximately true. A more accurate measure of this product at zero time can be achieved by multiplying r_1 and r_2 at the different times and extrapolating to zero time, as in Fig. 3. In a series of four such experiments, three using cyclooctene, *cis*-2-butene, and *cis*-4-octene, and one using cyclooctene, *trans*-2-butene, and *trans*-4-octene, the average value of $\overline{r_1 \times r_2}$ at zero time was measured as 4.05 ± 0.05 (*66*).

Now an analysis (*66*) of the kinetics of the metal-carbene chain mechanism shows that this product should be 4, and it also allows a precise definition of another notable feature of Fig. 1, the formation initially of more C_{12} than C_{16} although less 2-butene was present than 4-octene. 2-Butene must be more reactive than 4-octene, presumably for steric reasons, and the ratios of the rate constants defined by Eqs. (15)

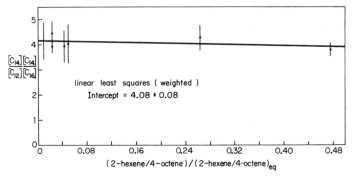

FIG. 3. Product $r_1 \times r_2$ as a function of extent of reaction. The experimental data are the same as in Fig. 1. Reprinted with permission from Katz and McGinnis (66); copyright by the American Chemical Society.

(here M is again the metal) can be

$$\overset{+}{R}\overset{-}{CH}\overset{-}{-}\overset{-}{M} + CH_3CH=CHCH_3 \xrightarrow{k_1} RCH=CHCH_3 + CH_3\overset{+}{CH}\overset{-}{-}\overset{-}{M} \qquad (15a)$$

$$\overset{+}{R}\overset{-}{CH}\overset{-}{-}\overset{-}{M} + C_3H_7CH=CHC_3H_7 \xrightarrow{k_2} RCH=CHC_3H_7 + C_3H_7\overset{+}{CH}\overset{-}{-}\overset{-}{M} \qquad (15b)$$

evaluated according to the kinetic analysis (66), from Eqs. (16) and (17) and the initial values of r_1 and r_2.

$$r_1 \text{ (initial)} = 2k_2[\text{octene}]/k_1[\text{butene}] \qquad (16)$$

$$r_2 \text{ (initial)} = 2k_1[\text{butene}]/k_2[\text{octene}] \qquad (17)$$

For cis-2-butene and cis-4-octene $k_1/k_2 = 6.46 \pm 0.9$; for trans-2-butene and trans-4-octene $k_1/k_2 = 3.08 \pm 0.2$ (66). Figures 1 and 3 also show that although $r_1 \times r_4 = 4$ at all times, the ratios of C_{12}, C_{14}, and C_{16} at zero time are not the equilibrium ratios, which is why they change with time.

Consider now what the product $r_1 \times r_2$ would be if the conventional mechanism were correct, but the olefin displacement reaction were rate determining. The kinetics of a mechanistic scheme much like that in Scheme 4 shows (66) that $r_1 \times r_2$ at zero time is a function of two ratios, that of the concentrations of butene and octene and that of their reactivities. However, for all reasonable values of these ratios, $r_1 \times r_2$ is never greater than 2.94. Accordingly, since in the double cross experiment the product was determined to be about 4 at zero time, the conventional mechanism is excluded no matter which step is rate-determining. The implication is that the carbene chain mechanism is correct.

The rigor with which either the diene cyclization (68) or the double cross (66) experiment disproves the conventional mechanism depends on the strengths of the arguments that can be brought to bear against the idea that the initial reaction in either experiment is a polymerization, followed by a depolymerization. For example, in the double cross experiments the conventional mechanism would account for the product distribution if cyclooctene first polymerized to polyoctenamer (13, 27) and then were cleaved by the acyclic olefins, but this possibility was rejected for two reasons. One was that, since, as Fig. 1 illustrates, cyclooctene was largely unconsumed when the first measurements of $[C_{12}]$, $[C_{14}]$, and $[C_{16}]$ were being made, for this hypothesis to account for the yields of C_{12}, C_{14}, and C_{16}, polyoctenamer would have to be more reactive than cyclooctene toward the acyclic olefins, and that appears unlikely. Two, this hypothesis would not account for why reactions of cyclic olefins with unsymmetrically substituted acyclic olefins give mainly the conventional products.

D. *Other Experimental Results*

The metal-carbene mechanism also accounts for a number of other experimental results.

The metatheses of acetylenes like 2-pentyne (91–94, 104) can be interpreted (65) as in Eq. (18) to proceed through the intermediacy of

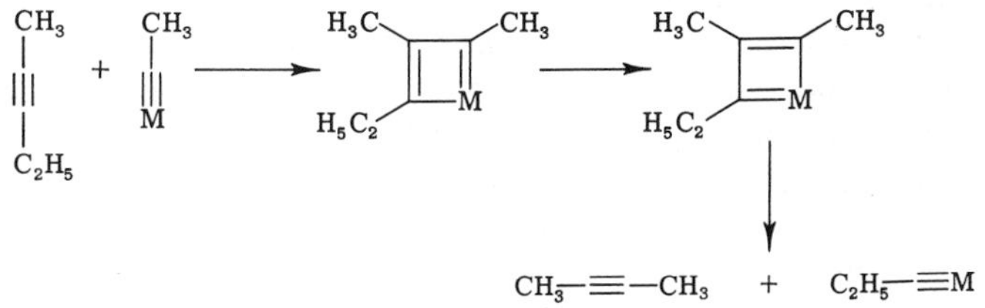

$$\tag{18}$$

metal-carbynes (41). The conventional interpretation of olefin metathesis on the other hand seems impossible to reconcile with these reactions, for the stability of cyclobutadiene–metal complexes (2, 35, 78) preclude their being unstable intermediates. It is hard to believe that the energy change upon complexing a cyclobutadiene to a metal is similar to that upon complexing a cyclobutane to a metal, and thus the metathesis of acetylenes more than any other single datum seems to show that the conventional mechanism must be wrong.

The formation of large rings from smaller cyclic olefins (53, 54, 117, 120) can be interpreted (36) to involve terminal carbenes cyclizing upon internal double bonds, as in Eq. (19), and the formation of catenanes (8,

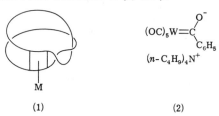

$$(19)$$

119) to involve the same reactions of intertwined chains (65) rather than twisted intermediates like structure 1 (8, 119).

$$(OC)_5W=C\begin{matrix} O^- \\ \diagdown \\ C_6H_5 \end{matrix}$$

$$(n\text{-}C_4H_9)_4N^+$$

(1)   (2)

The cyclization in Eq. (19) would also account (36) for why cyclodo-deca-1,5,9-triene or cycloocta-1,5-diene alone (36, 110) or with 2-pentene (52) give oligomers containing n C_4H_6-fragments where n's are even and odd, and why the composition of the oligomer mixture formed from either the octadiene or the dodecatriene is the same (110).

A chain reaction mechanism accounts naturally for the high molecular weight (4, 48) of the polymers formed early in the reactions of cycloolefins (36, 52).

The kinetics of the olefin metathesis of 2-pentene by (pyridine)$_2$Mo(NO)$_2$Cl$_2$ and organoaluminum halides have been measured (56) as first order in the metal and variable order in olefin (seemingly first order at high olefin concentration and up to order 1.7 at low olefin concentration) and were originally interpreted to support the conventional mechanism, but they now also seem in accord with the metal-carbene chain mechanism.

Both Cardin, Doyle, and Lappert (20) and Casey and Burkhardt (22) found support for the metal-carbene chain mechanism in their discoveries that stable metal-carbenes and olefins exchange their carbene moieties. The former group's olefins were tetraaminoethylenes, and a corresponding diaminocarbene–rhodium complex effected their metathesis. The latter group's olefins had one or no heteroatom substituents and the metal-carbene, (diphenylcarbene)pentacarbonyltungsten, reacted only stoichiometrically with them. Their outstanding example, α-methoxy-styrene, transferred one carbene moiety to the diphenylcarbene forming 1,1-diphenylethylene, and the other to the metal forming (phenylmethox-ycarbene)pentacarbonyltungsten. Other olefins—ethyl vinyl ether, iso-butylene, and trans-2-butene—gave isolable products in which a carbene moiety transferred to the diphenylcarbene, but none in which a carbene transferred to the metal. (Phenylmethoxycarbene)pentacarbonyl-chromium exchanges similarly with vinyl ethers (*43*) and with *N*-vinylamides (*42*).

The mechanism suggests that metal-carbenes and metal-containing four-membered rings will initiate olefin metathesis. Indeed, Kroll and Doyle showed that isolable metal-carbenes like **2**, when mixed with methyl aluminum dichloride do initiate metatheses (*4, 74*), and Dolgoplosk *et al.* showed that so do various diazocompounds, $RCH{=}N_2$ (where $R = C_6H_5$, C_2H_5, or $CO_2C_2H_5$), when mixed with WCl_6 (*36, 37, 45*). Both results might be consequences of the initiating mixtures generating metal-carbenes unstabilized by heteroatoms at the carbene center, but there is no proof that either mixture does generate metal-carbenes or that if formed they are catalysts. Initiation of olefin metatheses by metal-carbenes is considered again in Section IV.

III

SELECTIVITY IN OLEFIN METATHESES OF UNSYMMETRICALLY SUBSTITUTED ETHYLENES

A. *Monosubstituted Ethylenes*

Considered here is why when terminal olefins undergo the olefin metathesis reaction the yields using some catalysts are very low (*69, 83, 87, 115, 123, 125*), while with others the yields are high (*5, 9, 34, 38, 40, 60, 73, 74, 96, 125*). The reasons are neither that terminal olefins are unreactive toward olefin metathesis, nor that they are destroyed by side reactions with acids, for terminal olefins with internal olefins do give appreciable amounts of all products except those of their self-metathesis (*52, 59, 81*). The possible reasons seem to be either (a) that metal-carbenes combine with terminal olefins selectively according to Eq. (13a) rather than Eq. (13b), that this selectivity varies with the catalysts, and that only a limited number of metatheses take place before the chain reactions terminate or (b) that with some catalysts the chain reactions proceed through a greater number of cycles, this number rather than the selectivity determining effectiveness.

The selectivity for reaction (13a) over (13b) has been measured (*83*) by determining k/k', the ratios of the rate constants, for five catalyst preparations. This was done by measuring the rates at which products form when 1-octene-1,1-d_2, 1-hexene, and the catalysts are combined.

TABLE I

RATIOS OF RATE CONSTANTS k AND k' [Eqs. (13)][a]

Catalyst (temperature)[b]	k/k'
$Mo[(C_6H_5)_3P]_2Cl_2(NO)_2 + (CH_3)_3Al_2Cl_2$ (0°)	26.5 ± 6
$(C_6H_5)_2C{=}W(CO)_5$ (50°)	75.2 ± 4
$WCl_6 + n\text{-}C_4H_9Li$ (rt)	80.0 ± 8
$WCl_6 + (C_6H_5)_3SnC_2H_5$ (rt)	103 ± 20
$WCl_6 + C_2H_5OH + C_2H_5AlCl_2$ (rt)	155 ± 33

[a] From McGinnis et al. (83).
[b] rt = room temperature.

The figures in Table I mean that the main reaction that all the catalysts effect is the conversion of 1-octene and 1-hexene into 1-octene and 1-hexene, a reaction that would not have been noticed had it not been for the isotopic labels. Moreover, the molybdenum catalyst that is known to be most effective in bringing about metathesis of terminal olefins is the least selective, but even this catalyst brings about 26 ineffective metatheses, transforming starting materials into themselves, before it brings about a metathesis recognizable other than with isotopic labels. $WCl_6 + C_2H_5OH + C_2H_5AlCl_2$, which is notoriously ineffective in bringing about metatheses of terminal olefins, is much more selective, which may also be why it works so poorly.

That terminal olefins react selectively to interchange their unsubstituted methylene groups was also discovered (69) to occur when 1-pentene-d_{10} and 1-pentene were combined with a mixture of WCl_6, C_2H_5OH, and $C_2H_5AlCl_2$ (only traces of 4-octene were produced when the 1-pentenes were 70% scrambled), when 1-hexene was combined (87) either with 1-heptene-1,1-d_2, WCl_6, and $n\text{-}C_4H_9Li$ or with 1-pentene-d_{10}, WCl_6, and $C_2H_5AlCl_2$ (when the ratios of nonproductive and productive metatheses were about 1000), and when 1-pentene-d_{10} and 1-hexene were combined (87) with WCl_6 and $Zn(CH_3)_2$ (when the ratio was about 100). The rate constants were not analyzed in these studies.

A different way [Eq. (20), Tol is p-tolyl] to determine ratios like k/k'

$$(Tol)_2C{=}W(CO)_5 \quad + \quad CH_3(CH_2)_2CH{=}CH_2$$

$$\downarrow \tag{20}$$

$$(Tol)_2C{=}CH_2 \quad + \quad (Tol)_2C{=}CH(CH_2)_2CH_3 \quad + \quad Tol{\bigtriangleup}(CH_2)_2CH_3$$

$$35.9 \pm 0.3\% \qquad\qquad 0.06 \pm 0.02\% \qquad\qquad 26.9\%$$

was devised by Casey *et al.* (*23*). They had discovered that (diphenylcarbene)pentacarbonyltungsten, which they had synthesized earlier [Eq.
(21)] (*21*), reacted with olefins forming 1,1-diphenylethylenes (*22*), and
they measured the ratio, 600, of the olefin products, $CH_2{=}C(Tol)_2$ and
$CH_3(CH_2)_2CH{=}C(Tol)_2$, to determine rate-constant ratios like those in
Eqs. (13). That the ratio determined in this way is different from those in
Table I is not surprising because while the reacting carbene there was
monosubstituted, here it is disubstituted. A comparison of the two ex-

$$\underset{(3)}{\overset{\displaystyle C_6H_5}{\underset{\displaystyle CH_3O}{\diagdown\diagup}}C{=}W(CO)_5} \xrightarrow[\text{2. HCl}]{\text{1. } C_6H_5Li} \underset{(4)}{\overset{\displaystyle C_6H_5}{\underset{\displaystyle C_6H_5}{\diagdown\diagup}}C{=}W(CO)_5} \qquad (21)$$

periments indicates that the more highly substituted carbene is in fact
the more selective, as seems plausible. The Casey experiment has the
virtue of distinguishing experimentally whether metal-carbenes react
with olefins to form selectively the more-substituted or the less-substi-
tuted metal-carbene (the former forms), which the experiments analyzing the reaction kinetics of labeled and unlabeled olefins do not. [They
do not for the same reason that the ratio r in the single cross experiment
does not distinguish k_3/k_4 from its inverse, Eq. (10). This also means
that, in the metal-carbenes, if the charge on carbon were negative, rather
than positive as assumed above and as indicated by other experiments
(*26, 33, 72*), all observables but Casey's would be unaffected.] The latter
experiments, however, have the virtue of being applicable to a variety of
catalyst preparations.

Casey *et al.* (*23*) also measured the reactivities of different olefins by
combining their mixtures with $(Tol)_2C{=}W(CO)_5$ and measuring the
ratios of the olefin products. The reactivities of 1-pentene, isobutylene,
cis-2-butene, and 2-methyl-2-hexene were found in this way to be in the
ratios $49 \pm 5 : 10.4 \pm 0.5 : 1 :$ ca. 0.

Consider now the reactions of 1,1-disubstituted ethylenes. Since
monosubstituted ethylenes react selectively to interchange their more
highly substituted halves, 1,1-disubstituted ethylenes should react even
more selectively in the same way [Eq. (22)]. This increased selectivity
manifests itself in isobutylene's reaction with $(Tol)_2C{=}W(CO)_5$ yielding
even less than the 0.06% of more highly substituted 1,1-ditolylethylene
yielded by 1-pentene (*23*). But to demonstrate that metathesis catalysts
will induce the reaction in Eq. (22) selectively is not easy, for the

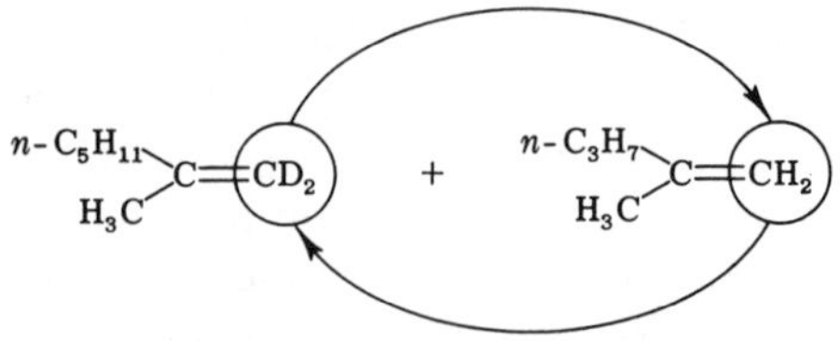

$$n\text{-}C_5H_{11}\diagdown C{=}CD_2 \quad + \quad n\text{-}C_3H_7\diagdown C{=}CH_2 \qquad (22)$$
$$\phantom{n\text{-}}H_3C\diagup \phantom{C{=}CD_2} \phantom{n\text{-}}H_3C\diagup$$

common catalyst mixtures containing WCl_6 or aluminum halides cannot be used since these acids oligomerize 1,1-disubstituted ethylenes.

For this purpose, a new catalyst was discovered (83), (diphenylcarbene)pentacarbonyltungsten (molecule 4), and its effectiveness is notable for two reasons: (a) because it supports the ideas about the mechanism of the olefin metathesis reaction, and (b) because the catalyst does not contain any metal halide or other Lewis acid and does not induce cationic oligomerizations even of sensitive olefins.

This catalyst does bring about the reaction in Eq. (22) selectively, and the ratio k/k' analogous to those in Table I was measured as being between 1400 and infinity, which was larger, as expected, than the ratio for the monosubstituted ethylenes.

B. Trisubstituted Ethylenes

If high selectivities also applied to trisubstituted olefins, a predictable consequence [Eq. (12)] would be that cycloalkenes unsymmetrically substituted on the double bond would yield polymers that, except possibly for geometrical isomerism, would be translationally invariant.

The metathesis of 1-methylcyclobutene [Eq. (23)] was the first reported (67) of a trisubstituted olefin, and the structure of the resulting

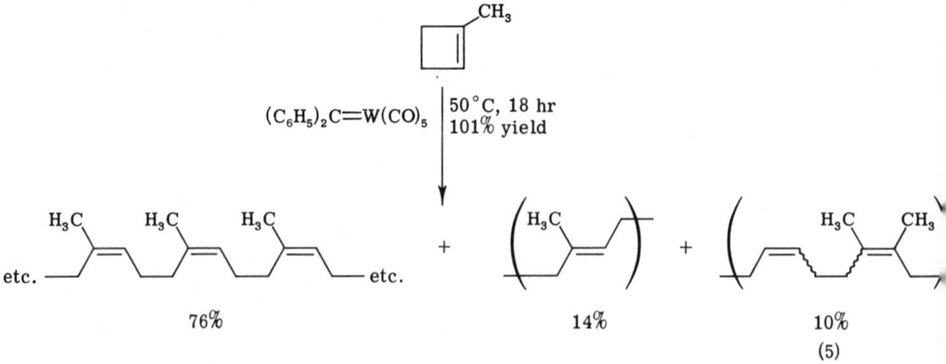

$$(C_6H_5)_2C{=}W(CO)_5 \quad \Big|\; \begin{array}{l} 50\,^\circ C,\ 18\ hr \\ 101\%\ yield \end{array}$$

etc. ———————— etc.   +   +

76%   14%   10%

(5)

(23)

polymer substantiated these expectations to a considerable extent.

Dall'Asta and Manetti (28) had tried many years ago to polymerize 1-methylcyclobutene with WCl_6 and either $(C_2H_5)_3Al$ or $C_2H_5AlCl_2$, but the polymers they obtained were largely saturated, presumably because under the acidic reaction conditions the olefin polymerized. The combination of $Mo[(C_6H_5)_3P]_2Cl_2(NO)_2$ and $(CH_3)_3Al_2Cl_3$ also gave largely saturated polymer, but $WCl_6 + (C_6H_5)_3SnC_2H_5$ gave more satisfactory material, which was 58% the expected product of olefin metathesis and 42% saturated hydrocarbon, while $WCl_6 + n\text{-}C_4H_9Li$ gave material that was better yet: only 17% of the resulting polymer's mass seemed to be saturated impurities and 8% aromatic impurities (67).

The best product, however, was produced by warming 1-methylcyclobutene with small amounts of (diphenylcarbene)pentacarbonyltungsten at 50°C (67). This consumed the 1-methylcyclobutene and gave in approximately quantitative yield a polymer containing no measurable saturated impurities. The polymer was identified as polyisoprene, 85% of whose double bonds had the Z-configuration ("cis") and 15% the E-configuration ("trans"), but once in every 10.1 ± 1 reactions, the selectivity rules were violated, weaving into the polymer chain occasional units with the structure 5, in which methylcyclobutenes were coupled head-to-head and tail-to-tail. The polyisoprene structure and stereochemistry were identified by proton and carbon nuclear magnetic resonance (NMR) spectroscopy, and the units of structure 5 were recognized by identifying a proton resonance at 5.3 δ as that of a 2-butene (like that in polybutadiene) and two ^{13}C resonances at 18.14 and 18.43 ppm as methyl resonances of a 2,3-dimethyl-2-butene.

The selectivity for forming head-to-tail linkages, while appreciable, is not tremendous. This may not be anomalous in that the directional specificity for other additions to substituted cyclobutenes is not known. In fact, the impetus to form 6 rather than 7 might be less in four-membered rings than in rings that are larger for the same reason that methyl substitution at the reacting center increases rates of solvolysis about 1000 times less in four-membered than in larger rings (118).

(6)   (7)

Unfortunately it is not easy to test the implication that more perfect translational invariance will be observed in polymers made from 1-methylcyclopentene and 1-methyl-cis-cyclooctene because metatheses of these and other simple trisubstituted olefins as now effected usually fail.

However, the metathesis of 1-methyl-trans-cyclooctene does succeed

[Eq. (24)], giving a polymer that is not perceivably saturated and that within the detection limits of a ^{13}C-NMR spectrometer (ca. 4%) is (except for geometric isomerism) translationally invariant (76). No units like **8** could be found. Thus to the extent that the reaction is a valid

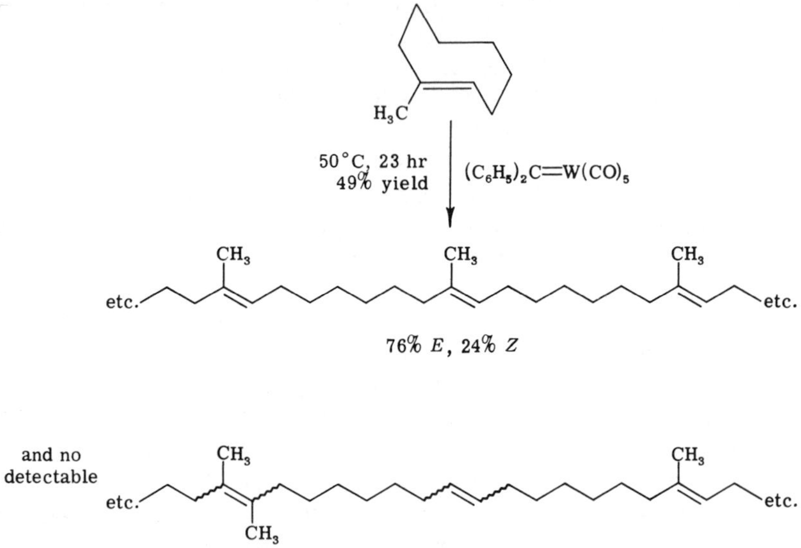

$$50°C, 23 \text{ hr} \quad 49\% \text{ yield} \quad | \quad (C_6H_5)_2C=W(CO)_5$$

76% E, 24% Z

and no detectable

(8)

(24)

measure, the selectivity for a disubstituted carbene-metal reacting with a trisubstituted olefin yielding a disubstituted carbene-metal is greater than 50 times that yielding a monosubstituted carbene-metal.

This metathesis provides an opportunity, complementary to that provided by the metathesis of 1-methylcyclobutene, to determine the stereochemistry of trisubstituted olefin metathesis. The stereochemistry of the polymer's double bonds could be measured as 76% E, 27% Z by the intensities of two methyl resonances (at 1.57 and 1.67 δ) in the proton NMR spectrum.

The stereochemical preference for the E precursor giving the E

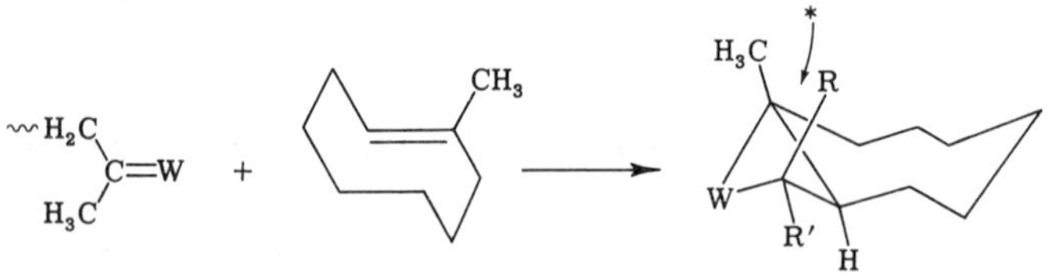

(25)

R = CH₃, R' = chain
or vice versa

product probably reflects the lesser steric interaction [indicated by the star in Eq. (25)] of a methyl with another that is least substituted. The stereochemistry of this reaction is considered further in Section V.

A significant characteristic of the polymer besides its translational invariance and its stereochemistry is its absorption in the ultraviolet spectrum (after purification by thin-layer chromatography) with a maximum at 245 nm, the same wavelength as that at which 1,1-diphenyl-1-propene exhibits its ultraviolet maximum. This should be expected if the diphenylcarbene moiety of the initiator [Eq. (24)] were attached to the beginning of the polymer chain, as it would be if the mechanism of initiation and metathesis involved the addition in Eq. (4) of a metal-carbene to an olefin.

Assuming the extinction coefficient of the polymer to be the same as that of 1,1-diphenyl-1-propene, the intensity of absorption indicated one diphenylethene for every 70 methylcyclooctenes. Since the number average molecular weight of the polymer measured by gel permeation chromatography was 6800, the average chain had, according to these figures, just under one diphenylethene attached to it $(6800/(70 \times 124) = 0.8)$. This is one of the best indications that the mechanism of the olefin metathesis reaction is the metal-carbene chain.

IV

METATHESES INDUCED BY METAL-CARBENES

A. Polymerizations of Cycloalkenes

(Diphenylcarbene)pentacarbonyltungsten initiates not only the metatheses of strained olefins, like 1-methylcyclobutene and 1-methyl-trans-cyclooctene discussed above, and of special olefins like those in Eq. (22), but also of simple olefins like cyclooctene, which it converts into a polyoctenamer [Eq. (2)] of high molecular weight (64). Although polyoctenamers have been made a number of times, the reaction is notable because while the double bonds in most samples have 35–95% of their double bonds trans (4, 11, 15, 16, 97, 108), in samples made with (diphenylcarbene)pentacarbonyltungsten they are >97% cis. The only other reports of cis-polyoctenamers are in the parent literature (29, 122).

Similarly cycloheptene yields a cis-polyheptenamer (64). Although previous attempts have given only very low yields, low molecular weights, and trans-stereochemistry (97, 108), the samples made using (diphenylcarbene)pentacarbonyltungsten are obtained in high yields, have high molecular weights, and have 98% or more of their double bonds cis.

Cyclobutene, cyclopentene, and norbornene also give their polyalkenamers (64). The molecular weights are all high, and the stereochemistries are largely cis. Of these, *cis*-polypentenamer has also been made with molybdenum, tungsten, and rhenium halide catalysts, and *cis*-polynorbornenamer (of lower molecular weight) with a molybdenum chloride catalyst, but polybutenamer made with metal halide catalysts (the metals tried were Ti, Mo, W, V, Cr, and Ru) has never been found to have more than 60% of its double bonds cis (64).

In one experiment with 150 g of norbornene the amount of (diphenylcarbene)pentacarbonyltungsten was decreased to 1/4000 the number of moles of olefin, but the yield was still good (67%) and the intrinsic viscosity high: $[\eta] = 2.56$ dl/g in benzene at 30°C. The double bonds were 95% cis.

Compared to conventional catalysts, (diphenylcarbene)pentacarbonyltungsten has three advantages: (a) it converts cycloalkenes stereoselectively into *cis*-polyalkenamers; (b) it contains no metal alkyl bonds, and therefore no acyclic olefins can form to decrease molecular weights by chain transfer; and (c) it contains no metal halide to induce side reactions or corrode equipment.

B. *Metatheses of Acyclic Olefins*

Isolated metal carbenes have been used in only few instances to initiate metatheses of acyclic olefins. One of the first examples was that of 2-hexene, which in *n*-heptane in the presence of 10 mmoles of (diphenylcarbene)pentacarbonyltungsten after 16 hours at 50°C was converted into a seemingly equilibrium mixture of 2-butene, 2-hexene, and 4-octene (82).

Similarly the 2-pentenes are converted into 2-butenes and 3-hexenes, and in these cases the stereochemistries of the transformations have been determined (63). Like the *cis*-cycloalkenes, *cis*-2-pentene selectively gives the cis products. Thus at 21°–25°C it gives 2-butene that is 94% cis and 3-hexene that is 92% cis.

The opposite isomer, *trans*-2-pentene, under the same conditions gives butenes and hexenes that have largely the opposite stereochemistry, but the stereoselectivity is considerably less than that of the *cis*-olefin metathesis: the butene is only 73% trans and the hexene 83% trans.

The stereochemistries of olefin metatheses are analyzed further in Section V.

C. Metatheses Induced by (Phenylmethoxycarbene)pentacarbonyltungsten

Although metal-carbenes should initiate olefin metatheses, whether or not they do has not been investigated extensively. Despite the large numbers of metal-carbenes stabilized by a heteroatom conjugated with the carbene center that have been synthesized (19) since Fischer and Maasböl (44) made the first recognized example, (phenylmethoxycarbene)pentacarbonyltungsten (structure 3), it was only recently discovered that even these stabilized metal-carbenes react with olefinic hydrocarbons (62). (Phenylmethoxycarbene)pentacarbonylchromium, -molybdenum, and -tungsten have been known to add to α,β-unsaturated esters and to vinyl ethers (and, as discussed above, the chromium compound exchanges its chromium carbonyl moiety for the methylene of ethyl vinyl ether or of N-vinylpyrrolidone), but for a number of years it was believed that with pure alkenes no reaction occurred. This is because the only alkenes studied, cyclohexene and tetramethylethylene, were chosen not for their propensity to undergo the olefin metathesis reaction, but for their bent to add carbenes. As substrates for olefin metathesis it is now known that they rank among the least responsive (62).

The more active cyclobutene and norbornene, in contrast, do react with such metal-carbenes yielding the corresponding polyalkenamers (62). Thus, when norbornene in heptane is warmed at 50°C for 6 days with 1 mmole of (phenylmethoxycarbene)pentacarbonyltungsten it gives in 53% yield a long-chain polynorbornenamer. The weight average molecular weight $\bar{M}_w = 1.1 \times 10^5$ and the number average molecular weight $\bar{M}_n = 3.8 \times 10^4$.

Cyclobutene behaves similarly. After 15 hours at 50°C with 5.5 mmoles of (phenylmethoxycarbene)pentacarbonyltungsten in toluene it gives in 60% yield a polybutenamer whose molecular weights are $\bar{M}_w = 6.0 \times 10^5$ and $\bar{M}_n = 1.7 \times 10^5$, and whose intrinsic viscosity in benzene at 30°C $[\eta] = 4.95$ dl/g.

This suggests that a variety of metal-carbenes will initiate metatheses, and in fact the isolable relatives of (phenylmethoxycarbene)pentacarbonyltungsten, in which the phenyl is replaced by methyl or a carbon monoxide is replaced by triphenylphosphine, also initiate metatheses of active olefins (76a). Since metal-carbenes are structurally homogeneous, it should be possible to use them to achieve a predictable control over metathesis reactions that has not been achieved with the structurally less well-defined initiating mixtures currently used.

These reactions also raise the following interesting mechanistic point. If the olefin metathesis reaction is, as supposed, the metal-carbene chain reaction, the structure of the polymer formed according to Eq. (26) might be expected to be exactly the same, except at the initiating end, when X is C_6H_5 as when X is CH_3O because after the initiation the local structure of the metal carbene center would be the same whichever the initiator. Accordingly, it is reasonable that the polybutenamer formed by the initiator (phenylmethoxycarbene)pentacarbonyltungsten (62) has the same, about 90% cis, stereochemistry as does that formed by the initiator (diphenylcarbene)pentacarbonyltungsten (64).

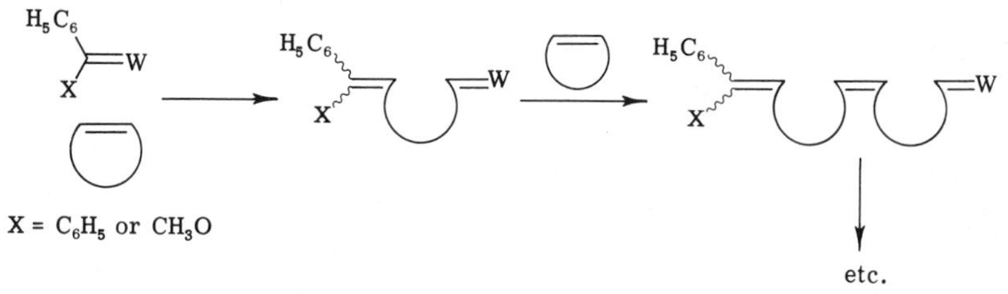

X = C_6H_5 or CH_3O

etc.

$$(26)$$

But while the polynorbornenamers formed by the two initiators should also be the same, they are not. That formed by (diphenylcarbene)pentacarbonyltungsten has about 95% of its double bonds cis (64), while that formed by (phenylmethoxycarbene)pentacarbonyltungsten has only about 72% (62). The reasons for this difference have yet to be determined.

D. Molecules with Functional Groups

Although very few catalysts are effective with molecules containing functional groups as well as double bonds, (diphenylcarbene)pentacarbonyltungsten and (phenylmethoxycarbene)pentacarbonyltungsten do initiate metatheses of ester derivatives of norbornene, such as the methyl norbornene-5-carboxylates (62a).

V

THE STEREOCHEMISTRIES OF OLEFIN METATHESES

A. *cis-1,2-Disubstituted Ethylenes*

When (diphenylcarbene)pentacarbonyltungsten induces the metathesis of *cis*-2-pentene (*63*) or of a variety of *cis*-cycloalkenes (*64*), the double bonds of the metathesis products are 92% or more cis. The stereochemistry can be accounted for if the interactions in the intermediate four-membered ring are minimized between juxtaposed axial groups and ring atoms (*63*). Thus, the energies of all cyclobutane conformations but the first in Scheme 5 must be raised by the opposition (indicated by a star) of an axial group and a ring carbon atom, and in the last conformation there is a 1,3-diaxial interaction (indicated by two stars) that must raise its energy even more. But in the first conformation the axial group is juxtaposed to the metal, and since the carbon–metal bond is longer than the carbon–carbon bond, the energy of this conformation should be the lowest. Thus the reaction through it should be the fastest, and the cis precursor should, as it does, give the cis product. That the interaction of

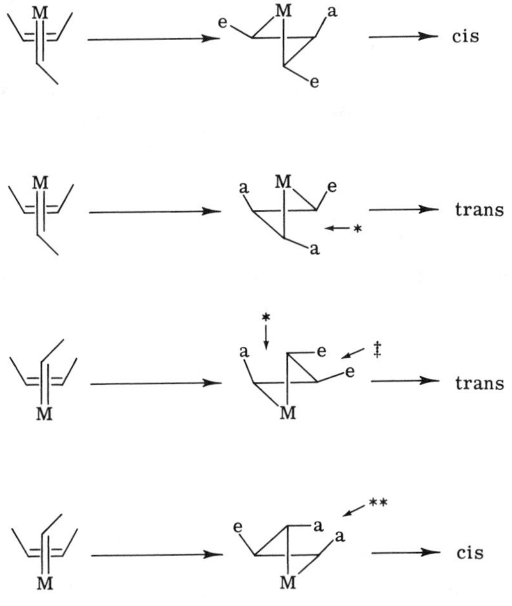

Scheme 5

the axial group would be less when it is opposed to the metal is indicated by the structure of a four-membered ring containing platinum, which as determined by X-ray diffraction shows the cross-ring distance from carbon to the metal to be 0.3 Å longer than to carbon (121). In the tungsten metallocycle this effect should be accentuated, as the carbon–tungsten bond (about 2.39 Å) should be longer (24).

B. trans-1,2-Disubstituted Ethylenes

When metathesis is effected with *trans*-2-pentene, rather than *cis*-, and (diphenylcarbene)pentacarbonyltungsten is the initiator, the 2-butene and 2-hexene products are largely trans. The stereospecificity (73–83% trans) is not as great as for *cis*-olefin metathesis, but it is appreciable (63). The ratios of the stereoisomers in the products are close to the equilibrium ratios, but they probably are not determined by the products equilibrating, for in the short time the metathesis was run to determine the stereochemistry of the initial product, the precursor, *trans*-2-pentene, underwent only negligible isomerization. The stereochemistries therefore are determined by the kinetics, which in turn should be affected by conformational factors similar to those in Scheme 5.

The analogous scheme for *trans* olefin precursors is shown in Scheme 6. Here all conformations of the four-membered ring except the third have an axial group interacting with a ring carbon atom. Accordingly, this conformation should be the one of lowest energy, and the trans precursor should give largely trans products.

The conformational analysis in Schemes 5 and 6 might also explain why the stereospecificity is less for the trans isomer's metathesis, for a steric interaction (indicated by a dagger) that occurs when one alkyl group is rotated past another is met on the predominant reaction path of only the trans precursor.

C. Trisubstituted Ethylenes

If the conformational analyses above are correct, similar analyses should also account for why 1-methylcyclobutene (67) gives polyisoprene that has largely the Z configuration (it is 85% Z, 15% E). The analysis for this trisubstituted olefin metathesis focuses on whether reaction proceeds according to Eq. (27) or Eq. (28) and whether R is the

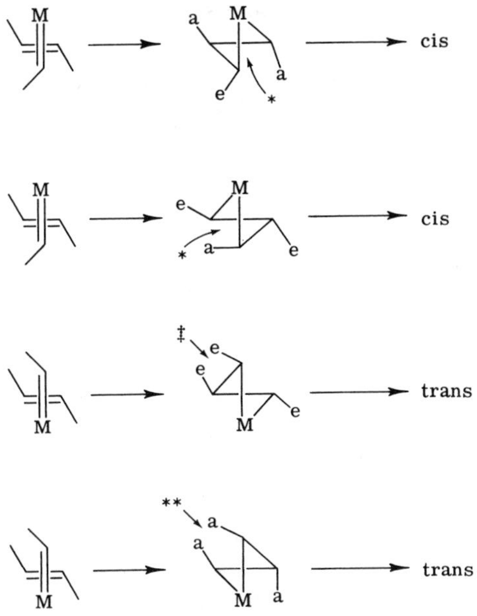

Scheme 6

methyl and R' the polymer chain or vice versa. The distinction among these paths seems to be that in conformation **B**, R' is almost eclipsed by a methylene group while in **A** it is not because, as indicated in structures **9** and **10**, the carbocyclic four-membered ring diminishes the interaction in **A** while increasing it in **B**. Thus the reaction proceeds through Eq. (27), and since the 1,3-diaxial interaction is less when a methyl is

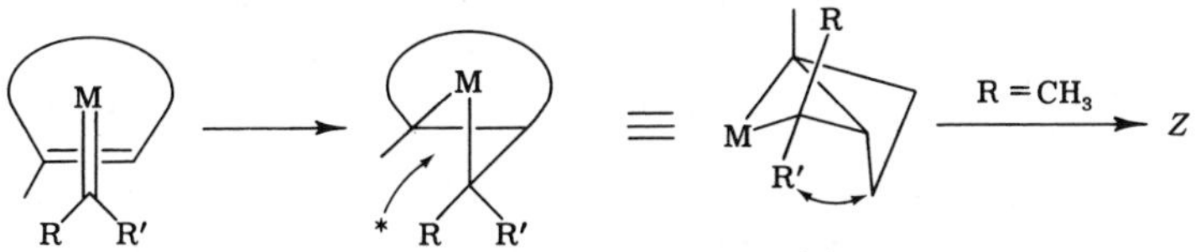

$$\xrightarrow{\hspace{2cm}} \equiv \xrightarrow{R=CH_3} Z \quad (27)$$

(A)

opposed to another that is unsubstituted, the major product should be, as it is, the Z isomer.

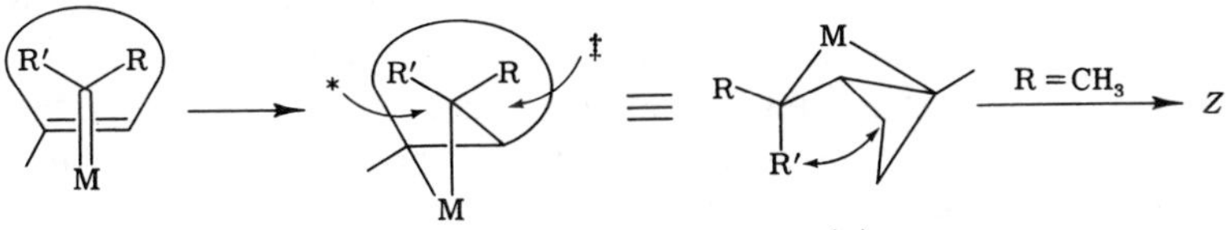

$$\xrightarrow{\hspace{2cm}} \equiv \xrightarrow{R=CH_3} Z \quad (28)$$

(B)

The metathesis of 1-methyl-*trans*-cyclooctene [Eqs. (24) and (25)] according to this analysis also proceeds through Eq. (27) [because the analog of structure **B** in Eq. (28) would have the eight-membered ring spanning diaxial positions], but now the carbocyclic ring is arranged as shown in structure **11**.

(9) = A   (10) = B   (11)

D. *Effects of Lewis Acids*

The stereospecificities of *cis*-1,2-disubstituted ethylene metathesis are very much higher when the initiator is (diphenylcarbene)pentacarbonyltungsten than when it is almost any other tungsten-containing initiator previously studied. With WCl_6—C_2H_5OH—$C_2H_5AlCl_2$, for example, *cis*-2-pentene gives 2-butene that is only 62% cis and 3-hexene that is 50% cis (*17*). Similar stereochemistries are reported for eleven other initiators containing tungsten and aluminum halides (*7*). (For the metal-carbene initiator, the figures were 94% and 93%.) One molybdenum-containing catalyst seems to give a similar stereochemistry (*38*), but two others give much more of the cis products (*57, 61*). These molybdenum-containing catalysts are considered again below. In cycloalkene metatheses, most initiators give materials with either mixtures of cis and trans double bonds or mainly trans double bonds, but some compositions of tungsten halides and organoaluminum or -tin compounds at low temperature do convert cyclopentene to *cis*-polypentenamer (*48, 85, 101*) and cyclooctene to *cis*-polyoctenamer (*29, 122*).

The reason the stereospecificities often are low may be that a carbon–metal bond in the metallocyclobutane cleaves heterolytically forming a 3-metallopropyl cation, the remaining bonds rotate, and the four-membered ring then reforms (*63*). If the heterolytic cleavage is facilitated by Lewis acids, their presence in quantity will diminish stereospecificity. In accord with this, stereospecificities do diminish with the Al:W ratio when (*48*) cyclopentene is metathesized at −30°C by WF_6 + $(C_2H_5)_3Al_2Cl_3$. The decrease in the amount of cis double bonds when the temperature is raised in two cyclopentene metatheses might also be a manifestation of this reaction (*85, 101*).

The stereospecificity of *trans*-1,2-disubstituted ethylene metathesis is higher when the initiator is (diphenylcarbene)pentacarbonyltungsten than when it is $WCl_6 + C_2H_5OH + C_2H_5AlCl_2$ (*17*) or $W(CO)_6 + CCl_4 + h\nu$ (*71*), but the same as when it is $WCl_6 + n\text{-}C_4H_9Li$ (*116*).

E. *Comparison of Molybdenum and Tungsten Catalysts*

It is interesting that two molybdenum-containing catalysts give high stereospecificities in trans olefin metatheses (*38, 57*), and molybdenum-containing catalysts give the highest stereospecificities observed in cyclopentene (*30*) and norbornene (*109*) polymerizations. The reason may be related to why a molybdenum-containing catalyst was found to be the least selective in terminal alkene metathesis (Section III,A): 3-metallopropyl cations are favored less when the metal is molybdenum than when it is tungsten.

ACKNOWLEDGMENTS

I have incorporated into this review ideas and experiments that are the products of my students' insights, labor, and skill. James McGinnis, Steven J. Lee, Nancy Acton, Robert Rothchild, and William Hersh did the bulk of the research with me, and Craig Altus, Samuel Hurwitz, Leslie Lobel, and David Treatman, as undergraduates, helped with the laboratory work and the algebra. I am grateful to all of them.

The National Science Foundation provided financial support for the early part of our work.[1]

REFERENCES

1. Adams, C. T., and Brandenberger, S. G., *J. Catal.* **13**, 360 (1969).
2. Amiet, R. G., Reeves, P. C., and Pettit, R., *J. Chem. Soc., Chem. Commun.* 1208 (1967).

[1] An evaluation used by the National Science Foundation to decide in January 1976 to discontinue its support:

> Apart from the fact that this is a perfectly competent technical piece of work, the main claim to glory seems to be that most of the workers in this field (including the P.I. [Principal Investigator]) seem to have been unaware of Chauvin's publications although these appeared in widely read journals and conveyed a very clear message. What would be an unforgivable demonstration of inadequacy in any other field of research seems to have been turned into a virtue in this case. I shudder at the implications for science and the cost to the taxpayer if every important scientific discovery had to be rediscovered independently after five years.

The references to Chauvin's publications in our first paper (*65*) are footnotes 11 and 16. This review summarizes progress up to about November 1976.

3. Anderson, A. W., and Merckling, N. G., U.S. Patent 2,721,189 (1955).
4. Arlie, J.-P., Chauvin, Y., Commereuc, D., and Soufflet, J.-P., *Makromol. Chem.* **175**, 861 (1974).
5. Bailey, G. C., *Catal. Rev.* **3**, 37 (1969).
6. Banks, R. L., *Top. Curr. Chem.* **25**, 39 (1972).
7. Basset, J. M., Bilhou, J. L., Mutin, R., and Theolier, A., *J. Am. Chem. Soc.* **97**, 7376 (1975).
8. Ben-Efraim, D. A., Batich, C., and Wasserman, E., *J. Am. Chem. Soc.* **92**, 2133 (1970).
9. Bespalova, N. B., Babich, E. D., Vdovin, V. N., and Nametkin, N. S., *Dokl. Chem. (Engl. Transl.)* 668 (1975).
10. Bradshaw, C. P. C., Howman, E. J., and Turner, L., *J. Catal.* **7**, 269 (1967).
11. *British Patent* 1,062,367 (1967).
12. Calderon, N., *Acc. Chem. Res.* **5**, 127 (1972).
13. Calderon, N., *J. Macromol. Sci. C7,* 105 (1972).
14. Calderon, N., Chen, H. Y., and Scott, K. W., *Tetrahedron Lett.* 3327 (1967).
15. Calderon, N., and Morris, M., *J. Polym. Sci., Part A-2* **5**, 1283 (1967).
16. Calderon, N., Ofstead, E. A., and Judy, W. A. *J. Polym. Sci., Part A-1* **5**, 2209 (1967).
16a. Calderon, N., Ofstead, E. A., and Judy, W. A., *Angew. Chem., Int. Ed. Engl.* **15**, 401 (1976).
17. Calderon, N., Ofstead, E. A., Ward, J. P., Judy, W. A., and Scott, K. W., *J. Am. Chem. Soc.* **90**, 4133 (1968).
18. Caldow, G. L., and MacGregor, R. A., *J. Chem. Soc. A* 1654 (1971).
19. Cardin, D. J., Cetinkaya, B., and Lappert, M. F., *Chem. Rev.* **72**, 545 (1972).
20. Cardin, D. J., Doyle, M. J., and Lappert, M. F., *J. Chem. Soc., Chem. Commun.* 927 (1972).
21. Casey, C. P., and Burkhardt, T. J., *J. Am. Chem. Soc.* **95**, 5833 (1973).
22. Casey, C. P., and Burkhardt, T. J., *J. Am. Chem. Soc.* **96**, 7808 (1974).
23. Casey, C. P., Tuinstra, H. E., and Saemen, M., *J. Am. Chem. Soc.* **98**, 608 (1976).
24. Churchill, M. R., *in* "Perspectives in Structural Chemistry" (J. D. Dunitz and J. A. Ibers, eds.), Vol. III, p. 91 ff. Wiley, New York, 1970.
25. Clark, A., and Cook, C., *J. Catal.* **15**, 420 (1969).
26. Connor, J. A., Jones, E. M., Randall, E. W., and Rosenberg, E., *J. Chem. Soc., Dalton Trans.* 2419 (1972).
27. Dall'Asta, G., *Rubber Chem. Technol.* **47**, 511 (1974).
28. Dall'Asta, G., and Manetti, R., *Atti Accad. Naz. Lincei, Rend., Cl. Sci. Fis. Mat. Nat.* **41**, 351 (1966).
29. Dall'Asta, G., and Manetti, R., *Italian Patent* 932,461 (referred to in Dall'Asta and Manetti (*28*), p. 551).
30. Dall'Asta, G., and Motroni, G., *Angew. Makromol. Chem.* **16/17**, 51 (1971).
31. Dall'Asta, G., and Motroni, G., *Eur. Polym. J.* **7**, 707 (1971).
32. Dall'Asta, G., Motroni, G., and Motta, L., *J. Polym. Sci., Part A-1* **10**, 1601 (1972).
33. Davison, A., and Reger, D. L., *J. Am. Chem. Soc.* **94**, 9237 (1972).
34. Descotes, G., Chevalier, P., and Sinou, D., *Synthesis* 364 (1974).
35. Dickson, R. C., and Fraser, P. J., *Adv. Organomet. Chem.* **12**, 323 (1974).
36. Dolgoplosk, B. A., Makovetsky, K. L., Golenko, T. G., Korshak, Yu. V., and Tinyakova, E. I., *Eur. Polym. J.* **10**, 901 (1974).
37. Dolgoplosk, B. A., Makovetskii, K. L., Tinykova, E. I., Golenko, T. G., and Oreshkin, I. O., *Izv. Akad. Nauk. SSSR, Ser. Khim.* 1084 (1976).

38. Doyle, G., *J. Catal.* **30**, 118 (1973).
39. Eleuterio, H. S., U.S. Patent 3,074,918 (1963).
40. Farona, M. F., and Greenlee, W. S., *J. Chem. Soc., Chem. Commun.* 759 (1975).
41. Fischer, E. O., *Adv. Organomet. Chem.* **14**, 1 (1976).
42. Fischer, E. O., and Dorrer, B., *Chem. Ber.* **107**, 1156 (1974).
43. Fischer, E. O., and Dötz, K. H., *Chem. Ber.* **105**, 3966 (1972).
44. Fischer, E. O., and Maasböl, A., *Angew. Chem., Int. Ed. Engl.* **3**, 580 (1964).
45. Golenko, T. G., Dolgoplosk, B. A., Makovetskii, K. L., and Ostrovskaya, I. Ya., *Dokl. Phys. Chem. (Engl. Transl.)* **220**, 80 (1975).
46. Grubbs, R. H., Burk, P. L., and Carr, D. D., *J. Am. Chem. Soc.* **97**, 3265 (1975).
47. Grubbs, R. H., Carr, D. D., Hoppin, C., and Burk, P. L., *J. Am. Chem. Soc.* **98**, 3478 (1976).
48. Günther, P., Haas, F., Marwede, G., Nützel, K., Oberkirch, W., Pampus, G., Schön, N., and Witte, J., *Angew. Makromol. Chem.* **14**, 87 (1970); **16/17**, 27 (1971).
49. Haas, F., Nützel, K., Pampus, G., and Thiesen, D., *Rubber Chem. Technol.* **43**, 116 (1970).
50. Haas, F., and Thiesen, D., *Kautschuk Gummi* **23**, 502 (1970).
51. Haines, R. J., and Leigh, G. J., *Chem. Soc. Rev.* **4**, 155 (1975).
52. Hérisson, J. L., and Chauvin, Y., *Makromol. Chem.* **141**, 161 (1970).
53. Höcker, H., and Musch, R., *Makromol. Chem.* **175**, 1395 (1974).
54. Höcker, H., Reimann, W., Riebel, K., and Szentivanyi, Z., *Makromol. Chem.* **177**, 1707 (1976).
55. Hocks, L., *Bull. Soc. Chim. Fr.* 1893 (1975).
56. Hughes, W. B., *J. Am. Chem. Soc.* **92**, 532 (1970).
57. Hughes, W. B., *J. Chem. Soc., Chem. Commun.* 431 (1969).
58. Hughes, W. B., *Chem. Technol.* **5**, 486 (1975).
59. Hughes, W. B., *Organomet. Chem. Syn.* **1**, 341 (1972).
60. Ichikawa, K., and Fukuzumi, K., *J. Org. Chem.* **41**, 2633 (1976).
61. Ismael-Milanovic, A., Basset, J. M., Praliaud, H., Dufaux, M., and de Morgues, L., *J. Catal.* **31**, 408 (1973).
62. Katz, T. J., and Acton, N., *Tetrahedron Lett.* 4241 (1976).
62a. Katz, T. J., and Acton, N., in preparation.
63. Katz, T. J., and Hersh, W. H., *Tetrahedron Lett.* 585 (1977).
64. Katz, T. J., Lee, S. J., and Acton, N., *Tetrahedron Lett.* 4247 (1976).
65. Katz, T. J., and McGinnis, J., *J. Am. Chem. Soc.* **97**, 1592 (1975).
66. Katz, T. J., and McGinnis, J., *J. Am. Chem. Soc.* **99**, 1903 (1977).
67. Katz, T. J., McGinnis, J., and Altus, C., *J. Am. Chem. Soc.* **98**, 606 (1976).
68. Katz, T. J., and Rothchild, R., *J. Am. Chem. Soc.* **98**, 2519 (1976).
69. Kelly, W. J., and Calderon, N., *J. Macromol. Sci., Chem.* **9**, 911 (1975).
70. Khidekel', M. L., Shebaldova, A. D., and Kalechits, I. V., *Russ. Chem. Rev. (Engl. Transl.)* **40**, 669 (1971).
71. Krausz, P., Garnier, F., and Dubois, J. E., *J. Am. Chem. Soc.* **97**, 437 (1975).
72. Kreiter, C. G., and Formacek, V., *Angew. Chem., Int. Ed. Engl.* **11**, 141 (1972).
73. Kroll, W. R., and Doyle, G., *J. Catal.* **24**, 356 (1972).
74. Kroll, W. R., and Doyle, G., *J. Chem. Soc., Chem. Commun.* 839 (1971).
75. Lal, J., and Smith, R. R., *J. Org. Chem.* **40**, 775 (1975).
76. Lee, S. J., McGinnis, J., and Katz, T. J., *J. Am. Chem. Soc.* **98**, 7818 (1976).
76a. Lee, S. J., and Katz, T. J., in preparation.
77. Lewandos, G. S., and Pettit, R., *Tetrahedron Lett.* 789 (1971).
78. Maitlis, P. M., *Adv. Organomet. Chem.* **4**, 95 (1966).

79. Mango, F. D., *Adv. Catal.* **20**, 291 (1969).
80. Mango, F. D., and Schachtschneider, J. H., *J. Am. Chem. Soc.* **93**, 1123 (1971).
80a. Mango, F. D., *Coord. Chem. Rev.* **15**, 109 (1975).
81. Mar'in, V. I., Shebaldova, A. D., Bol'shinskova, T. A., Khidekel', M. L., and Kalechits, I. V., *Kinet. Catal. (USSR)* **14**, 528 (1973).
82. McGinnis, J., Dissertation, Columbia University (1976).
83. McGinnis, J., Katz, T. J., and Hurwitz, S., *J. Am. Chem. Soc.* **98**, 605 (1976).
84. Michelotti, F. W., and Keaveney, W. P., *J. Polym. Sci. Part A* **3**, 895 (1965).
85. Minchak, R. J., and Tucker, H., *Polym. Prepr., Am. Chem. Soc. Div. Polym. Chem.* **13**, 885 (1970).
86. Minchak, R. J., Tucker, H., and Macey, J. H., *Soc. Plastic Eng., Tech. Papers* **20**, 666 (1974).
87. Mocella, M. T., Busch, M. A., and Muetterties, E. L., *J. Am. Chem. Soc.* **98**, 1283 (1976).
88. Mol, J. C., and Moulijn, J. A., *Adv. Catal.* **24**, 131 (1975).
89. Mol, J. C., Moulijn, J. A., and Boelhouwer, C., *J. Chem. Soc., Chem. Commun.* 633 (1968).
90. Mol, J. C., Visser, F. R., and Boelhouwer, C., *J. Catal.* **17**, 114 (1970).
91. Mortreux, A., and Blanchard, M., *Bull. Soc. Chim. Fr.*, 1641 (1972).
92. Mortreux, A., and Blanchard, M., *J. Chem. Soc., Chem. Commun.* 786 (1974).
93. Mortreux, A., Dy, N., and Blanchard, M., *J. Mol. Catal.* **1**, 101 (1975/1976).
94. Moulijn, J. A., Reitsma, H. J., and Boelhouwer, C., *J. Catal.* **25**, 434 (1972).
95. Muetterties, E. L., and Busch, M. A., *J. Chem. Soc., Chem. Commun.* 754 (1974).
96. Nametkin, N. S., Vdovin, V. M., Babich, E. D., Karel'skii, V. N., and Kacharmin, B. V., *Dokl. Chem. (Engl. Transl.)* **213**, 872 (1973).
97. Natta, G., Dall'Asta, G., Bassi, I. W., and Carella, G., *Makromol. Chem.* **91**, 87 (1966).
98. Natta, G., Dall'Asta, G., and Mazzanti, G., *Angew. Chem., Int. Ed. Engl.* **3**, 723 (1964).
99. Natta, G., Dall'Asta, G., Mazzanti, G., and Motroni, G., *Makromol. Chem.* **69**, 163 (1963).
100. Natta, G., Dall'Asta, G., and Porri, L., *Makromol. Chem.* **81**, 253 (1965).
101. Pampus, G., and Lehnart, G., *Makromol. Chem.* **175**, 2605 (1974).
102. Pearson, R. G., *J. Am. Chem. Soc.* **94**, 8287 (1972).
103. Pennella, F., and Banks, R. L., *J. Catal.* **31**, 304 (1973).
104. Pennella, F., Banks, R. L., and Bailey, G. C., *J. Chem. Soc., Chem. Commun.* 1548 (1968).
105. Pennella, F., Regier, R. B., and Banks, R. L., *J. Catal.* **34**, 52 (1974).
106. Peters, E. F., and Evering, B. L., U.S. Patent 2,963,447 (1960).
107. Pettit, R., Sugahara, H., Wristers, J., and Merk, W., *Disc. Faraday Soc.* **47**, 71 (1969).
108. Porri, L., Diversi, P., Lucherini, A., and Rossi, R., *Makromol. Chem.* **176**, 3121 (1975).
109. Sartori, G., Ciampelli, F., and Cameli, N., *Chim. Ind. (Milan)* **45**, 1478 (1963).
110. Scott, K. W., Calderon, N., Ofstead, E. A., Judy, W. A., and Ward, J. P., *Adv. Chem. Ser.* **91**, 399 (1969).
111. Soufflet, J.-P., Commereuc, D., and Chauvin, Y., *C.R. Acad. Sci. Ser. C.* **276**, 169 (1973).
112. Streck, R., *Chem. Ztg.* 397 (1975).
113. Taube, R., and Seyferth, K., *Wiss. Z. Tech. Hoch. Chem. "Carl. Schorlemmer" Leune-Merseburg* **17**, 305 (1975).

114. Truett, W. L., Johnson, D. R., Robinson, I. M., and Montague, B. A., *J. Am. Chem. Soc.* **82**, 2337 (1960).
115. Uchida, A., Hamano, Y., Mukai, Y., and Matsuda, S., *Ind. Eng. Chem. Prod. Res. Dev.* **10**, 372 (1971).
116. Wang, J. L., and Menapace, H. R., *J. Org. Chem.* **33**, 3794 (1968).
117. Wasserman, E., Ben-Efraim, D. A., and Wolovsky, R., *J. Am. Chem. Soc.* **90**, 3286 (1968).
118. Wiberg, K. B., and Chen, W., *J. Am. Chem. Soc.* **96**, 3900 (1974).
119. Wolovsky, R., *J. Am. Chem. Soc.* **92**, 2132 (1970).
120. Wolovsky, R., and Nir, Z., *Synthesis* **4**, 134 (1972).
121. Yarrow, D. J., Ibers, J. A., Lenarda, M., and Graziani, M., *J. Organomet. Chem.* **70**, 133 (1974).
122. Zimmermann, M., Lehnert, G., Maertens, D., and Pampus, G., Ger. Offen. 23 34 604 (1975).
123. Zowade, T., and Höcker, H., *Makromol. Chem.* **165**, 31 (1973).
124. Zuech, E. A., *J. Chem. Soc., Chem. Commun.* 1182 (1968).
125. Zuech, E. A., Hughes, W. B., Kubicek, D. H., and Kittleman, E. T., *J. Am. Chem. Soc.* **92**, 528 (1970).

Molecular Rearrangements in Polynuclear Transition Metal Complexes

JOHN EVANS

Department of Chemistry
The University
Southampton, United Kingdom

I

INTRODUCTION

Stereochemical nonrigidity of mononuclear coordination complexes of transition metals (90) and of organometallic complexes (27) has been the subject of considerable interest for the past ten years. However, extensive studies of corresponding polynuclear complexes awaited the availability of commercial Fourier transform nuclear magnetic resonance (NMR) spectrometers. Although the first publication of a variable temperature ^{13}C NMR spectrum on such a complex, $[(\eta^5\text{-}C_5H_5)Fe(CO)_2]_2$, was as late as 1972 (63), use of this technique is now standard in the characterization of new metal cluster compounds (8, 51). This work has yielded a considerable number of examples of molecular rearrangements within polynuclear transition metal complexes. Primarily, this review is concerned with the migration of hydrido, carbonyl, and unsaturated organic ligands around the vertices, edges, and faces denoted by the transition metal framework. Finally, rearrangements of that framework itself will be considered. Only essential aspects of the NMR spectra discussed will be described in the text; the details are contained in Table I.

TABLE I

NMR Spectra of the Polynuclear Transition Metal Complexes Discussed

Compound	Temperature (K)	Chemical shifts (multiplicity, J Hz, intensity)[a]	Reference
$(C_5H_5)_2Fe_2(CO)_4$	208	^{13}CO: 272.9, 211.0 (cis isomer)	65
	200	^{13}CO: 272.9, 211.0 (cis); 242.2 (trans)	63
	328	^{13}CO: 242.2 (av)	
$(C_5H_5)_2Fe_2 \cdot (CO)_3$-P(OPh)$_3$	242	^1H: (C_5H_5): τ 6.7 (d, J_{PH} 1.1), 5.65 (s) (cis); 6.05 (br), 5.43 (s) (trans)	38
	315	^1H: (C_5H_5): τ 6.1 (d), 5.5 (s)	
	233	^{13}CO: 277.6 (d, J_{PC} ~ 25), 215.1 (s) (cis); ~ 213 (s) (trans)	
	360	^{13}CO: 255.2 (br) (av)	
$(C_5H_5)_2Fe_2 \cdot (CO)_3$-P(OEt)$_3$	242	^{13}C: (C_5H_5) 87.6 (s), 86.3 (s) (cis); 89.3 (trans) ^{13}CO: 282.2 (d, J_{PC} 23), 217.7 (s) (cis); 213.6 (trans)	65
	400	^{13}CO: 256.6 (br) (av)	
$(C_5H_5)_2Rh_2 \cdot (CO)_3$	303	^1H (τ 4.48 (d, J_{RhH} 2)	55
	193	^{13}CO: 231.8 (t, J_{RhC} 45, 1), 191.8 (d, J_{RhC} 83, 2)	56
	293	$^{13}C[^1H]$: (C_5H_5) 89.3, (CO) 203.8 (t, J_{RhC} 43)	
$(C_5H_5)_2Rh_2 \cdot (CO)_3$-P(OPh)$_3$	213	^1H: $\overline{\tau}$ 2.70 (m, 15), 4.41 (s, 5), 5.32 (d, J 2, 5)	53
	168	^{13}CO: 239.4 (m, J_{RhC} 41, J_{RhC} 50, J_{PC} 19), 190.4 (d, J_{RhC} 85)	
	298	^{13}CO: 218.5 (m, J_{RhC} 64, J_{RhC} 24, J_{PC} 10)	
$(C_5H_5)_2Cr_2(CO)_6$	212	^1H: τ 5.36 (s), 5.55 (s)	3
	274	^1H: τ 5.4 (br)	
$Rh_2[P(OMe)_3]_8$	305	$^{31}P[^1H]$: A_3BX pattern (J_{AB} 122, J_{AX} 200, $J_{BX} \simeq J_{AB}$ 200)	
	373	$^{31}P[^1H]$: s, d (J 122)	
$Fe_3(CO)_{12}$	253	^{13}C: 212.9 (s)	62
$RuFe_2(CO)_{12}$	243	^{13}C: 206.4 (s)	62
$[MnFe_2(CO)_{12}]^-$	188	^{13}C: 223.3 (s)	62
$Ru_3(CO)_{12}$	223	^{13}C: 199.7 (s)	62
$Os_3(CO)_{12}$	283	^{13}C: 182.3 (s), 170.4 (s)	62
	373	^{13}C: 176.4 (s)	
$Rh_4(CO)_{12}$	208	^{13}C: 228.8 (t, J_{RhC} 35, 1), 183.4 (d, J_{RhC} 75, 1), 181.8 (d, J_{RhC} 64, 1), 175.5 (d, J_{RhC} 62)	57
	336	^{13}C: 190.3 (quintet, J_{RhC} 17.1)	37
$Co_4(CO)_{12}$	203	^{13}C: 243.1 (s), 195.9 (s), 191.9 (s)	52
$Co_3Rh(CO)_{12}$	188	^{13}C: 251.2 (s, 2), 238.3 (d, J_{RhC} 38, 2), 201.1 (s, 2), 200.1 (s, 2), 195.5 (s, 3), 188.2 (d, J_{RhC} 78, 1), 183.1 (d, J_{RhC} 51, 1)	73
	263	^{13}C: 208 (br), 186.2 (d, J_{RhC} 72)	
	303	^{13}C: 201.3 (br)	
$[Rh_6(CO)_{15}]^{2-}$	203	^{13}C: 209.2 (septet, J_{RhC} 13.9)	66
$Rh_6(CO)_{16}$	343	^{13}C: 231.5 (quartet, J_{RhC} 24, 4), 180.1 (d, J_{RhC} 70, 12)	66

TABLE I—(*Continued*)

Compound	Temperature (K)	Chemical shifts (multiplicity, J Hz, intensity)^a	Reference
$[Rh_7(CO)_{16}]^{3-}$	203	^{13}C: 254.3 (br, 3), 229.5 (t, J_{RhC} 40, 3), 218.1 (t, J_{RhC} 40, 3), 206.4 (d, J_{RhC} 93, 3), 205.7 (d, J_{RhC} 104, 1), 198.2 (d, J_{RhC} 81, 3)	66
$[Rh_7(CO)_{16}I]^{2-}$	242	^{13}C: 218.5 (octet)	21
$Os_6(CO)_{18}$	150	^{13}C: 194.4 (s, 1), 183.0 (s, 2), 179.2 (s, 1), 176.7 (s, 2), 171.1 (s, 2), 170.8 (s, 1)	50
	373	^{13}C: 181.9 (s, 3), 179.4 (s, 3), 175.0 (s, 3)	
$H_2Os_3(CO)_{11}$	191	^1H: τ 29.97 (d, J_{HH} 3.6, 1), 20.24 (d, J_{HH} 3.6, 1)	44
$H_2Ru_3(CO)_9S$	165	^{13}C: 197.7 (s, 2), 193.3 (d, J_{HC} 8, 2), 187.9 (s, 1), 186.1 (s, 2), 181.8 (d, J_{HC} 13, 2)	61
	273	$^{13}C[^1H]$: 193.9 (s, 1), 187.3 (s, 2)	
	363	$^{13}C[^1H]$: 189.3 (s)	
$[HFe_3(CO)_{11}]^-$	243	$^{13}C[^1H]$: 259.8 (s, 1), 214.8 (s, 10)	62
	313	$^{13}C[^1H]$: 221.0 (s)	
$H_4Ru_4(CO)_{11}\cdot P(OMe)_3$	~303	^1H: τ 6.31 (d, J_{PH} 12.2, 9), 27.72 (d, J_{PH} 2.65, 4)	76
$H_4Ru_4(CO)_{10}\cdot [P(OMe)_3]_2$	~303	^1H: τ 6.31 (d, J_{PH} 12.0, 18), 27.61 (t, J_{PH} 6.63, 4)	76
$H_4Ru_4(CO)_9\cdot [P(OMe)_3]_3$	~303	^1H: τ 6.23 (m, 27), 27.8 (quartet, J_{PH} 7.70, 4)	76
$H_4Ru_4(CO)_8\cdot [P(OMe)_3]_4$	~303	^1H: τ 6.26 (t, J_{sum} 12.2, 36), 27.83 (quintet, J_{PH} 7.95, 4)	76
$[H_3Ru_4(CO)_{12}]^-$	178	^1H: τ 25.95 (d, J_{HH} 2.5, 2), 27.44 (s, 3.9), 29.05 (t, J_{HH} 2.5, 1)	78
	327	^1H: τ 26.9 (s)	
$H_2Ru_3Fe(CO)_{13}$	181	$^{13}C[^1H]$: 231.9 (s, 2), 212.0 (s, 1), 204.4 (s, 2) 195.3 (s, 2), 191.1 (s, 2), 190.4 (s, 2), 188.2 (s, 2), 187.7 (s, 2)	89
	308	$^{13}C[^1H]$: 199.5 (s)	
$[HOs_5(CO)_{15}]^-$	303	^1H: τ 32.66	49
	165	^{13}C: 189.2 (d, J_{HC} ~2, 4), 189.4 (s, 3), 182.3 (s, 2), 180.3 (s, 4), 169.3 (d, J_{HC} ~6, 2)	
	328	^{13}C: 182.4 (s, 6), 180.3 (s, 4)	
$[HOs_6(CO)_{18}]^-$	313	^1H: τ 21.26 (s)	51
	161	$^{13}C[^1H]$: 191.5 (s), 184.6 (s)	
	273	^{13}C: 187.6 (s)	
$C_7H_8Fe_2(CO)_6$	193	^{13}CO: 223.7 (s, 1), 209.3 (s, 1), 207.0 (s, 1)	35
	313	^{13}CO: 212.6 (s)	
$C_8H_{10}Fe_2(CO)_6$	208	$^{13}C[^1H]$: 90.6 (s, 1), 74.8 (s, 1), 72.6 (s, 1), 71.3 (s, 1), 55.8 (s, 1), 47.8 (s, 1), 24.4 (s, 1), 23.6 (s, 1)	36
	363	$^{13}C[^1H]$: 73.6 (s, 2), 69.2 (s, 2), 48.0 (s, 2), 36.1 (s, 2)	
	198	^{13}CO: 216.3 (s), 216.1 (s), 215.8 (s), 209.8 (s), 208.6 (s), 205.0 (s)	
	301	^{13}CO: 215.8 (s, 1), 211.2 (s, 3), 210.4 (s, 2)	
	383	^{13}CO: 212.1 (s, 3), 211.2 (s, 3)	

(Continued)

TABLE I—*(Continued)*

Compound	Temperature (K)	Chemical shifts (multiplicity, J Hz, intensity)[a]	Reference
$C_8H_{10}Rh_2 \cdot (C_5H_5)_2$	223	$^{13}C[^1H]$: 87.0 (d, J_{RhC} 2), 83.1 (d, J_{RhC} 4), 72.2 (d, J_{RhC} 6), 62.5 (d, J_{RhC} 15), 53.5 (d, J_{RhC} 8), 49.5 (d, J_{RhC} 10), 48.4 (d, J_{RhC} 7), 45.3 (s), 24.2 (s), 17.9 (d, J_{RhC} 19)	58
	418	$^{13}C[^1H]$: 86.6 (d, J_{RhC} 4), 82.6 (d, J_{RhC} 5), 60.2 (d, J_{RhC} 5), 50.4 (t, 'J_{sum}' 8), 39.1 (d, J_{RhC} 16), 34.3 (s)	
$C_7H_8Rh_2 \cdot (C_5H_5)_2$	199	$^{13}C[^1H]$: 86.3 (d, J_{RhC} 4), 83.3 (d, J_{RhC} 4), 67.5 (d, J_{RhC} 7), 58.0 (d, J_{RhC} 5), 51.7 (d, J_{RhC} 9), 50.0 (d, J_{RhC} 10), 39.3 (d, J_{RhC} 3), 25.1 (d, J_{RhC} 7), −20.5 (d, J_{RhC} 11)	58
	308	$^{13}C[^1H]$: 86.2 (d, J_{RhC} 4), 82.9 (d, J_{RhC} 4), 58.2 (d, J_{RhC} 9), 54.7 (t, 'J_{sum}' 7), 39.3 (d, J_{RhC} 5), 2.0 (d, J_{RhC} 10)	
$C_{10}H_{12}Fe_2 \cdot (CO)_6$	153	^{13}CO: 218.6 (s), 216.5 (s), 215.8 (s), 209.7 (s), 208.6 (s), 204.8 (s)	34
	243	^{13}CO: 218.5 (s), 211.0 (s), 210.3 (s)	
	375	^{13}CO: 212.6 (s), 210.3 (s)	
$C_8H_{10}Fe_2 \cdot (CO)_5$-$PEt_3$	151	^{13}CO: 222.2 (d, J_{PC} 19), 220.9 (s), 220.2 (s), 213.1 (d, J_{PC} ~19), 207.3 (s)	34
	223	^{13}CO: 220.9 (s), 217.2 (d, J_{PC} ~19), 213.2 (s)	
	311	^{13}CO: 217.2 (d, J_{PC} ~19), 215.1 (s)	
$H_2Os_3(CO)_9 \cdot C_2H_3$	206	$^{13}CO[^1H]$: 184.5 (s), 181.0 (s), 179.7 (s), 176.6 (s), 173.9 (s), 172.6 (s), 172.2 (s), 165.6 (s)	96
	301	^{13}CO: 185.2 (s), 181.5 (s), 181.5 (d, J_{HC} 1, 1), 178.5 (d, J_{CH} 2.5, 2), 174.2 (s, 2), 173.4 (d, J_{CH} 2, 2), 169.8 (d, J_{CH} 12, 2)	
$H_2Os_3(CO)_9C_2H_2$	300	^1H: τ 3.90 (s, 1), 4.43 (s, 1), 28.13 (d, J_{HH} 1.4, 1), 31.83 (d, J_{HH} 1.4, 1)	47
$(C_5H_5)_3Rh_3 \cdot (PhCCPh)(CO)$	183	^{13}CO: 241.6 (m, J_{RhC} 43.7, J_{RhC} 43.7, J_{RhC} 28.4)	102
	303	^{13}CO: 236.0 (quartet, J_{RhC} 38.7)	
	186	$^{13}C[^1H]$: (Ph^{13}CCPh) 149.8 (m, J_{RhC} 7, J_{RhC} 15.4, J_{RhC} 25.2)	98
	303	$^{13}C[^1H]$: (Ph^{13}CCPh) 150.8 (quartet, J_{RhC} 11)	
$(C_5H_5)_3Rh_3 \cdot (PhCCPh)$	303	$^{13}C[^1H]$: (Ph^{13}CCPh) 301.6 (quartet, J_{RhC} 36.6)	98
$Os_3(CO)_7 \cdot (C_6H_4)(PMe_2)_2$	213	^1H: τ 8.21 (t, 'J_{sum}' 8.5, 6), 7.63 (t, 'J_{sum}' 8.5, 6), 3.24 (m, 2), 2.72 (d, J_{HH} 7, 1), 2.43 (d, J_{HH} 7, 1)	45
	333	^1H: τ 8.21 (t, 6), 7.63 (t, 6), 3.38 (m, 2), 2.62 (m, 2)	
$Os_3(CO)_7 \cdot (C_6H_4)$-$(AsMe_2)_2$	183	^1H: τ 8.46 (s, 6), 7.88 (s, 6), 3.50 (m, 2), 2.66 (d, J_{HH} 7, 1), 2.86 (d, J_{HH} 7, 1)	45
	333	^1H: τ 8.49 (s, 6), 7.88 (s, 6), 3.51 (m, 2), 2.77 (m, 2)	
$(C_8H_8)_2Ru_3(CO)_4$	233	^1H: τ 6.26 (s)	28

TABLE I—(*Continued*)

Compound	Temperature (K)	Chemical shifts (multiplicity, J Hz, intensity)[a]	Reference
$Ni_3(CO)_3(C_2F_6)$ (C_8H_8)	183	^{1}H: τ 5.5 (s)	42
$(C_7H_7)(C_7H_9)$ $Ru_3(CO)_6$	179	^{1}H: τ 3.0 (m, 2), 3.4 (m, 1), 4.8 (m, 1), 5.1 (m, 1), 5.9 (br, 8)	101
	252	^{1}H: τ 3.4 (t, 1), 3.9 (m, 2), 5.1 (m, 2), 5.9 (s, 7)	
$(C_8H_6)Ru_3(CO)_8$	213	^{1}H: τ 3.7 (t, J_{HH} 2.5, 2), 5.4 (d, J_{HH} 2.5, 2), 6.9 (d, J_{HH} 2.5, 2)	68
	333	^{1}H: τ 3.7 (t, J_{HH} 2.5, 2), 6.2 (d, J_{HH} 2.5, 4)	
$[C_8H_3(SiMe_3)_3]\cdot$ $Ru_3(CO)_8$	303	^{1}H: major isomer: τ 4.05 (s, 1), 6.23 (s, 2), 9.62 (s, 9), 9.89 (s, 18); minor isomer: τ 4.54 (s, 1), 5.97 (s, 2), 9.70 (s, 18), 9.78 (s, 9)	67
$(Me_2C_2B_7H_9)\cdot$ $Pt(PEt_3)_2$	~303	^{1}H: τ 8.02 (m, 12), 8.03 (s, 6), 8.98 (m, 18)	64
$(C_2B_7H_9)Pt\cdot$ $(PEt_3)_2$	~303	^{1}H: τ 6.46 (m, 2), 8.04 (m, 12), 9.0 (m, 18); ^{11}B: (wrt $BF_3\cdot OEt_2$) δ −1.6 (m, 3), 10.5 (m, 2) 28.4 (d, J_{BN} 140, 2)	64

[a] s = Singlet, d = doublet, t = triplet, br = broad, m = multiplet.

II

DINUCLEAR METAL CARBONYLS AND THEIR DERIVATIVES

$[(\eta^5\text{-}C_5H_5)Fe(CO)_2]_2$ has been characterized in the crystal in both trans (*87, 15*) and cis (*16*) bridged forms. Interpretation of the solution infrared spectrum of the dimer and its derivatives has required the presence, in smaller proportions, of nonbridged isomers (*81, 79*), probably in the staggered anti and gauche conformations (*17*). Cis–trans isomerism was demonstrated to occur above 213 K ($\Delta G^{\ddagger} \sim 44$ kJ mole^{-1}) (*17*). Gansow's ^{13}C NMR data (*63*), however, have clearly demonstrated two carbonyl exchange steps. Although it seems that the limiting low-temperature spectrum was not recorded (*65*), it is apparent that bridge-terminal carbonyl site exchange is rapid ($k > 10^3$ s^{-1}) above 200 K solely for the trans isomer. It is only by 238 K that the bridging (δ 272.9) and terminal (δ 211.0) carbonyl resonances of the cis isomer start to broaden, as does the averaged ^{13}CO resonance of the trans isomer (δ 242.2). By 328 K, a single ^{13}CO resonance is observed, indicating rapid bridge-terminal ^{13}CO exchange and cis–trans isomerization.

By a careful consideration of the stereochemistry involved, Adams and Cotton (5) rationalized these data using two proposals: (a) that carbonyl bridge opening and closing occurs in a pairwise, concerted manner and (b) that rotations by $\pm 2\pi n/3 (n = 1$ or 2) of unbridged isomers allow interconversions of the various staggered conformers. Consequently (Fig. 1) synchronous opening of the bridges of the trans isomer produces an unbridged species in an anti conformation. Both the original bridging and terminal groups are in the correct orientation for closure to reform the trans isomer, thus allowing bridge-terminal exchange. However, on opening the two bridges of the cis isomer, to produce a gauche conformer, only the original bridging CO groups B and T′ are in a position to reform bridges. Rotation about the metal—metal bond is required to yield another gauche conformer before exchange can be effected. Once rotation is allowed, gauche and anti conformers can interchange, permitting cis–trans isomerization. So it is apparent that an extra energy barrier, viz. rotation about the iron—iron bond, is required for the cis isomer to undergo carbonyl site exchange, and once this can be achieved it will also isomerize.

This mechanism has successfully accounted for the behavior of many related binuclear complexes for which bridge–terminal ligand exchange has been observed, including $[(\eta^5\text{-}C_5H_5)Cr(NO)_2]_2$ (74), $[(\eta^5\text{-}C_5H_5)Mn(NO)(CO)]_2$ (74), $(\eta^5\text{-}C_5H_5)_2Fe_2(CO)_3(CNR)$ (R = ^{t}Bu, Ph) (4, 6, 69), $(\eta^5\text{-}C_5H_5)_2Fe_2(CO)_3(GeMe_2)$ (1), and $Co_2(CO)_8$ (97)—all of which have two bridging ligands. One pair of derivatives in particular $\{(\eta^5\text{-}C_5H_5)_2Fe_2(CO)_3[P(OR)_3]\}$, where R is Ph (32, 38) or Et (65), provided a test of this mechanism. The phosphite ligand is likely to remain in a terminal

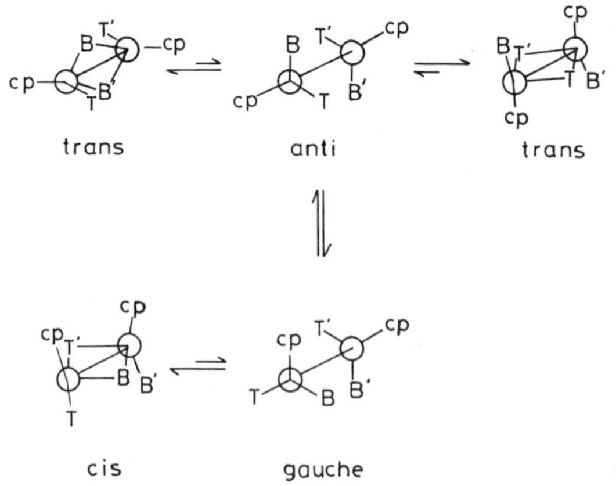

FIG. 1.　Part of the Adams–Cotton scheme for ligand exchange in $(C_5H_5)_2Fe_2(CO)_4$ and its derivatives by pairwise bridge opening and closing.

position (T in Fig. 1). Pairwise bridge opening of both isomers generates unbridged species in which only the original bridging ligands are in a position to reclose, precluding exchange without rotation about the iron—iron bond. Indeed, by following the permitted permutations, it is evident that bridge–terminal carbonyl exchange will occur only on cis–trans isomerization. Two groups (32, 38, 65) have verified this experimentally; they also found that introducing the bulky phosphite ligand increased the energy barrier to site exchange—by ca. 33 kJ mole^{-1} when R = Ph. This demonstrates the increased steric inhibition of rotation about the metal—metal bond.

$(\eta^5\text{-}C_5H_5)_2Rh_2(CO)_3$ possesses one bridging and two terminal carbonyl groups, the latter in trans orientation, in the crystal (88). At 193°K, the ^{13}CO NMR absorptions (at δ 231.8 and 191.8) show that this structure is maintained in solution (56). However, by 293 K, rapid, intramolecular carbonyl site exchange is apparent from the averaged ^{13}CO resonance observed. A proposal that this exchange process also involved pairwise bridge closure (1), via a triple bridging intermediate of similar structure to $Fe_2(CO)_9$, was substantiated by the isolation of a related complex $[(\eta^5\text{-}C_5H_5)_2Rh_2(CO)(CPh_2)_2]$ (103) with three bridging ligands—one carbonyl and two carbenes. On this basis, the substituted complex $(\eta^5\text{-}C_5H_5)_2Rh_2(CO)_2P(OPh)_3$ would not be expected to undergo site exchange of its single bridging and terminal carbonyl groups (Fig. 2). However, the $\Delta G^\ddagger$ (37 kJ mole^{-1} at 208 K) barrier to such exchange is very similar to that of the parent tricarbonyl (53). Furthermore, the eight-line pattern of the fast-exchange ^{13}CO spectrum demonstrates that a simple one for one carbonyl site exchange process is occurring, the phosphite ligand remaining bound to one rhodium atom; this pathway apparently involves racemization at the substituted rhodium atom, Rh_2 (Fig. 2) (55).

The free energies of activation for rotational interconversion of anti and gauche isomers in the series $[(\eta^5\text{-}C_5H_5)M(CO)_3]_2$ (Fig. 3: L' = η^5—C_5H_5, L = CO) increases from Cr (51 kJ mole^{-1}) through Mo (63 kJ mole^{-1}) to W (68 kJ mole^{-1}) (2); this is the inverse order to that for the M—M distances, indicating particularly that the long Cr—Cr bond is flexible and torsion about it is more ready. Indeed, above ca. 274 K, a

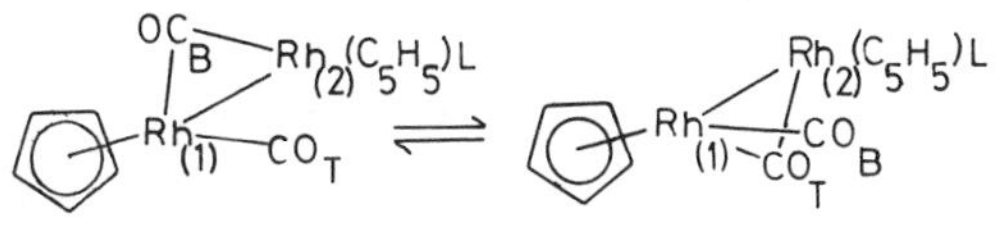

FIG. 2. Proposed synchronous mechanism for the exchange of the bridging and terminal carbonyl ligands of $(C_5H_5)_2Rh_2(CO)_2L$ derivatives.

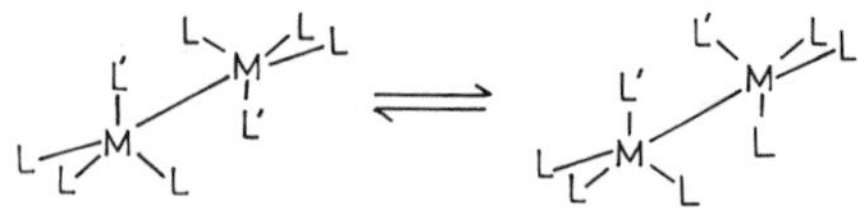

anti   gauche

Fig. 3. Conformers of an $M_2L_6L_2$ molecule with trigonal bipyramidal metal coordination geometry.

second process is apparent which involves the cleavage of this bond (3) (Eq. 1) to form radical monomers.

$$[(\eta^5\text{-}C_5H_5)Cr(CO)_3]_2 \rightleftharpoons 2[(\eta^5\text{-}C_5H_5)Cr(CO)_3] \tag{1}$$

On the other hand, in one example, [Rh{P(OMe)$_3$}$_4$]$_2$ (85), steric inhibition to intramolecular exchange is so great that, at 373 K, a specific intermolecular exchange of only the equatorial ligands is observed [Fig. 3; L = L' = P(OMe)$_3$].

In summary, four types of ligand exchange processes have been characterized: (i) rotation about the metal–metal bond, (ii) cleavage of the metal–metal bond, (iii) two for two synchronous bridge opening and closing, and (iv) one for one bridge opening and closing.

III

METAL CARBONYL AND ISONITRILE CLUSTERS

A. Carbonyls

1. Trinuclear Clusters

The ^{13}C NMR spectrum of Fe$_3$(CO)$_{12}$ is a sharp singlet (62, 63) down to 123 K (33), demonstrating a very low barrier to equilibration of the five carbonyl environments of the molecular structure in the crystal (99), if that structure (Fig. 4) were maintained in solution. Even though that may not be the case (33, 72), rapid equilibration must still be concluded. Mixed metal derivatives of the same structural type, viz. Fe$_2$Ru(CO)$_{12}$ (104, 75) and [MnFe$_2$(CO)$_{12}$]$^-$ (11), are also nonrigid (62).

Exchange mechanisms can be postulated for these clusters involving the types of mechanisms reported for the binuclear complexes. However, it is more informative to consider the shape of the ligand polyhedron, as well as that of the circumscribed metal skeleton (71). For

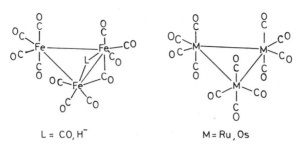

L = CO, H⁻   M = Ru, Os

FIG. 4.   Structures of the $M_3(CO)_{12}$ clusters.

our purposes, this allows a better description of the topology of any postulated mechanism than the normal line diagram. The permutative character of site exchange on clusters is inherently a more complicated problem than for mononuclear complexes. In the latter class of compounds there are only the relative motions of the ligand polyhedron around a spherical core. However, in cluster compounds this core is nonspherical, so its orientation within the ligand envelope is important. (Reorientations can be simple rotations or involve migration of the atoms contained in cluster core—Section VII). Fortunately, not all of these permutations are operating for each cluster (or at least they cannot be detected). The example of $Fe_3(CO)_{12}$ has been described in these terms (72), and it was observed that the triangular iron core resides within an icosahedron of carbonyl groups (Fig. 5). Different relative

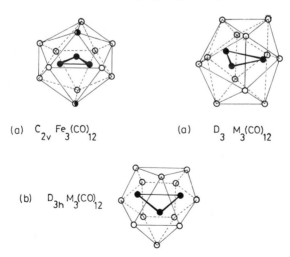

(a) C_{2v} $Fe_3(CO)_{12}$ (a) D_3 $M_3(CO)_{12}$

(b) D_{3h} $M_3(CO)_{12}$

FIG. 5.   Relative orientations of the metal triangle and CO polyhedron in some $M_3(CO)_{12}$ structures: (a) CO icosahedron, (b) CO tetrakaidecahedron. ●, M; ○, terminal CO; ◑, edge-bridging CO.

orientations of these two shapes produce different formal bonding representations, which may coexist in equilibrium in solution. All points on the ligand polyhedron are potentially identical, so reorientation of the metal triangle within it will give total ligand equilibration. ^{13}C NMR data (see above) show the barrier to reorientation within this high-symmetry ligand field to be evidently very small.

Both $Ru_3(CO)_{12}$ (83) and $Os_3(CO)_{12}$ (24) adopt an alternative structure in the solid state (Fig. 4) with a different carbonyl polyhedron (Fig. 5), a tetrakaidecahedron; there are intrinsically two ligand environments in this polyhedron. Down to the lowest temperature studied (173 K), the ruthenium complex exhibits an averaged [13]CO resonance (62, 89). This situation is reached only at 373 K for $Os_3(CO)_{12}$, the frozen ^{13}C NMR spectrum being observed at 283 K. Total equilibration of carbonyl groups is precluded if their polyhedral structure remains undisturbed. However, rearrangement to an icosahedral ligand envelope in either an intermediate or transition state would allow averaging. If the metal triangle is oriented to produce the structure of D_3 symmetry (Fig. 5), the carbonyl groups will remain terminally bound (72).

2. Tetranuclear Clusters

At 208 K, the ^{13}C NMR spectrum of $Rh_4(CO)_{12}$ demonstrates that the structure in the crystal (100) is maintained in solution (57) (Fig. 6). As the temperature is raised, all resonances broaden uniformly, eventually producing a sharp quintet at 336 K (37). A random ligand exchange process is apparent with the twelve carbonyl groups remaining associated with the four rhodium atoms. This was interpreted in terms of a previously postulated mechanism (26) involving an unbridged intermediate of T_d symmetry (the $Ir_4(CO)_{12}$ structure). However, since the idealized ligand polyhedron has only one ligand environment (I_h symme-

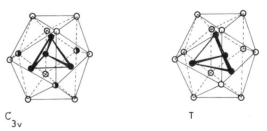

C_{3v} T

FIG. 6.   Alternative orientations of an M_4(M=Co, Rh) tetrahedron in a CO icosahedron; the C_{3v} form is found in the crystal and in solution. ●, M; ○, terminal CO; ◑, edge-bridging CO.

try), only a rotation of the metal tetrahedron is required to cause random equilibration. A small rotation about the molecule's C_3 axis will produce a structure (Fig. 6) possessing twelve equivalent terminal carbonyl groups, and an intermediate of this type may be involved. A distortion of the carbonyl envelope is required to produce the cuboctahedral geometry in $Ir_4(CO)_{12}$ (Fig. 7) (71). These two mechanisms (with or without ligand–ligand motion) are indistinguishable in this case. The ^{13}C NMR spectrum of the isostructural $Co_4(CO)_{12}$ (25, 99) remains problematical (52, 23). Three, not four as expected, resonances of equal intensity are observable at 203 K and below. This intensity anomaly may result from relaxation caused by the ^{59}Co nuclei. As the temperature is raised, these resonances broaden, but it is not possible to categorically determine whether this is due to a site exchange process or to interactions with the ^{59}Co nuclei.

The mixed metal complex $Co_3Rh(CO)_{12}$ (82) has the basic structure as its homometallic parents, the rhodium atom being in the basal, carbonyl-bridged, metal triangle; this was verified by the cluster's ^{13}C NMR spectrum at 188 K (73). By 303 K, an averaged ^{13}C NMR signal is observable; this can be rationalized either on the basis of random jumps of the metal tetrahedron to equivalent positions within the icosahedral carbonyl envelope or of a redistribution of carbonyl groups via an $Ir_4(CO)_{12}$-type intermediate. However, there is a lower-energy process which at 243 K averages the two carbonyl groups terminally bound to the rhodium atom on one hand and the remaining ten on the other. A rotation of the metal tetrahedron through 180° maintaining the binding of the two terminal rhodium atoms within a fixed icosahedral carbonyl envelope (Fig. 6) would still yield four apparent ligand environments (of populations 2:2:4:2:2), instead of the two observed. By maintaining this binding and allowing the ligands to distort to the cuboctahedral form, either as an intermediate or transition state, and relax back into their favored icosahedral geometry, then one would produce the observed effect. At higher temperatures, this orientation preference of the rhodium atom is overcome and so total equilibration is allowed. In this

FIG. 7. $Ir_4(CO)_{12}$ is an Ir_4 tetrahedron in a CO cuboctahedron. ●, Ir; ○, terminal CO.

case at least, ligand rearrangement does involve a redistribution of carbonyl groups relative to each other as well as to the metals, probably via the initially prepared $Ir_4(CO)_{12}$-type intermediate (26). It seems probable, therefore, that this process is also that giving rise to the stereochemical nonrigidity of $Rh_4(CO)_{12}$.

3. Larger Clusters

Hexa- and heptanuclear clusters with 15, 16, 17, and 18 ligands have been studied. $[Rh_6(CO)_{15}]^{2-}$ readily undergoes total ligand equilibration, as evidenced by a rhodium split septet ^{13}CO NMR spectrum at 203 K (66), while $Rh_6(CO)_{16}$ and its derivatives are rigid up to or above 293 K (66, 55, 22). The coordination polyhedron of $Rh_6(CO)_{16}$ is the fully triangulated tetracapped truncated tetrahedron (T_d symmetry), but that of the $[M_6(CO)_{15}]^{2-}$ anions (C_{3v} symmetry) is more open, possessing triangular, square, and hexagonal faces (21); the latter may thus have a lower energy barrier to distortion. Relative carbonyl migration is necessary to produce the total ligand equilibration observed for the anion. This probably involves an intermediate with a different ligand envelope. Such an intermediate, of D_3 symmetry, has been proposed (66) (Fig. 8).

Another 16-hedron is described by the ligands in the anion $[Rh_7(CO)_{16}]^{3-}$ (7), possessing C_{3v} symmetry (Fig. 9). At 203 K, a ^{13}C NMR spectrum appropriate to a rigid structure was observed (66), but by 298 K, partial bridge–terminal ligand exchange occurs, probably involving a rotation of the bottom hexagon of carbonyl groups, with the remaining carbonyl ligands and the metal atoms fixed. Addition of the extra ligand in $[Rh_7(CO)_{16}I]^{2-}$ (10) (Fig. 9) apparently promotes ligand mobility since the anion's ^{13}C NMR spectrum consists of a single rhodium split octet at 242 K (21).

$Os_6(CO)_{18}$ has a bicapped tetrahedral metal framework in the crystal (84) with three different metal environments (Fig. 10). Three terminal

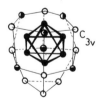 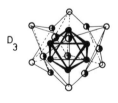

FIG. 8. $Rh_6(CO)_{15}^{2-}$. The C_{3v} structure is predominant; the D_3 form is proposed as an intermediate in the CO site exchange. ●, Rh; ○, terminal CO; ◐, edge-bridging CO; ◒, face-bridging CO.

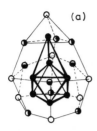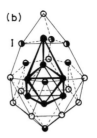

FIG. 9.   Structures of (a) $Rh_7(CO)_{16}^{3-}$ and (b) $Rh_7(CO)_{16}I^{2-}$. ●, Rh; ○, terminal CO; ◑, edge-bridging CO or I; ◒, face-bridging CO.

carbonyl groups are bound to each metal, two of which are related by a reflection plane. This gives a total of six CO environments, and all are observed in the cluster's solution ^{13}C NMR spectrum at 150 K (*50*). As the temperature is raised, three pairs of resonances (of intensity ratio 1:2) collapse (at different temperatures) and reform three averaged signals by 373 K. In this case, the metal core is definitely not rotating within the ligand cage, but rather localized sections of the cage are rearranging. These sections correspond to the triangle of ligands on each metal. It is evident that the energy barrier to rotation of the ligand triangle of an $M(CO)_3$ group is affected by the metal atom environment and the position of the triangle in the ligand polyhedron.

4. *Summary*

The various types of carbonyl site exchange mechanisms can be placed in three classes.

i. Rotation of the metal core within a fixed carbonyl envelope. The ready equilibration of carbonyl groups in $Fe_3(CO)_{12}$ is possibly by this means.

ii. Rotation of the metal core accompanying a permutation within the ligand envelope, via the intermediary of an alternative carbonyl polyhe-

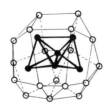

FIG. 10.   Structure of $Os_6(CO)_{18}$. ●, Os; ○, terminal CO.

dral structure. Both $M_3(CO)_{12}$ (M = Ru, Os) and $M_4(CO)_{12}$ (M = Rh, Co) are examples of this class.

iii. Local rearrangements within sections of the ligand envelope with a fixed orientation of the metal core with respect to the geometry of the carbonyl polyhedron. The exchange processes of $[Rh_7(Co)_{16}]^{3-}$ and $Os_6(CO)_{18}$ are of this type.

All of these three classes can produce bridge–terminal and/or terminal–terminal exchange, depending upon the particular example in question. In comparison with binuclear complexes, the dissociation pathway of $[(\eta^5\text{-}C_5H_5)Cr(CO)_3]_2$ has not been identified for any cluster. Since most of the dinuclear complexes studied contain cyclopentadienyl ligands which, whenever evidence can determine this, stay bound to one metal, it seems more appropriate to place their exchange processes in the third classification.

B. Isonitriles

In the $Ni_4(CNBu^6)_7$ structure (C_{3v} symmetry), there are four terminally bound isonitrile ligands and three bridging groups which are both carbon and nitrogen bound (43). The bridging groups are associated with either three edges or three faces of the nickel tetrahedron. At low temperatures, the ^1H NMR spectrum of this complex exhibits the expected three methyl group environments (in the ratio 3:3:1). On raising the temperature, the two high-field resonances, assigned to the terminal ligands, coalesce, and by 373 K total equilibration is evident. Migration of the bridging groups over all edges and face of the Ni_4 core provides the lower-energy process for equilibrating the terminal ligands. Bridge–terminal exchange ensues at higher temperatures.

IV

HYDRIDO CARBONYL CLUSTERS

The introduction of hydrido ligands to a metal carbonyl cluster adds still further modes of rearrangement to those already possible in the binary carbonyls. Migration of the hydrogen atoms relative to the frame described by the auxiliary ligands, as found in mononuclear hydrido complexes (70, 90) is possible. The interchange of bridging and terminal hydrido ligands in $H_2Os_3(CO)_{11}$ (44, 95) and its derivatives, by ca. 323 K, could be an example of this type of behavior. Since the ^{13}C NMR spectra of these complexes are hitherto unreported, then a permutation of carbonyl ligands cannot be excluded. Both hydride and carbonyl migration is evident from the temperature variation of the ^{13}C NMR spectrum of $H_2Ru_3(CO)_9S$ (61) (Fig. 11). At 165 K, the proton coupled spectrum, consisting of five resonances, is in accord with the illustrated structure. By 213 K all these signals broaden, and at 273 K two sharp singlets (proton decoupled) of relative intensity 2:1 are observable. A higher-energy process allows equilibration of these two resonances to one, which is narrow by 373 K. Although alternatives can be presented, a probable sequence of events is that the lower-energy process involves the migration of a hydrido ligand to below the vacant Ru—Ru edge. With respect to the carbonyl frame, this involves a hydride jump from one square face to another. Above 273 K a second process involving the rotation of one (or more) of the two types of $Ru(CO)_3$ unit becomes sufficiently rapid to affect the observed ^{13}C NMR spectrum. A combination of these two events will equilibrate the five carbonyl environments.

Migration of the hydride position about the icosahedral frame in $[HFe_3(CO)_{11}]^-$ is also apparent (62). Rotation of the metal triangle within a fixed ligand envelope would yield three CO environments (in the ratio 5:5:1), ortho, meta, and para with respect to the hydride's position (Fig. 5). However at 243 K, two resonances in the ratio 10:1 are observed. Even an accidental degeneracy of the ortho and para resonances will not resolve the discrepancy. The less intense line has a chemical shift typical of a bridging carbonyl (δ 259.8), but if the molecular structure in the crystal (41) were maintained in solution and the triangle were rapidly hopping to equivalent positions inside the ligand envelope, this para carbonyl group would be a terminal one. Within the structure of the C_2 isomer proposed for $Fe_3(CO)_{12}$ (72), this para position is in a face-bridging environment. So possibly, with the restrictions imposed above, a similar structure is adopted by this anion. Alternatively, a selective carbonyl exchange process involving all save this one bridging group may be operating (62). At 313 K, the two resonances have coalesced to one, so site exchange within the ligand polyhedron is occurring.

Hydride migration has also been observed on tetranuclear clusters. A

FIG. 11.   Structure of $H_2Ru_3(CO)_9S$. ●, Ru; ◕, S; ○, terminal CO; ⊕, edge-bridging H.

fast-exchange proton NMR spectrum is observed at ambient temperatures for phosphite-substituted derivatives of $H_4Ru_4(CO)_{12}$, viz. $H_4Ru_4(CO)_{12-n}[P(OMe)]_n$ (n = 1, 2, 3, 4) (76). The hydrogen atoms are considered to reside on four of the six edges of the Ru_4 tetrahedron, corresponding to four of the six square faces of the idealized cuboctahedron described by the other ligands (Fig. 12). These are thus vacant sites, with respect to both the metal and ligand polyhedra, to which the hydrido ligands may jump. The rearrangement could involve an isomeric intermediate in which the two vacant square faces of the ligand cuboctahedron were cis, not trans. Two isomers have been identified in the related anion $[H_3Ru_4(CO)_{12}]^-$ (78). One of these isomers can be attributed with C_{3v} symmetry, in which 3 fac-square faces of the carbonyl cuboctahedron are occupied by hydrogen atoms (two alternative arrangements of hydrogen atoms on the Ru—Ru edges are allowed—with the "base" of the tetrahedron having either three or none of its edges bridged). Occupation of mer-square faces would give rise to another form with C_2 symmetry, in which there are two hydride environments. At 178 K, distinct proton resonances due to two isomers were observed, but at 327 K, site exchange of the second isomer and isomerization are sufficiently rapid to yield one averaged hydride resonance. Spin saturation studies indicate that both these pathways for the C_2 isomer were operating. The difference lies only in which hydride migrates and to which adjacent square carbonyl envelope face it hops (Fig. 12).

Since only ^1H NMR was used in these examples, redistribution of the carbonyl ligands would not have been detected. Such rearrangements have been found for $H_2Ru_2Fe(CO)_{13}$ using ^{13}C NMR (89). At 210 K, the resonances due to the four carbonyl groups bound to the iron atom broaden (Fig. 12), while the remainder are unaffected. A local interchange of these four ligands, involving cleavage of the longer side of the asymmetric carbonyl groups, is evident. By 228 K, the carbonyl groups

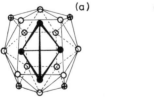

FIG. 12.   (a) Idealized structure of $H_4Ru_4(CO)_{12}$ and its derivatives. (b) Structure of $H_2Ru_3Fe(CO)_{13}$. ●, Ru; ◑, Fe; ○, terminal CO; ◐, edge-briding CO; ⊕, edge-bridging H.

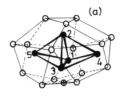 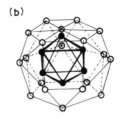

Fig. 13. Structures of (a) $HOs_5(CO)_{15}^-$ and (b) $HOs_6(CO)_{18}^-$. ●, Os; ○, terminal CO; ⊕, edge-bridging H; ⊗, face-bridging H.

bound to the ruthenium atoms only are averaged to one line. This second process involves another carbonyl migration, with or without hydride exchange. At higher temperatures the two groups of carbonyl ligands interchange, a sharp, averaged resonance being observed at 368 K. Scrambling between these two sections of the ligand envelope, or rotation of the metal polyhedron, is now occurring.

The anion $[HOs_5(CO)_{15}]^-$ (49) appears to have its hydride ligand spanning an equatorial osmium–osmium edge on a carbonyl-denoted square face (Fig. 13), as confirmed by the cluster's ^{13}C NMR spectrum at 165 K. At that temperature, rotation of the $M(CO)_3$ group on Os (2) is occurring. Collapse of the two resonances coupled to the proton at 223 K demonstrates that at this temperature the other two equatorial $Os(CO)_3$ groups are rotating. A simple two-resonance spectrum is observed at 328 K. In addition to the localized carbonyl scrambling, a hydride shift mechanism about the three equatorial Os–Os edges would produce this effect. The carbonyl environments around the edges, Os(1)–Os(2) and Os(3)–Os(2), do not offer square faces for the hydride ligand. However, these carbonyl groups are already mobile at the elevated temperatures necessary for the hydride shift, so appropriate accommodation for the hydride can be made. ^{13}C NMR has also been used to detect hydride migration in $[HOs_6(CO)_{18}]^-$ (51) (Fig. 13). Two resonances of equal intensity are observable at 161 K; if local scrambling of each $Os(CO)_3$ group is rapid, this spectrum is consistent with the hydride bridging, a face of the Os octahedron corresponding to the center of a hexagonal face of the carbonyl envelope (80). At 188 K, these two resonances have formed a broad averaged one, which sharpens on warming. The hydride is moving over all the faces of the Os octahedron; carbonyl mobility will again assist in this by generating alternative hexagonal faces in the carbonyl envelope necessary for a degenerate rearrangement.

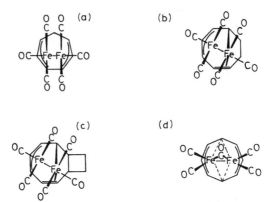

FIG. 14.   Structures of (a) $C_7H_8Fe_2(CO)_6$, (b) $C_8H_{10}Fe_2(CO)_6$, (c) $C_{10}H_{12}Fe_2(CO)_6$, and (d) $C_8H_8Fe_2(CO)_5$.

V

DINUCLEAR OLEFIN COMPLEXES

The mirror plane in $C_7H_8Fe_2(CO)_6$ (Fig. 14) reduces the number of carbonyl environments to three (29), and accordingly three ^{13}CO resonances have been observed (35) at 193 K. As the temperature is raised, all three broaden uniformly, coalescing to a singlet at 313 K. Of several possible mechanisms, rotation of the $Fe(CO)_3$ groups was favored. Complexes of the structural type of $C_8H_{10}Fe_2(CO)_6$ (30) are unsymmetrical. This structure of the cyclo-octatriene complex has been shown to persist in solution by proton NMR (39, 12). At room temperature, racemization of the two enantiomers (Fig. 15) gives rise to an apparent mirror plane through the organic ligand. As expected, the low-temperature ^{13}C NMR spectrum of this complex (below 208 K) consists of fourteen resonances (36)—due to the eight ring and six carbonyl carbon environments. Above 208 K, the ring carbon resonances broaden uniformly, giving rise to four averaged ones by 363 K, in accord with the

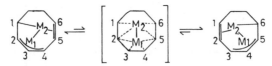

FIG. 15.   The twitch mechanism for exchange between the enantiomers of $C_8H_{10}Fe_2(CO)_6$ and related complexes.

proton NMR data. The temperature dependence of the ^{13}CO resonances gives a more detailed picture of the molecule's fluxionality. Up to 238 K, five of the six ^{13}CO NMR signals collapse at the same rate, and by 273 K, new resonances of intensities 3 and 2 start to sharpen. At about 301 K, the resonance of intensity 3 is sharp, but the other two (of ratio 2:1) broaden. Sample decomposition prevented the recording of spectra above 383 K, at which point a coalesced signal, of intensity 3, from these two was starting to sharpen. At low temperatures two processes are occurring at similar rates (a) the "twitch" mechanism (Fig. 15) and (b) rotation of one of the $Fe(CO)_3$ groups. Rotation about the second $Fe(CO)_3$ moiety is more hindered but is also fast by 383 K. Since the carbonyl groups on each iron atom are not interconverted, the iron atoms do not switch ends, with respect to the organic ligand, at any time in these pathways. This eliminated an alternative mechanism for the ring oscillation, viz. the "glide" (36) involving an intermediate of similar structure to $C_7H_8Fe_2(CO)_6$. The "twitch" mechanism of ring oscillation has been demonstrated for $(C_8H_{10})Rh_2(\eta^5-C_5H_5)_2$ and $(C_7H_8)Rh_2(\eta^5-C_5H_5)_2$, which both adopt the $(C_8H_{10})Fe_2(CO)_6$ bonding mode (58). In these complexes the two $(\eta^5-C_5H_5)$ resonances, for both ^1H and ^{13}C nuclei, remain separate and sharp up to the highest temperatures studied. (For the cyclooctatriene complex this was 418 K and the ring oscillation is rapid by 338 K.) Furthermore, in the fast-exchange ^{13}C NMR spectrum only carbon atoms C_2 and C_5 (Fig. 14) were coupled with both ^{103}Rh nuclei, with C_1, C_3, C_4, and C_6 coupled to only one. This unambiguously determines the ring oscillation as being a twitch process.

The bicyclo[6.2.0]deca-1,3,5-triene derivative $(C_{10}H_{12})Fe_2(CO)_6$ also adopts the $C_8H_{10}Fe_2(CO)_6$ bonding type (31), but by 203 K the ring oscillation is sufficiently rapid to produce an averaged mirror plane down the bicyclic ligand. All six ^{13}CO environments have been frozen out, though, at 153 K (34). In this complex, the three exchange processes, viz. the ring twitch and rotation about each $M(CO)_3$ group, are more separated in activation energy than in $(C_8H_{10})Fe_2(CO)_6$. The twitch causes two pairs of ^{13}CO resonances to average between 153 K and 185 K, at which point rotation about one $Fe(CO)_3$ group starts to occur. Rotation about the second iron atom is only evident above 223 K. Again at the highest temperature recorded, 375 K, two ^{13}CO resonances were observed, indicating that the iron atoms do not interchange. A PEt$_3$-substituted complex was also characterized; in the crystal the phosphine resides on the iron atom bound to the η^3-alkyl group, trans to the Fe—Fe bond. A more complete spectral assignment was possible for the ^{13}CO resonances, and this complex provided evidence that rotation about the η^3-allyl-bound metal has a lower energy barrier than about the other $M(CO)_3$ group.

Although three proton environments would be anticipated for $(C_8H_8)Fe_2(CO)_5$ if the molecular structure in the crystal were maintained in solution (60) (Fig. 14), only one sharp proton NMR line is observed down to 193 K (the lowest temperature studied). Evidently this complex is also fluxional, and, indeed, ring rotation is rapid even in the solid state at 77 K (19).

VI

OLEFIN AND ACETYLENE CLUSTER COMPLEXES

NMR characterization of $HOs_3(CO)_9(C_2H_3)$ has positioned the vinyl group above an Os–Os edge (Fig. 16) (96); the ^{13}C NMR spectrum in the carbonyl region at 206 K demonstrates the absence of symmetry. However by 301 K, the ^{13}CO NMR pattern is indicative of an averaged mirror plane perpendicular to the Os_3 triangle. This is readily explained in terms of a ligand oscillation over the metal–metal edge, involving the interchange of formal σ and π bonds. Since the vinyl protons H(1) and H(2) are not interconverted by this process, the sides of the vinyl group are still differentiated; hence the plane of the vinyl ligand must be perpendicular to the Os_3 plane in the transition state.

Unsaturated ligands also binds to the faces of metal clusters, and many such complexes exhibit nonrigidity. The complex $H_2Os_3(CO)_9$ (CCH_2) (Fig. 17) is unsymmetrical, but by 345 K, the resonances due to the two inequivalent olefinic protons coalesce (46). At this temperature, eight of the nine ^{13}CO NMR signals are broadened (54), and it seems that an averaged mirror plane, again perpendicular to the Os_3 plane, is being produced by the migration of H(a) to the vacant Os–Os edge. Exchange between the two hydride environments H(a) and H(b) is much slower;

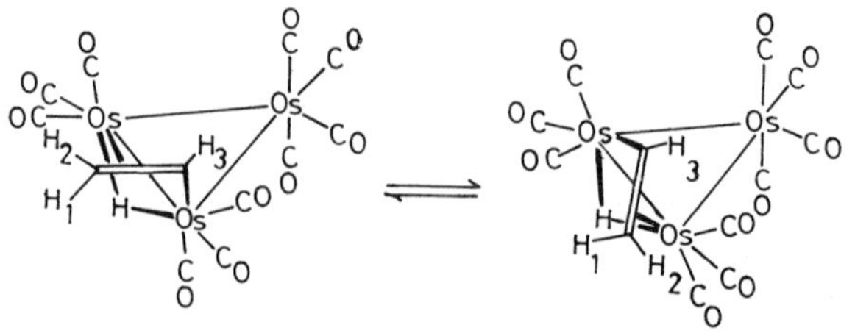

FIG. 16.   Oscillation of vinyl group in $HOs_3(C_2H_3)(CO)_{10}$.

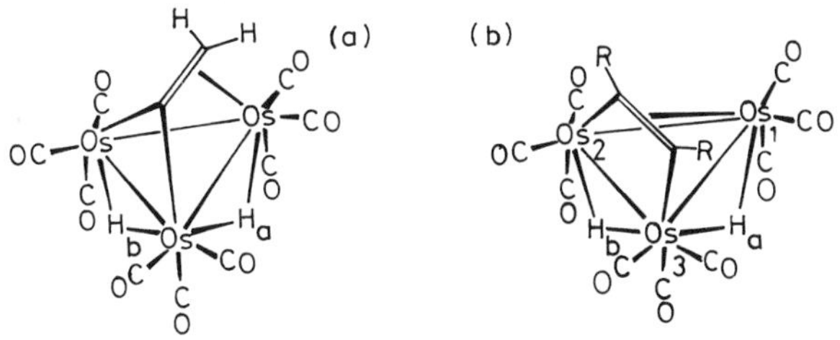

FIG. 17. Two bonding modes of a C_2 unit to an M_3 face. (a) $H_2Os_3(C_2H_2)(CO)_9$; (b) $H_2Os_3(C_2R_2)(CO)_9$ e.g., $R_2 = C_6H_{12}$.

only at 388 K do the hydride resonances themselves start to suffer broadening (47).

An alternative bonding made to a triangular metal face is exemplified by $H_2Os_3(CO)_9(C_8H_{12})$ (Fig. 17). In the limiting low-temperature ^{53}CO NMR spectrum, below 173 K, all nine carbonyl environments are resolved (54). As the temperature is raised, four pairs of these resonances broaden and coalesce to give a 2:2:2:2:1 spectrum at 213 K. By 253 K, the spectrum further simplifies to 2:2:2:3 and only one, averaged, resonance is observed at 373 K. The two low-energy processes appear to be, first, the degenerate hydride migration to the vacant Os–Os edge and, second, rotation of the carbonyl ligands about Os(1). Since the free energy of activation for the third step is experimentally indistinguishable from that measured for the averaging of the two hydride resonances due to H(a) and H(b) (20), it is probable that one mechanism causes both. With the other two, processes are already occurring rapidly, a degenerate ligand migration through 120° to become parallel with the other hydride bridged Os–Os edge will cause total carbonyl and hydride equilibration; this involves a similar bonding change to that observed for $HOs_3(CO)_{10}(C_2H_3)$ (Fig. 16).

The diphenylacetylene ligand in $(\eta^5-C_5H_5)_3Rh_3(PhCCPh)(CO)$ is also bonded in this second manner, being parallel with one of the metal–metal edges (102); the carbonyl ligand is in a face-bridging position. At 186 K, the ^{13}C NMR spectrum due to $Ph^{13}CCPh$ exhibits one rhodium split eight-line multiplet, consistent with this structure. On raising the temperature to 303 K, rotation of the ligand is demonstrated by this pattern simplifying to a rhodium split quartet (98). Rotation of the acetylene ligand in the decarbonylated complex is rapid even at 146 K.

Migration of the benzyne ligand in $Os_3(CO)_7(C_6H_4)(EMe_2)_2$ (E = As, Ph) is also necessary to explain the temperature dependence of these complexes' proton NMR spectra (45). Four methyl resonances would be expected from the unsymmetrical structure (93), but only two are observed down to the lowest temperatures studied (E = As, 153 K; E = P, 213 K). By 300 K an apparent mirror plane, perpendicular to the C_6 plane, is introduced into the benzene ligand. These two averaging mechanisms were considered to involve intermediates with one bridging carbonyl group and different orientations of the benzyne ligand.

The larger type of organic ligands, some of which form nonrigid binuclear complexes, also migrate about metal clusters. $(C_8H_8)_2Ru_3(CO)_4$ possesses two cyclo-octatetraene ligands bridging Ru–Ru edges (14) (Fig. 18). Down to 233 K, the rings are rotating rapidly in solution (28). This rotation is frozen at 135 K in the solid state, but above this temperature line-narrowing indicates that the exchange process is occurring (40). At higher temperatures, the observed second moment of the proton line fitted independent rotation of the two rings; 1,2 or 1,3 shift mechanisms could not be distinguished. Rapid rotation of a planar cyclo-octatetraene ring relative to the face of an Ni_3 triangle is evident in the complex $Ni_3(CO)_3(C_8H_8)(CF_3CCCF_3)$ (42). The complex $(\eta^7-C_7H_7)(\eta^5-C_7H_9)Ru_3(CO)_6$ (Fig. 18) has two types of cyclic ligands, one bound to an edge of the Ru_3 triangle and the other to one Ru atom (13). Even at 173 K, the tropyllium ligand gives rise to a sharp averaged proton resonance (101). Oscillation of the $(\eta^5-C_7H_9)$ ring about the Ru atom is apparent above 179 K, that ligand having an averaged mirror plane by 252 K.

Complexed pentalene derivatives (77) have been shown to migrate over all the faces and edges of an Ru_3 triangle (67). The predominant isomer of $Ru_3(CO)_8(C_8H_6)$ contains a pentalene ligand bridging two ruthenium atoms (Fig. 19), with the plane of the triangle at 50° to that of the ligand (68). A minor isomer is in dynamic equilibrium with this one, and its structure, with the bicyclic ligand bound to an Ru_3 face, has been characterized for a derivative $Ru_3(CO)_8[1,3,5-(Me_3Si)_3C_8H_3]$ (67). The proton NMR spectra of the major, edge-bonded, isomers are temperature dependent. In the high-temperature limit an averaged mirror plane is introduced through the pentalene ligand, which is removed in the static structure by the orientation of the Ru_3 triangle. Two processes in

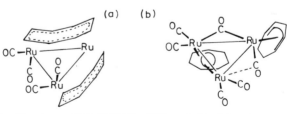

FIG. 18.   Structures of (a) $(C_8H_8)_2Ru_3(CO)_4$ and (b) $(C_7H_7)(C_7H_9)Ru_3(CO)_6$.

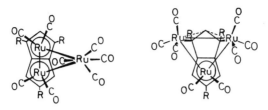

FIG. 19.   Structures of the two isomers of $(C_8H_3R_3)Ru_3(CO)_8(R=H, SiMe_3)$.

combination allow the eventual ligand migration over the faces and edges of this triangle (Fig. 19). The lower-energy one, giving rise to the temperature-dependent proton NMR spectrum, involves the swinging of the pentalene ligand, and also the carbonyl groups of the $Ru(CO)_2$ moieties, with respect to the Ru_3 triangle. This allows the ligand to change sides of the cluster. In addition, when the isomers are interconverted, the ligand is transferred between edge- and face-bridging positions, and these processes allow the pentalene complete mobility.

VII

SKELETAL ATOM MIGRATION

Polytopal rearrangements of the cage atoms in class-polyhedral boranes, e.g., $(B_8H_8)^{2-}$ and $(B_{11}H_{11})^{2-}$, have been discussed (91). In these ions, skeletal atom migration is sufficient to average the ^{11}B NMR environments. Rearrangements of metal atoms in the thermal isomerization of metallocarboranes are well known (18, 48). Several of these include the relative migrations of two or more transition metals (59, 94). The seven-vertexed metallocarboranes $[(\eta^5\text{-}C_5H_5)_2Co_2C_2B_3H_5]$ have a relatively high transition metal content (86). Above 573 K, the 1,7,2,3 isomer (Fig. 20) undergoes an apparent carbon atom migration to the 1,7,2,4 species. However, at lower temperatures (474–523 K) two sequential intermediates—the 1,2,4,5 and 1,2,3,5 isomers—were detected. Metal migration was apparent at two stages, the first one being reversible. The authors favored a rotating-triangle mechanism (92), fitting other patterns previously observed in monometallic species (48). There is also an example of a stereochemically nonrigid metallocarborane in which a rapid oscillation about the metal is occurring at room temperature (64). Even though the structure of $(Me_2C_2B_7H_9)Pt(PEt_3)_2$ in the crystal is unsymmetrical, room   temperature   NMR   spectra   of   it   and

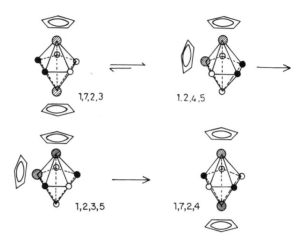

Fig. 20.   Isomerization sequence for $(C_5H_5)_2Co_2C_2B_3H_5$. ⊘, Co; ●, CH; ○, BH.

$(C_2B_7H_9)Pt(PEt_3)_2$ demonstrate an apparent mirror plane. This can be produced if the two enantiomers of the structure (Fig. 21) are interconverting rapidly. Although the relationship of the platinum atom to the rest of the cage is altered during this process, the metal atom is still localized.

At the time of writing, there are no proven examples of skeletal atom migrations in transition metal carbonyl clusters. In a number of such complexes, e.g., $Os_6(CO)_{18}$ (50) and $[Os_5(CO)_{15}]^{2-}$ (49), such migrations are evidently not sufficiently rapid to affect the reported ^{13}C NMR spectra. In part, there is a problem of detection; NMR using transition metal nuclei has not yet been widely applied to metal carbonyl clusters. Within mixed metal clusters, isomerizations as well as degenerate rearrangements can be envisaged. With this in mind, the random distributions of cobalt and nickel atoms in $[Ni_2Co_4(CO)_{14}]^{2-}$ (9), and possibly of the two metals in $Co_2Rh_4(CO)_{16}$ (82), allow the possibility of the coexistence of cis and trans substituted metal polyhedra. Intercon-

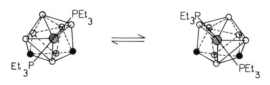

Fig. 21.   Oscillatory exchange process of $Pt(PEt_3)_2C_2R_2B_7H_7(R=H, Me)$. ⊘, Pt; ●, CR; ○, BH.

versions between these two isomers would involve skeletal atom migration and could be studied by ^{59}Co NMR.

ACKNOWLEDGMENTS

I wish to thank the Royal Society for support in the form of a Pickering Research Fellowship and also my colleagues, particularly Professor J. Lewis and Dr. B. F. G. Johnson, for many helpful discussions.

REFERENCES

1. Adams, R. D., Brice, M. D., and Cotton, F. A., *Inorg. Chem.* **13,** 1080 (1974).
2. Adams, R. D., Collins, D. M., and Cotton, F. A., *Inorg. Chem.* **13,** 1086 (1974).
3. Adams, R. D., Collins, D. M., and Cotton, F. A., *J. Am. Chem. Soc.* **96,** 749 (1974).
4. Adams, R. D., and Cotton, F. A., *J. Am. Chem. Soc.* **94,** 6193 (1972).
5. Adams, R. D., and Cotton, F. A., *J. Am. Chem. Soc.* **95,** 6589 (1973).
6. Adams, R. D., Cotton, F. A., and Troup, J. M., *Inorg. Chem.* **13,** 257 (1974).
7. Albano, V. G., Bellon, P. L., and Ciani, G. F., *Chem. Commun.* 1024 (1969).
8. Albano, V. G., Chini, P., Ciani, G., Sansoni, M., Strumolo, D., Heaton, B. T., and Martinengo, S., *J. Am. Chem. Soc.* **98,** 5027 (1976).
9. Albano, V. G., Ciani, G., and Chini, P., *J. Chem. Soc., Dalton Trans.* 432 (1974).
10. Albano, V. G., Ciani, G., Martinengo, S., Chini, P., and Giordano, G., *J. Organometal. Chem.* **88,** 381 (1975).
11. Anders, U., and Graham, W. A. G., *Chem. Commun.* 186 (1967).
12. Aumann, R., and Lohmann, B., *J. Organometal. Chem.* **44,** C51 (1972).
13. Bau, R., Burt, J. C., Knox, S. A. R., Laine, R. M., Phillips, R. P., and Stone, F. G. A., *J. Chem. Soc., Chem. Commun.* 726 (1973).
14. Bennett, M. J., Cotton, F. A., and Legzdins, P., *J. Am. Chem. Soc.* **90,** 6335 (1968).
15. Bryan, R. F., and Greene, P. T., *J. Chem. Soc. A* 3064 (1970).
16. Bryan, R. F., Greene, P. T., Field, D. S., and Newlands, M. J., *Chem. Commun.* 1477 (1969).
17. Bullitt, J. G., Cotton, F. A., and Marks, T. J., *Inorg. Chem.* **11,** 671 (1972).
18. Callahan, K. P., and Hawthorne, M. F., *Advan. Organometal. Chem.* **14,** 145 (1976).
19. Campbell, A. J., Fyfe, C. A., and Maslowsky, Jun., E., *Chem. Commun.* 1032 (1971).
20. Canty, A. J., Johnson, B. F. G., and Lewis, J., *J. Organometal. Chem.* **43,** C35 (1972).
21. Chini, P., Longoni, G., and Albano, *Advan. Organometal. Chem.* **14,** 285 (1976).
22. Chini, P., Martinengo, S., McCaffrey, D. J. A., and Heaton, B. T., *J. Chem. Soc., Chem. Commun.* 310 (1974).
23. Cohen, M. A., Kidd, D. R., and Brown, T. L., *J. Am. Chem. Soc.* **97,** 4408 (1975).
24. Corey, E. R., and Dahl, L. F., *Inorg. Chem.* **1,** 521 (1962).
25. Corradini, P., *J. Chem. Phys.* **31,** 1676 (1959).
26. Cotton, F. A., *Inorg. Chem.* **5,** 1083 (1966).
27. Cotton, F. A., *Acc. Chem. Res.* **1,** 257 (1968).
28. Cotton, F. A., Davison, A., Marks, T. J., and Musco, A., *J. Am. Chem. Soc.* **90,** 6335 (1968).
29. Cotton, F. A., De Boer, B. G., and Marks, T. J., *J. Am. Chem. Soc.* **93,** 5069 (1971).
30. Cotton, F. A., and Edwards, W. T., *J. Am. Chem. Soc.* **91,** 843 (1969).

31. Cotton, F. A., Frenz, B. A., Deganello, G., and Shaver, A., *J. Organometal. Chem.* **50**, 227 (1973).
32. Cotton, F. A., Frenz, B. A., and White, A. J., *Inorg. Chem.* **13**, 1407 (1974).
33. Cotton, F. A., and Hunter, D. L., *Inorg. Chim. Acta* **11**, L9 (1974).
34. Cotton, F. A., and Hunter, D. L., *J. Am. Chem. Soc.* **97**, 5739 (1975).
35. Cotton, F. A., Hunter, D. L., and Lahuerta, P., *Inorg. Chem.* **14**, 511 (1975).
36. Cotton, F. A., Hunter, D. L., and Lahuerta, P., *J. Am. Chem. Soc.* **97**, 1046 (1975).
37. Cotton, F. A., Kruczynski, L., Shapiro, B. L., and Johnson, L. F., *J. Am. Chem. Soc.* **94**, 6191 (1972).
38. Cotton, F. A., Kruczynski, L., and White, A. J., *Inorg. Chem.* **13**, 1402 (1974).
39. Cotton, F. A., and Marks, T. J., *J. Organometal. Chem.* **19**, 237 (1969).
40. Cottrell, C. E., Fyfe, C. A., and Senoff, C. V., *J. Organometal. Chem.* **43**, 203 (1972).
41. Dahl, L. F., and Blount, J. F., *Inorg. Chem.* **4**, 1373 (1965).
42. Davison, J. L., Green, M., Stone, F. G. A., and Welch, A. J., *J. Am. Chem. Soc.* **97**, 7490 (1975).
43. Day, V. W., Day, R. O., Kristoff, J. S., Hirsckorn, F. J., and Muetterties, E. L., *J. Am. Chem. Soc.* **97**, 2573 (1975).
44. Deeming, A. J., and Hasso, S., *J. Organometal. Chem.* **88**, C21 (1975).
45. Deeming, A. J., Kimber, R. E., and Underhill, M., *J. Chem. Soc., Dalton Trans.* 2589 (1973).
46. Deeming, A. J., and Underhill, M., *J. Chem. Soc., Chem. Commun.* 277 (1973).
47. Deeming, A. J., and Underhill, M., *J. Chem. Soc., Dalton Trans.* 1415 (1974).
48. Dustin, D. F., Evans, W. J., Jones, C. J., Wiersema, R. J., Gong, H., Chan, S., and Hawthorne, M. F., *J. Am. Chem. Soc.* **96**, 3085 (1974).
49. Eady, C. R., Guy, J. J., Johnson, B. F. G., Lewis, J., Malatesha, M. C., and Sheldrick, G. M., *J. Chem. Soc., Chem. Commun.* 807 (1976).
50. Eady, C. R., Jackson, W. G., Johnson, B. F. G., Lewis, J., and Matheson, T. W., *J. Chem. Soc., Chem. Commun.* 958 (1975).
51. Eady, C. R., Johnson, B. F. G., and Lewis, J., *J. Chem. Soc., Chem. Commun.* 302 (1976).
52. Evans, J., Johnson, B. F. G., Lewis, J., and Matheson, T. W., *J. Am. Chem. Soc.* **97**, 1245 (1975).
53. Evans, J., Johnson, B. F. G., Lewis, J., and Matheson, T. W., *J. Chem. Soc., Chem. Commun.* 576 (1975).
54. Evans, J., Johnson, B. F. G., Lewis, J., and Matheson, T. W., *J. Organometal. Chem.* **97**, C16 (1975).
55. Evans, J., Johnson, B. F. G., Lewis, J., Matheson, T. W., and Norton, J. R., *J. Chem. Soc., Dalton Trans.*, to be submitted.
56. Evans, J., Johnson, B. F. G., Lewis, J., and Norton, J. R., *J. Chem. Soc., Chem. Commun.* 79 (1973).
57. Evans, J., Johnson, B. F. G., Lewis, J., Norton, J. R., and Cotton, F. A., *J. Chem. Soc., Chem. Commun.* 807 (1973).
58. Evans, J., Johnson, B. F. G., Lewis, J., and Watt, R., *J. Chem. Soc., Dalton Trans.* 2368 (1974).
59. Evans, J., Jones, C. J., Stibr, B., Grey, R. A., and Hawthorne, M. F., *J. Am. Chem. Soc.* **96**, 7405 (1974).
60. Fleischer, E. B., Stone, A. L., Dewar, R. B. K., Wright, J. D., Keller, C. E., and Petit, R., *J. Am. Chem. Soc.* **88**, 3158 (1966).
61. Forster, A., Johnson, B. F. G., Lewis, J., and Matheson, T. W., *J. Organometal. Chem.* **104**, 225 (1976).

62. Forster, A., Johnson, B. F. G., Lewis, J., Matheson, T. W., Robinson, B. H., and Jackson, W. G., *J. Chem. Soc., Chem. Commun.* 1042 (1974).

63. Gansow, O. A., Burke, A. R., and Vernon, W. D., *J. Am. Chem. Soc.* **94,** 2550 (1972).

64. Green, M., Spencer, J. L., Stone, F. G. A., and Welch, A. J., *J. Chem. Soc., Chem. Commun.* 571 (1974).

65. Harris, D. C., Rosenberg, E., and Roberts, J. D., *J. Chem. Soc., Dalton Trans.* 2399 (1974).

66. Heaton, B. T., Towl, A. D. C., Chini, P., Fumagilli, A., McCaffrey, D. J. A., and Martinengo, S., *J. Chem. Soc., Chem. Commun.* 523 (1975).

67. Howard, J. A. K., Knox, S. A. R., McKinney, R. J., Stansfield, R. F. D., Stone, F. G. A., and Woodward, P., *J. Chem. Soc., Chem. Commun.* 557 (1976).

68. Howard, J. A. K., Knox, S. A. R., Riera, V., Stone, F. G. A., and Woodward, P., *J. Chem. Soc., Chem. Commun.* 452 (1974).

69. Howell, J. A. S., Matheson, T. W., and Mays, M. J., *J. Chem. Soc., Chem. Commun.* 865 (1975).

70. Jesson, J. P., *in* "Transition Metal Hydrides." (E. L. Muetterties, ed.), p. 75 ff. Marcel Dekker, 1971.

71. Johnson, B. F. G., *J. Chem. Soc., Chem. Commun.,* 211 (1976).

72. Johnson, B. F. G., *J. Chem. Soc., Chem. Commun.,* 703 (1976).

73. Johnson, B. F. G., Lewis, J., and Matheson, T. W., *J. Chem. Soc., Chem. Commun.,* 442 (1974).

74. Kirchner, R. M., Marks, T. J., Kristoff, J. S., and Ibers, J. A., *J. Am. Chem. Soc.* **95,** 6602 (1973).

75. Knight, J., and Mays, M. J., *Chem. Ind. (London)* 115 (1968).

76. Knox, S. A. R., and Kaesz, H. D., *J. Am. Chem. Soc.* **93,** 4594 (1971).

77. Knox, S. A. R., and Stone, F. G. A., *Acc. Chem. Res.* **7,** 331 (1974).

78. Koepke, J. W., Johnson, J. R., Knox, S. A. R., and Kaesz, H. D., *J. Am. Chem. Soc.* **97,** 3947 (1975).

79. McArdle, P., and Manning, A. R., *J. Chem. Soc. A.* 2128 (1970).

80. McPartlin, M., Eady, C. R., Johnson, B. F. G., and Lewis, J., *J. Chem. Soc., Chem. Commun.* 883 (1976).

81. Manning, A. R., *J. Chem. Soc. A* 1319 (1968).

82. Martinengo, S., Chini, P., Albano, V. G., Cariati, F., and Salvatori, T., *J. Organometal. Chem.* **59,** 379 (1973).

83. Mason, R., and Rae, A. J. M., *J. Chem. Soc. A* 778 (1968).

84. Mason, R., Thomas, K. M., and Mingos, D. M. P., *J. Am. Chem. Soc.* **95,** 3802 (1973).

85. Mathieu, R., and Nixon, J. F., *J. Chem. Soc., Chem. Commun.* 147 (1974).

86. Miller, V. R., and Grimes, R. N., *J. Am. Chem. Soc.* **97,** 4213 (1975).

87. Mills, O. S., *Acta Crystallogr.* **11,** 620 (1958).

88. Mills, O. S., and Nice, J. P., *J. Organometal. Chem.* **10,** 337 (1967).

89. Milone, L., Aime, S., Randall, E. W., and Rosenberg, E., *J. Chem. Soc., Chem. Commun.* 452 (1975).

90. Muetterties, E. L., *Acc. Chem. Res.* **3,** 266 (1970).

91. Muetterties, E. L., Hoel, E. L., Salentine, C. G., and Hawthorne, M. F., *Inorg. Chem.* **14,** 950 (1975).

92. Muetterties, E. L., and Knoth, W. H., "Polyhedral Boranes." Dekker, New York, 1968.

93. Nyholm, R. S., Gainsford, G. J., Guss, J. M., Ireland, P. R., and Mason, R., *J. Chem. Soc., Chem. Commun.* 87 (1972).

94. Salentine, C. G., and Hawthorne, M. F., *J. Am. Chem. Soc.* **97**, 6382 (1975).
95. Shapley, J. R., Keister, J. B., Churchill, M. R., and De Boer, B. G., *J. Am. Chem. Soc.* **97**, 4145 (1975).
96. Shapley, J. R., Richter, S. I., Tachikawa, M., and Keister, J. B., *J. Organometal. Chem.* **94**, C43 (1975).
97. Todd, L. J., and Wilkinson, J. R., *J. Organometal. Chem.* **77**, 1 (1974).
98. Todd, L. J., Wilkinson, J. R., Rausch, S. A., and Dickson, R. S., *J. Organometal. Chem.* **101**, 133 (1975).
99. Wei, C. H., and Dahl, L. F., *J. Am. Chem. Soc.* **88**, 1821 (1966).
100. Wei, C. H., Wilkes, G. R., and Dahl, L. F., *J. Am. Chem. Soc.* **89**, 4793 (1967).
101. Whitesides, T. H., and Budnik, R., *J. Chem. Soc., Chem. Commun.* 302 (1974).
102. Yamamoto, T., Garber, A. R., Bodner, G. M., Todd, L. J., Rausch, M. D., and Gardner, S. A., *J. Organometal. Chem.* **56**, C23 (1973).
103. Yamamoto, T., Garber, A. R., Wilkinson, I. R., Boss, C. B., Streib, W. E., and Todd, L. J., *J. Chem. Soc., Chem. Commun.* 354 (1974).
104. Yawney, D. B. W., and Stone, F. G. A., *J. Chem. Soc. Chem. Commun.* 186 (1968); *J. Chem. Soc. A* 502 (1969).

POSTSCRIPT

Since writing this article, the carbonyl exchange in $Co_3Rh(CO)_{12}$ has been reconsidered (Section III, A, 2). The spectrum observed at 243 K (a 2:10 pattern) would not be produced by keeping two carbonyl groups terminally bound to the rhodium atom and allowing the ligands to distort to a cuboctahedral form and then relax back to an icosahedron. This process instead yields a 2:8:2 spectrum since the two ligands para to the $Rh(CO)_2$ pair (one of these is edge bridging and the other terminally bound to the apical cobalt) also only interchange with each other. However, at 188 K the ligands of the $Co(CO)_3$ group appear as one, not two, resonances. If this is due to a rapid rotation of this unit at 188 K, then the two para ligands could become averaged with the other eight by the icosahedral-cuboctahedral-icosahedral process at 243 K. So unless the ligand motions are totally uncorrelated, three processes are necessary to explain the temperature variation of the ^{13}C NMR spectra of $Co_3Rh(CO)_{12}$. It must also be mentioned that while all ligands in the cuboctahedral intermediate for $Rh_4(CO)_{12}$ exchange would be terminally bound, having the full T_d symmetry of $Ir_4(CO)_{12}$, this cannot be the case for the mixed metal complex.

A reinvestigation of the ^{13}C NMR of $[HFe_3(CO)_{11}]^-$ has yielded a more satisfactory description of this anion's nonrigidity(*P3*). At 166 K, the ^{13}C NMR spectrum is consistent with the structure in the crystal being maintained in solution (Fig. 4). At ca. 246 K, the intermediate spectrum is a 2:9 not a 1:10 pattern, and one of the terminal carbonyls is specifically exchanging with the bridging one. A concerted distortion of the icosahedral ligand frame (cf. Fig. 5) to a tetrakaidecahedron with all ligands terminal, and then relaxing back, causes this effect; the bridging carbonyl then specifically exchanges with one axial group of the $Fe(CO)_4$ moiety and also all the other nine exchange as a set. At higher temperatures, the hydrido ligand is probably moving with respect to the rest of the icosahedron.

Two other recent papers also merit mention at this, the proof stage. The first one is a multinuclear investigation of the $[Rh_{13}(CO)_{24}H_{5-n}]^{n-}$ (n is 2 or 3) anions (*P2*). Analysis of the proton spectra in particular indicated a rapid migration (at 298 K) of the hydrido ligands within the hexagonal close packed Rh_{13} array. It was evident that the hydrides spent time in both tetrahedral and octahedral holes. In the second paper (*P1*) skeletal atom migration was detected in $[Pt_9(CO)_{18}]^{2-}$ using ^{195}Pt NMR. The magnetic equivalence of the

Pt nuclei of the outer triangles requires the rapid rotation of the outer Pt_3 triangles with respect to the central one even at 188 K.

P1. Brown, C., Heaton, B. T., Chini, P., Fumagalli, A., and Longoni, G., *J. Chem. Soc., Chem. Commun.* 309 (1977).
P2. Martinengo, S., Heaton, B.T., Goodfellow, R. J., and Chini, P., *J. Chem. Soc., Chem. Commun.* 39 (1977).
P3. Wilkinson, J. R., and Todd, L. J., *J. Organometal. Chem.* **118,** 199 (1977).

Subject Index

Cumulative List of Contributors

Abel, E. W., **5**, 1; **8**, 117
Aguilo, A., **5**, 321
Albano, V. G., **14**, 285
Armitage, D. A., **5**, 1
Atwell, W. H., **4**, 1
Bennett, M. A., **4**, 353
Birmingham, J., **2**, 365
Brook, A. G., **7**, 95
Brown, H. C., **11**, 1
Brown, T. L., **3**, 365
Bruce, M. I., **6**, 273; **10**, 273; **11**, 447; **12**, 379
Cais, M., **8**, 211
Callahan, K. P., **14**, 145
Cartledge, F. K., **4**, 1
Chalk, A. J., **6**, 119
Chatt, J., **12**, 1
Chini, P., **14**, 285
Churchill, M. R., **5**, 93
Coates, G. E., **9**, 195
Collman, J. P., **7**, 53
Corey, J. Y., **13**, 139
Courtney, A., **16**, 241
Coutts, R. S. P., **9**, 135
Coyle, T. D., **10**, 237
Craig, P. J., **11**, 331
Cullen, W. R., **4**, 145
Cundy, C. S., **11**, 253
de Boer, E., **2**, 115
Dessy, R. E., **4**, 267
Dickson, R. S., **12**, 323
Eisch, J. J., **16**, 67
Emerson, G. F., **1**, 1
Ernst, C. R., **10**, 79
Evans, J., **16**, 319
Faller, J. W., **16**, 211
Fischer, E. O., **14**, 1
Fraser, P. J., **12**, 323
Fritz, H. P., **1**, 239
Furukawa, J., **12**, 83
Fuson, R. C., **1**, 221
Gilman, H., **1**, 89; **4**, 1; **7**, 1
Green, M. L. H., **2**, 325
Griffith, W. P., **7**, 211
Grovenstein, Jr., E., **16**, 167
Gubin, S. P., **10**, 347
Gysling, H., **9**, 361

Haiduc, I., **15**, 113
Harrod, J. F., **6**, 119
Hartley, F. H., **15**, 189
Hawthorne, M. F., **14**, 145
Heck, R. F., **4**, 243
Heimbach, P., **8**, 29
Henry, P. M., **13**, 363
Hieber, W., **8**, 1
Hill, E. A., **16**, 131
Ibers, J. A., **14**, 33
Ittel, S. A., **14**, 33
Jolly, P. W., **8**, 29
Jones, P. R., **15**, 273
Jukes, A. E., **12**, 215
Kaesz, H. D., **3**, 1
Katz, T. J., **16**, 283
Kawabata, N., **12**, 83
Kettle, S. F. A., **10**, 199
Kilner, M., **10**, 115
King, R. B., **2**, 157
Kingston, B. M., **11**, 253
Kitching, W., **4**, 267
Köster, R., **2**, 257
Kühlein, K., **7**, 241
Kuivila, H. G., **1**, 47
Kumada, M., **6**, 19
Lappert, M. F., **5**, 225; **9**, 397; **11**, 25; **14**, 345
Lednor, P. W., **14**, 345
Longoni, G., **14**, 285
Luijten, J. G. A., **3**, 397
Lupin, M. S., **8**, 211
McKillop, A., **11**, 147
Maddox, M. L., **3**, 1
Maitlis, P. M., **4**, 95
Mann, B. E., **12**, 135
Manuel, T. A., **3**, 181
Mason, R., **5**, 93
Matsumura, Y., **14**, 187
Mingos, D. M. P., **15**, 1
Moedritzer, K., **6**, 171
Morgan, G. L., **9**, 195
Mrowca, J. J., **7**, 157
Nagy, P. L. I., **2**, 325
Nakamura, A., **14**, 245
Nesmeyanov, A. N., **10**, 1
Neumann, W. P., **7**, 241

358

Cumulative List of Titles

Acetylene and Allene Complexes: Their Implication in Homogeneous Catalysis, **14,** 245

Activation of Alkanes by Transition Metal Compounds, **15,** 147

Alkali Metal Derivatives of Metal Carbonyls, **2,** 157

Alkyl and Aryl Derivatives of Transition Metals, **7,** 157

Alkylcobalt and Acylcobalt Tetracarbonyls, **4,** 243

Allyl Metal Complexes, **2,** 235

π-Allylnickel Intermediates in Organic Synthesis, **8,** 29

1,2-Anionic Rearrangements of Organosilicon and Germanium Compounds, **16,** 1

Applications of ^{119m}Sn Mössbauer Spectroscopy to the Study of Organotin Compounds, **9,** 21

Arene Transition Metal Chemistry, **13,** 47

Aryl Migrations in Organometallic Compounds of the Alkali Metals, **16,** 167

Boranes in Organic Chemistry, **11,** 1

Carbene and Carbyne Complexes, On the Way to, **14,** 1

Carboranes and Organoboranes, **3,** 263

Catalysis by Cobalt Carbonyls, **6,** 119

Catenated Organic Compounds of the Group IV Elements, **4,** 1

Chemistry of Carbon-Functional Alkylidynetricobalt Nonacarbonyl Cluster Complexes, **14,** 97

^{13}C NMR Chemical Shifts and Coupling Constants of Organometallic Compounds, **12,** 135

Compounds Derived from Alkynes and Carbonyl Complexes of Cobalt, **12,** 323

Conjugate Addition of Grignard Reagents to Aromatic Systems, **1,** 221

Coordination of Unsaturated Molecules to Transition Metals, **14,** 33

Cyclobutadiene Metal Complexes, **4,** 95

Cyclopentadienyl Metal Compounds, **2,** 365

Diene-Iron Carbonyl Complexes, **1,** 1

Dyotropic Rearrangements and Related σ–σ Exchange Processes, **16,** 33

Electronic Effects in Metallocenes and Certain Related Systems, **10,** 79

Electronic Structure of Alkali Metal Adducts of Aromatic Hydrocarbons, **2,** 115

Fast Exchange Reactions of Group I, II, and III Organometallic Compounds, **8,** 167

Flurocarbon Derivatives of Metals, **1,** 143

Fluxional and Nonrigid Behavior of Transition Metal Organometallic π-Complexes, **16,** 211

Free Radicals in Organometallic Chemistry, **14,** 345

Heterocyclic Organoboranes, **2,** 257

α-Heterodiazoalkanes and the Reactions of Diazoalkanes with Derivatives of Metals and Metalloids, **9,** 397

High Nuclearity Metal Carbonyl Clusters, **14,** 285

Infrared Intensities of Metal Carbonyl Stretching Vibrations, **10,** 199

Infrared and Raman Studies of π-Complexes, **1,** 239

Insertion Reactions of Compounds of Metals and Metalloids, **5,** 225

Insertion Reactions of Transition Metal-Carbon σ-Bonded Compounds I. Carbon Monoxide Insertion, **11,** 87

Insertion Reactions of Transition Metal-Carbon σ-Bonded Compounds II. Sulfur Dioxide and Other Molecules, **12,** 31

Isoelectronic Species in the Organophosphorus, Organosilicon, and Organoaluminum Series, **9,** 259

A 7
B 8
C 9
D 0
E 1
F 2
G 3
H 4
I 5
J 6